PRACTICAL MATH

Fourth Edition

0 2 6 5 3 9 1 4 8 7

ratio remainder exponent volume graph

AMERICAN TECHNICAL PUBLISHERS
Orland Park, Illinois 60467-5756

American Technical Publishers, Inc., Editorial Staff

Editor in Chief:
Jonathan F. Gosse
Vice President—Production:
Peter A. Zurlis
Digital Media Manager:
Carl R. Hansen
Assistant Production Manager:
Nicole D. Bigos
Technical Editor:
Cathy A. Scruggs
Copy Editor:
James R. Hein
Editorial Assistant:
Alex C. Tulik
Cover Design:
Robert M. McCarthy
Illustration/Layout:
Bethany J. Fisher
Nicole S. Polak
Nick W. Basham
Robert M. McCarthy
Digital Resources:
Robert E. Stickley
Cory S. Butler

The publisher acknowledges Sprecher + Schuh as well as the Air Conditioning Contractors of America for providing technical images and information.

4 5 6 7 8 9 – 15 – 9 8 7 6 5 4

Printed in the United States of America

ISBN 978-0-8269-2250-2

 This book is printed on recycled paper.

Contents

Chapter 3 Working with Complex Fractions 53

Digital Resources

- Quick Quizzes®
- Illustrated Glossary
- Flash Cards
- Master Math® Tutorials
- Reference Tables
- Media Library
- ATPeResources.com

Book Features

Practical Math covers essential math concepts and how they are applied on the job in agriculture, alternative energy, culinary arts, boiler operation, construction, electrical, HVAC, manufacturing, maintenance, mechanics, pipefitting, plumbing, and welding. Each chapter presents math concepts with visual steps and includes both math exercises and practical applications that reinforce learning. Selected answers to math exercises and practical applications are listed at the back of the book.

Features of this edition include the following:

- Each chapter is divided into several sections to maximize both learning and instructional flexibility.
- Over 300 practical applications provide opportunities for learners to apply essential math concepts in 13 different occupational fields.
- Online learner resources reinforce math concepts and encourage practice in a variety of formats.
- Each chapter ends with a formative review and a summative test to assess math comprehension achieved.
- The expanded appendix includes occupational reference tables as well as essential math tables.

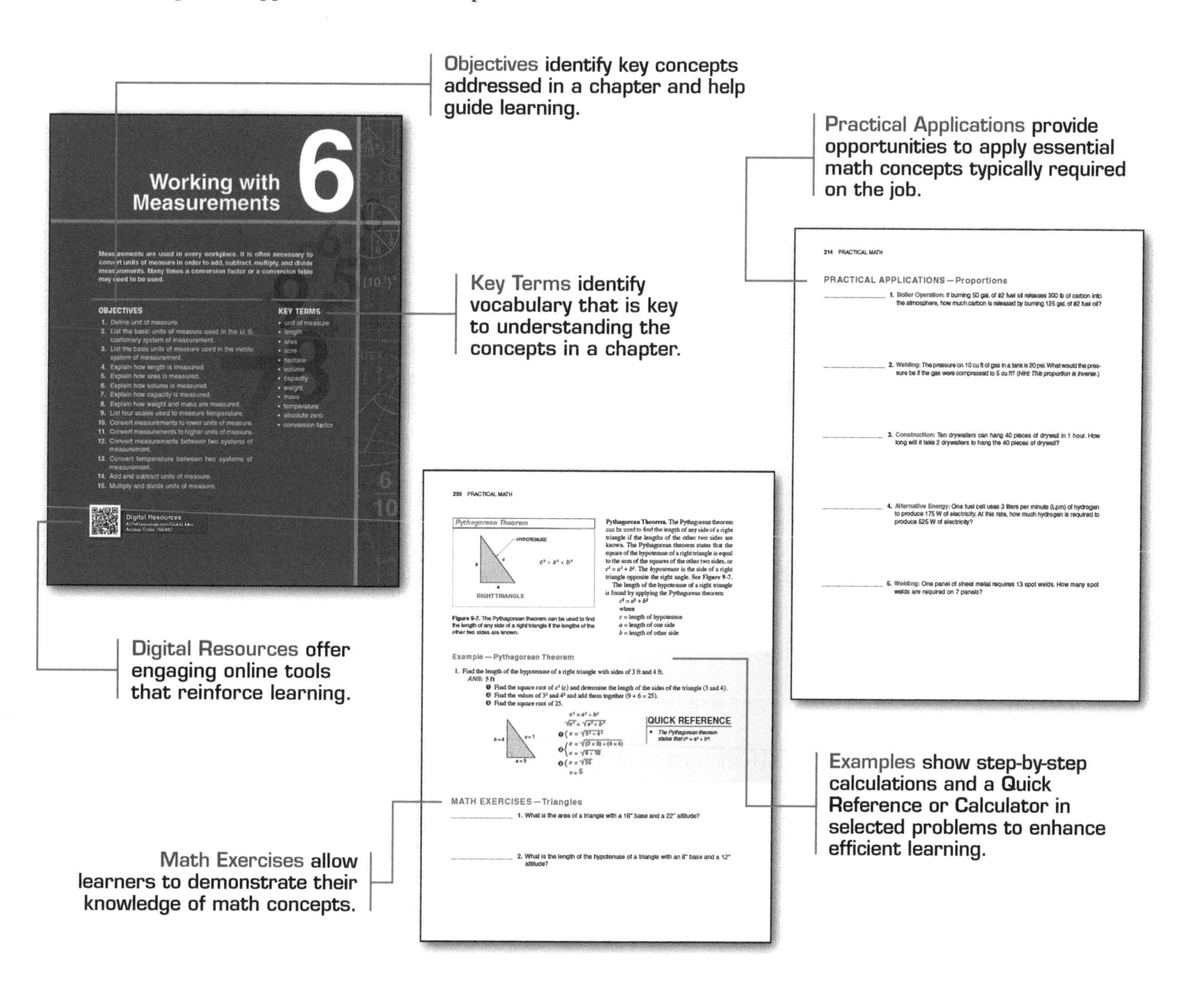

Digital Resources

This edition of *Practical Math* includes access to digital resources that enhance and reinforce learning. These online Learner Resources can be accessed using either of the following methods.

- Key ATPeResources.com/QuickLinks into a web browser and enter QuickLink™ code **764460**.
- Use a Quick Response (QR) reader app on a mobile device to scan the QR code located on the opening page of each chapter.

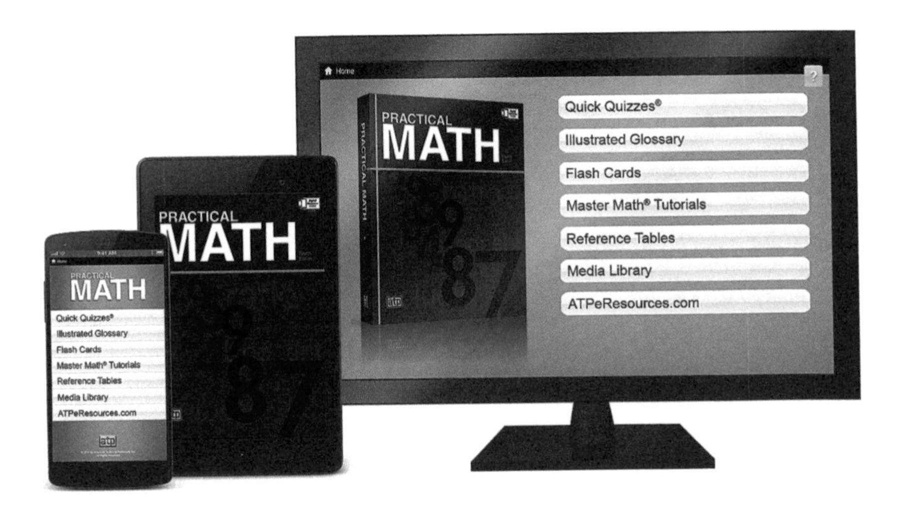

Learner Resources

The online Learner Resources provide the following interactive tools to maximize both comprehension and retention.

- **Quick Quizzes®** provide 10 interactive questions for each chapter and include reference links to the textbook and Illustrated Glossary.
- The **Illustrated Glossary** serves as a helpful reference to math terms and definitions, with selected terms linked to illustrations.
- **Flash Cards** serve as a review tool for key terms and definitions and for practicing a variety of math exercises that reinforce learning.
- **Master Math® Tutorials** demonstrate how common math calculations are performed step-by-step.
- **Reference Tables** are printable tables of both math and occupational reference information.
- The **Media Library** provides videos that demonstrate the application of key math concepts.
- **ATPeResources.com** provides access to online reference materials that support continued learning.

Digital Resources

This edition of *Practical Math* also offers a set of digital resources that empower instructors to tailor their presentations, demonstrations, and assessments within a specific course or program.

Instructor Resources

The online *Practical Math Instructor Resources* provide the following set of dynamic teaching and learning tools to maximize instructional effectiveness.

- The **Instructional Guide** provides an overview of how to structure a course and includes a step-by-step instructional plan for each chapter. The Instructional Plans are divided by section and correlated to learning objectives.
- **Premium PowerPoint® Presentations** include objectives and key content in each chapter of the book. Section breaks are denoted, and presentation notes are provided.
- The **Image Library** provides all numbered figures in a format that can be resized by the instructor for maximum instructional flexibility.
- **Assessments** include a pretest, a posttest, and test banks based on the objectives for each chapter that can be used to document knowledge and skills attained. Each test bank is provided in several formats to facilitate use in testing software or a learning management system (LMS).
- **Answer Keys** list answers to the pretest, the posttest, and all of the math exercises and practical applications in the textbook.
- **Learner Resources** (Quick Quizzes, Illustrated Glossary, Flash Cards, Master Math® Tutorials, Reference Tables, Media Library, and a link to ATPeResources.com) are included.

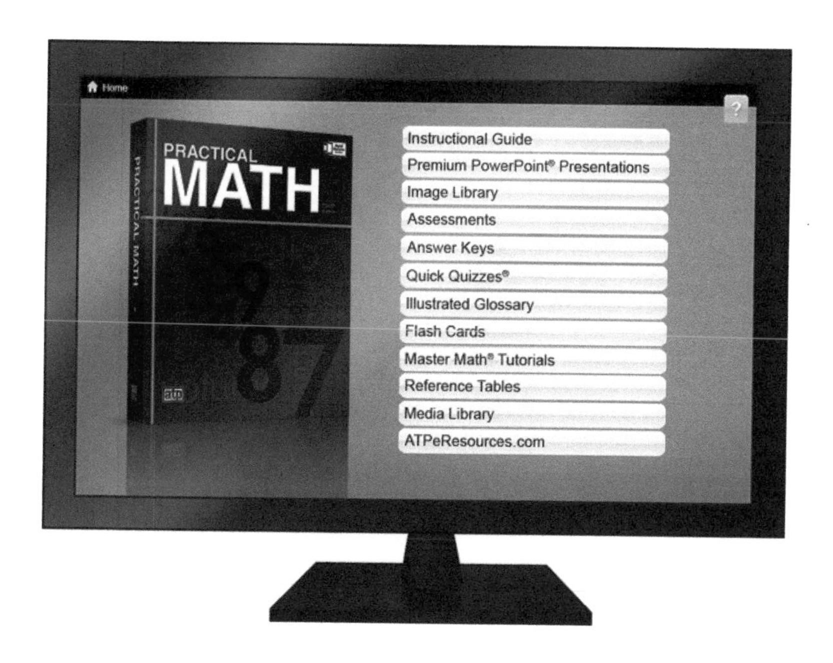

To obtain information on other related training materials, visit www.atplearning.com.

The Publisher

1

Working with Whole Numbers

Whole numbers are positive numbers that have no fractional or decimal parts. Operations used for calculating whole numbers include adding, subtracting, multiplying, and dividing. Carrying and borrowing are two processes used when performing these operations.

OBJECTIVES

1. Differentiate between whole numbers and integers.
2. Explain how Arabic numerals are expressed.
3. Explain how Roman numerals are expressed.
4. Add whole numbers.
5. Subtract whole numbers.
6. Multiply whole numbers.
7. Divide whole numbers.

KEY TERMS

- whole number
- odd number
- even number
- integer
- Arabic numerals
- period
- Roman numerals
- addition
- sum
- carrying
- unit of measure
- subtraction
- difference
- borrowing
- multiplication
- factor
- product
- division
- dividend
- divisor
- quotient
- remainder

Digital Resources
ATPeResources.com/QuickLinks
Access Code: 764460

SECTION 1-1 UNDERSTANDING WHOLE NUMBERS

A *whole number* is any positive number that has no fractional or decimal parts, such as 1, 10, or 100. Whole numbers are used for counting. Whole numbers include odd numbers such as 31 and 999 and even numbers such as 68 and 734. **See Figure 1-1.** An *odd number* is any number that ends in 1, 3, 5, 7 or 9 and cannot be divided by 2 an exact number of times. An *even number* is any number, except 0, that ends in 2, 4, 6, 8 or 0 and can be divided by 2 an exact number of times.

Similar to whole numbers, integers do not have fractional parts, but they do include negative numbers such as −32°F. An *integer* is a negative whole number, a positive whole number, or 0. Whole numbers are written as either Arabic numerals or Roman numerals.

Whole Numbers

EVEN NUMBERS
CAN BE DIVIDED BY 2 AN EXACT NUMBER OF TIMES

WHOLE NUMBERS
HAVE NO FRACTIONAL OR DECIMAL PARTS

1 2 19 46 67 328

ODD NUMBERS
CANNOT BE DIVIDED BY 2 AN EXACT NUMBER OF TIMES

Figure 1-1. Whole numbers have no fractional or decimal parts and are either odd or even.

Arabic Numerals

Arabic numerals are numerals expressed using ten digits—0, 1, 2, 3, 4, 5, 6, 7, 8, and 9. Large numbers with many digits are made easier to read by the use of periods. **See Figure 1-2.** A *period* is a group of three places in a number that is separated from other periods by a comma. The units period (000 through 999) is the first period. The thousands period (1000 through 999,999) is the second period. The millions period (1,000,000 through 999,999,999) is the third period, and so on. A digit can occupy one of three places in each period—the units, tens, or hundreds place.

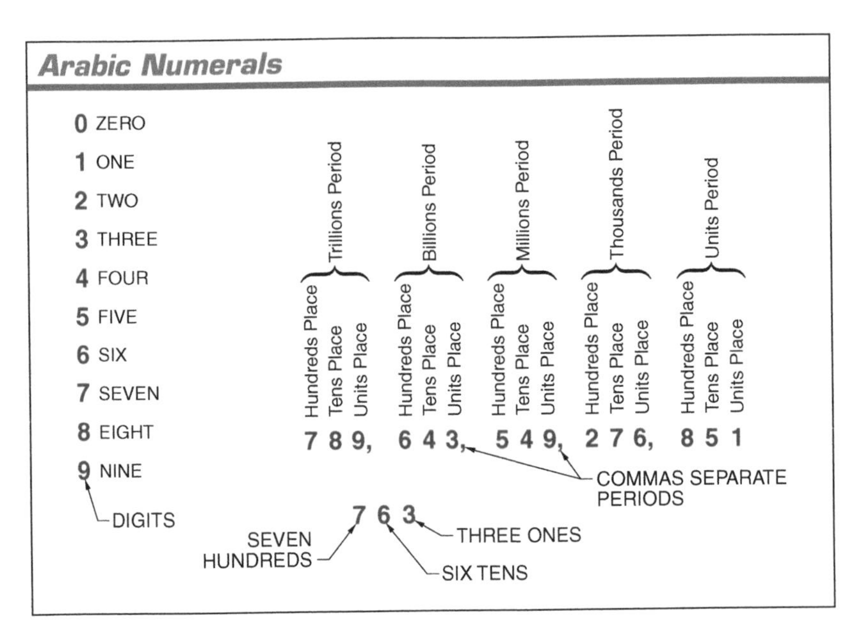

Figure 1-2. Arabic numerals are expressed using ten digits grouped in periods and seperated by commas.

Each period is read or written out using the name of the period at the end, except for the units period. For example, 125,000,300,057 is read as one hundred twenty-five billion, three hundred thousand fifty-seven.

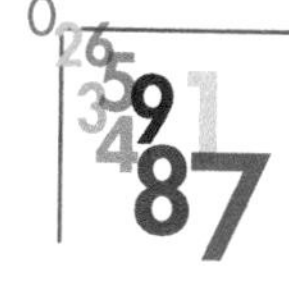

The comma can be omitted after a digit that occupies the units place of the thousands period. For example, 7,222 can also be written as 7222.

Roman Numerals

Roman numerals are numerals expressed by the letters I, V, X, L, C, D, and M. Roman numerals are occasionally used on clock faces, on public buildings such as libraries and museums, and for numbering chapters in books. **See Figure 1-3.**

In the Roman numbering system, when a letter is followed by the same letter, or one lower in value, the values of the letters are added together. For example, XX = 20 and XV = 15. When a letter is followed by a letter greater in value, the letter of lesser value is subtracted from the letter of greater value. For example, IV = 4.

When a letter is placed between two letters both greater in value, the letter of lesser value is subtracted from the sum of the other two. For example, XIV = 14 and CVL = 145. When a rule (line) is placed over a letter, the value of the letter increases a thousand times. For example, the Roman numeral C has a value of 100, but $\overline{\text{C}}$ has a value of 100,000.

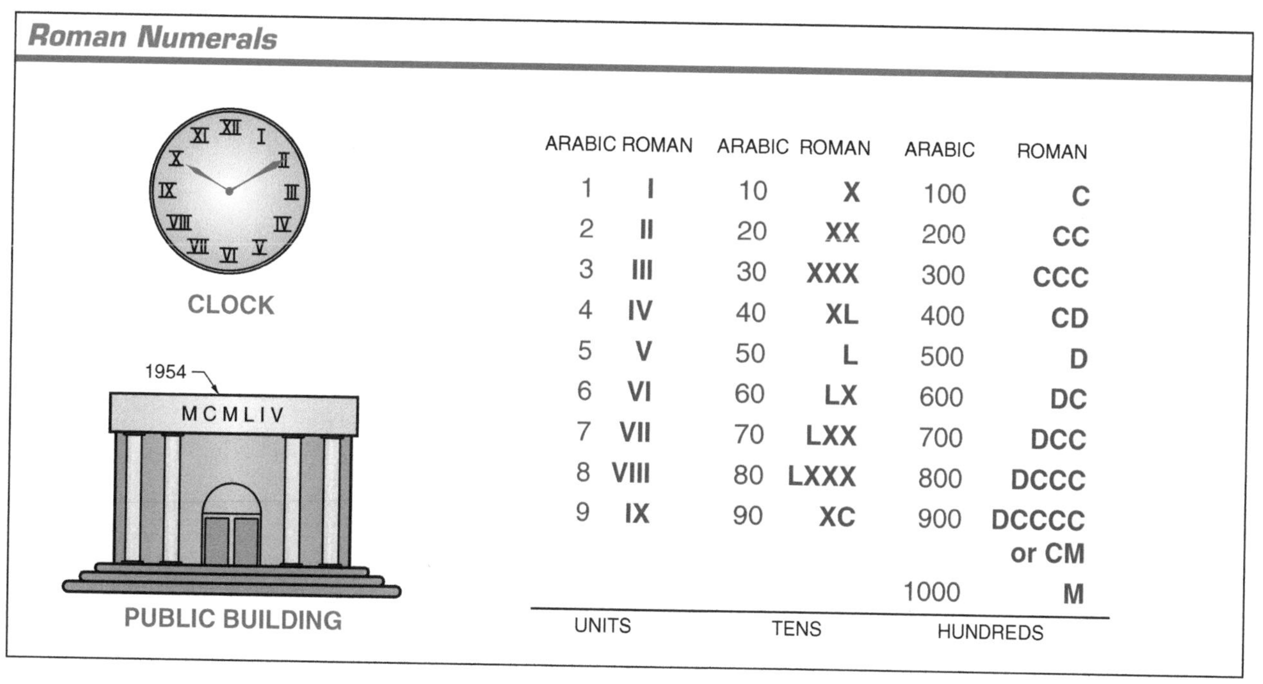

ARABIC	ROMAN	ARABIC	ROMAN	ARABIC	ROMAN
1	I	10	X	100	C
2	II	20	XX	200	CC
3	III	30	XXX	300	CCC
4	IV	40	XL	400	CD
5	V	50	L	500	D
6	VI	60	LX	600	DC
7	VII	70	LXX	700	DCC
8	VIII	80	LXXX	800	DCCC
9	IX	90	XC	900	DCCCC or CM
				1000	M
UNITS		TENS		HUNDREDS	

Figure 1-3. Roman numerals are expressed using seven letters in various combinations.

MATH EXERCISES—Expressing Whole Numbers

Write numbers as words and words as numbers in the blanks provided.

1. 123,507 ____________________

2. 55,017 ____________________

3. 6300 ____________________

____________________ **4.** Seventy-eight million, forty-one thousand, seven

____________________ **5.** Five hundred six thousand, nine hundred twenty-five

Write Roman numerals as Arabic numerals and Arabic numerals as Roman numerals.

____________________ **6.** CCLXIX

____________________ **7.** 21

____________________ **8.** LXXI

____________________ **9.** 1993

____________________ **10.** IX

SECTION 1-2 ADDING WHOLE NUMBERS

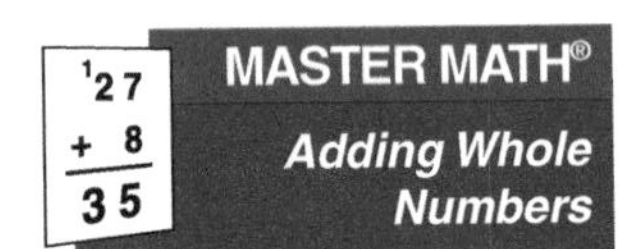

Adding whole numbers requires an understanding of addition. *Addition* is the process of combining two or more numbers into a single sum. The *sum* is the number that is produced as a result of addition. Addition is the most common mathematical operation and is often used in daily tasks. For example, numbers are added together during quantity takeoff for estimating, calculating the total cost of a job, or calculating the number of linear feet in a kitchen countertop.

Numbers can be added vertically or horizontally. When two numbers are added vertically, the plus sign (+) is used. When more than two numbers are added vertically the operation is apparent, and no sign is required.

To add whole numbers vertically, stack the numbers and align the digits in columns by place. **See Figure 1-4.** Add the columns from top to bottom, beginning with the units column and working from right to left. When the sum of the units column is 0 to 9, record the sum and add the tens column. Follow this procedure for the remaining columns.

When the sum of any column is greater than 10, record the last digit and carry the remaining digit to the next column. *Carrying* is the process of moving a digit from one column into the column to its left. **See Figure 1-5.**

Numbers can also be added horizontally, but this is more difficult than adding numbers vertically. For example, 15 + 120 + 37 + 9 = 181 shows whole numbers added horizontally. However, since the numbers are not aligned in columns, mistakes can occur more easily when adding these numbers.

A *unit of measure* is a standard by which a quantity is measured. For example, minutes, feet, and cubic inches are all units of measure. If two quantities are expressed using different units of measure, they cannot be added. For example, the addition of $205.00 and 110 gal. yields the number 315, but since the units of measure are different, the sum cannot be expressed in either dollars or gallons.

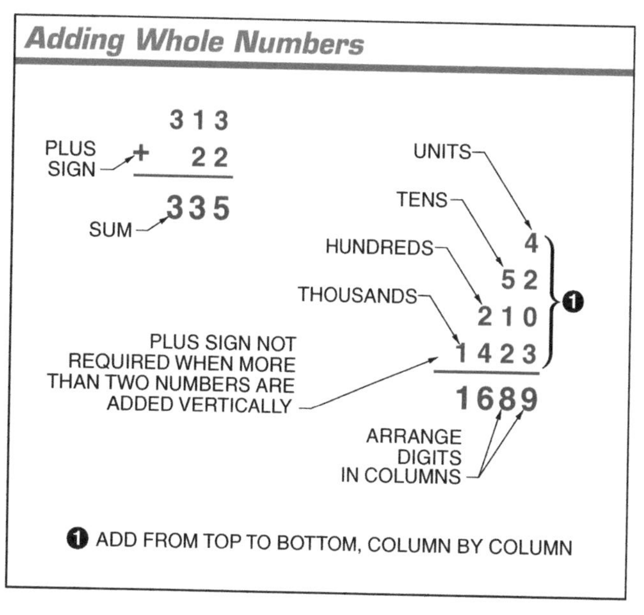

Figure 1-4. Addition is the process of combining two or more numbers into a single number.

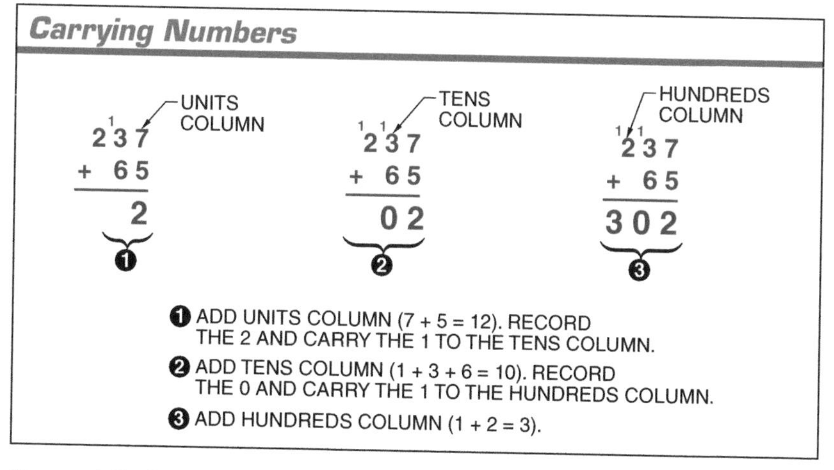

Figure 1-5. Carrying is the process of moving one digit from one column into the column to its left.

Examples—Adding Whole Numbers

1. Find the sum of 25, 20, 19, 12, 32, and 8.

ANS: **116**

❶ Add the digits in the units column first (5 + 0 + 9 + 2 + 2 + 8 = 26). Record the 6 and carry the 2 to the tens column.

❷ Add the digits in the tens column (2 + 2 + 2 + 1 + 1 + 3 = 11). Record the 11.

$$\begin{array}{rrr} & ❷ & ❶ \\ & {}^{2}2 & 5 \\ & 2 & 0 \\ & 1 & 9 \\ & 1 & 2 \\ & 3 & 2 \\ & & 8 \\ \hline 1 & 1 & 6 \end{array}$$

QUICK REFERENCE

- *Add columns starting with the units column and moving left until all columns are added.*
- *Carry as required.*

2. Find the sum of 4, 52, 268, and 1873.

ANS: **2197**

❶ Add the digits in the units column first (4 + 2 + 8 + 3 = 17). Record the 7 and carry the 1 to the tens column.

❷ Add the tens column, including the 1 from the units column (1 + 5 + 6 + 7 = 19). Record the 9 and carry the 1 to the hundreds column.

❸ Add the hundreds column, including the 1 from the tens column (1 + 2 + 8 = 11). Record the 1 and carry the 1 to the thousands column.

❹ Add the thousands column, including the 1 from the hundreds column (1 + 1 = 2).

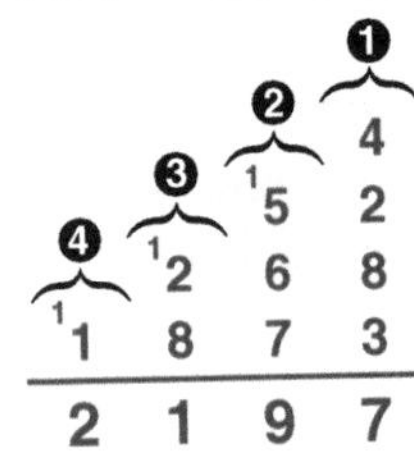

3. Find the sum of 102, 55, 190, 145, 80, and 131.

ANS: **703**

❶ Add the units column and carry the 1.

❷ Add the tens column and carry the 3.

❸ Add the hundreds column.

$$\begin{array}{rrr} ❸ & ❷ & ❶ \\ {}^{3}1 & {}^{1}0 & 2 \\ & 5 & 5 \\ 1 & 9 & 0 \\ 1 & 4 & 5 \\ & 8 & 0 \\ 1 & 3 & 1 \\ \hline 7 & 0 & 3 \end{array}$$

To check addition problems in which the numbers were added vertically, add the columns from bottom to top. The same sum should occur as when adding the columns from top to bottom.

To check addition problems in which the numbers were added horizontally, add the numbers from right to left. The same sum should occur as when adding the numbers from left to right.

MATH EXERCISES — Adding Whole Numbers

Write each sum in the blank provided.

_______________ **1.** 56 + 49 + 17 + 36 + 21

_______________ **2.** 467 + 536 + 84 + 705

_______________ **3.** 8950 + 15,765 + 7732

_______________ **4.** 14,005 + 1204 + 350 + 9786 + 43

_______________ **5.** 12 + 15,045 + 159,056 + 179

PRACTICAL APPLICATIONS — Adding Whole Numbers

_______________ **6. Manufacturing:** What is the total length of the spacing gauge?

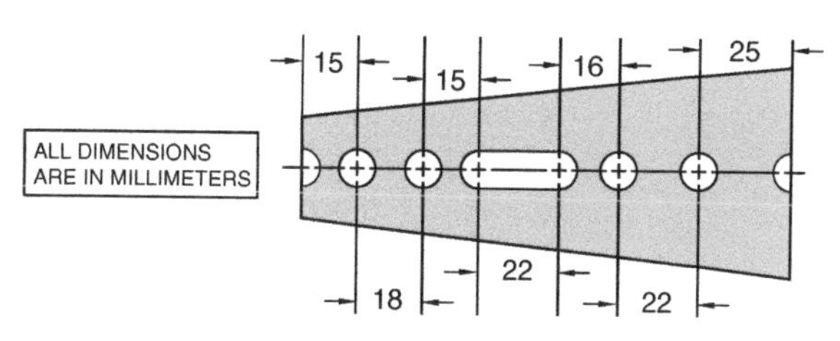

SPACING GAUGE

_______________ **7. Construction:** A contractor estimates the cost of remodeling a house as follows: $800.00 for masonry, $950.00 for lumber, $110.00 for hardware, $75.00 for trim, $120.00 for paint, and $3500.00 for labor. What is the total estimated cost?

______________ **8. Electrical:** What is the total number of watts in circuit A with all lamps turned on?

200 W LAMP
200 W LAMP
75 W LAMP
100 W LAMP

CIRCUIT A

SECTION 1-3 SUBTRACTING WHOLE NUMBERS

Subtracting whole numbers requires an understanding of subtraction. *Subtraction* is the process of taking one number away from another number. It is the opposite of addition. The *difference* is the number produced as a result of subtraction.

As in addition, numbers can be subtracted vertically or horizontally. When subtracted vertically, numbers are aligned in columns that have different values, and subtraction is indicated using a minus (–) sign.

To subtract whole numbers vertically, align the columns and subtract the second number from the first number beginning with the units column and working from right to left. Follow this procedure for the remaining columns. **See Figure 1-6.**

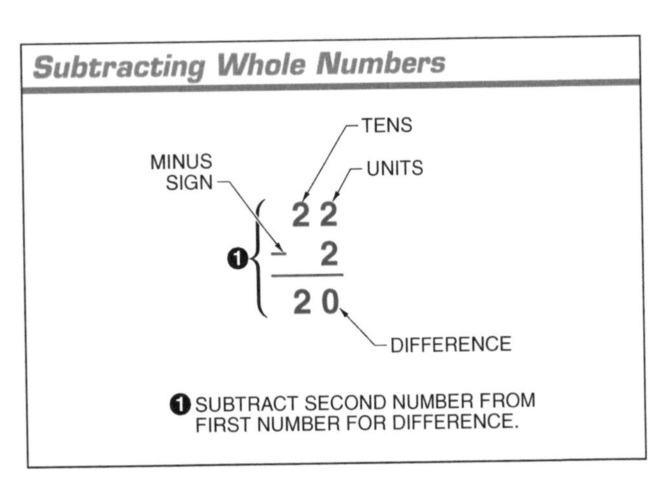

Figure 1-6. Subtraction is the process of taking one number away from another number.

When the second number is larger than the first number, borrow from the column immediately to the left and continue the operation. *Borrowing* is the process of moving "10" from the next higher column so that the difference will be positive. **See Figure 1-7.**

As in addition, the quantities to be subtracted must use the same unit of measure. For example, pounds and feet cannot simply be subtracted. The subtraction of 35 lb from 110′ has a difference of 75 but not in pounds or feet.

To check subtraction in which numbers were subtracted vertically, add the difference (answer) to the second number. If the sum equals the first number, subtraction was performed correctly.

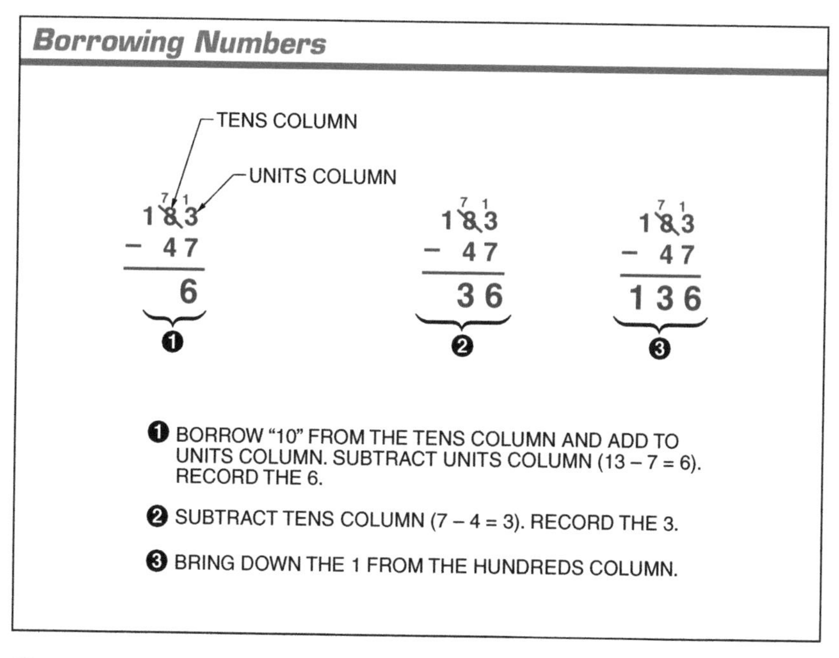

Figure 1-7. Borrowing is the process of moving "10" from the next higher column.

Examples — Subtracting Whole Numbers

1. Subtract 15 from 97.

ANS: **82**

❶ Subtract the units column.

❷ Subtract the tens column.

$$\begin{array}{r} 9\ 7 \\ -\ 1\ 5 \\ \hline 8\ 2 \end{array}$$

QUICK REFERENCE

- *Subtract columns starting with the units column and moving left until all columns are added.*
- *Borrow as required.*

2. Subtract 34 from 70.

ANS: **36**

❶ Borrow from the tens column to get 10 in the units column. Record the 6.

❷ Subtract 3 from 6 in the tens column (6 – 3 = 3). Record the 3.

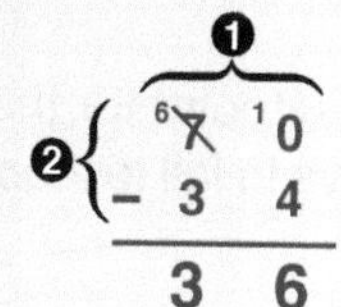

3. Subtract 364 from 942.

ANS: **578**

❶ Borrow from the tens column to get 12 in the units column. Subtract 4 from 12 (12 – 4 = 8). Record the 8.

❷ Borrow from the hundreds column to get 13 in the tens column. Subtract 6 from 13 (13 – 6 = 7). Record the 7.

❸ Subtract 3 from 8 (8 – 3 = 5).

❶
$$\begin{array}{rrr} 9 & \overset{3}{\not{4}} & {}^{1}2 \\ -3 & 6 & 4 \\ \hline & & 8 \end{array}$$

❷
$$\begin{array}{rrr} \overset{8}{\not{9}} & \overset{13}{\not{4}} & {}^{1}2 \\ -3 & 6 & 4 \\ \hline & 7 & 8 \end{array}$$

❸
$$\begin{array}{rrr} \overset{8}{\not{9}} & \overset{13}{\not{4}} & {}^{1}2 \\ -3 & 6 & 4 \\ \hline 5 & 7 & 8 \end{array}$$

MATH EXERCISES—Subtracting Whole Numbers

Write each difference in the blank provided.

________________ **1.** 467 – 84

________________ **2.** 9786 – 43

________________ **3.** 8950 – 7732

________________ **4.** 159,056 – 9179

________________ **5.** 560,894 – 30,101

PRACTICAL APPLICATIONS—Subtracting Whole Numbers

________________ **6. Manufacturing:** A conveyor gudgeon requires 2 holes to be drilled. The farthest hole is 6″ from the end of the gudgeon. The second hole is 2″ from the end. How far apart are the holes?

________________ **7. HVAC:** The temperature in a room is 84°F. The setpoint for the thermostat is 68°F. How many degrees does the temperature have to fall to reach the setpoint?

______________ **8. Maintenance:** A storage shed contains 8579 tons (t) of bar stock from which 3243 t are removed. How many tons remain?

______________ **9. Construction:** What is the length of A on the centering template?

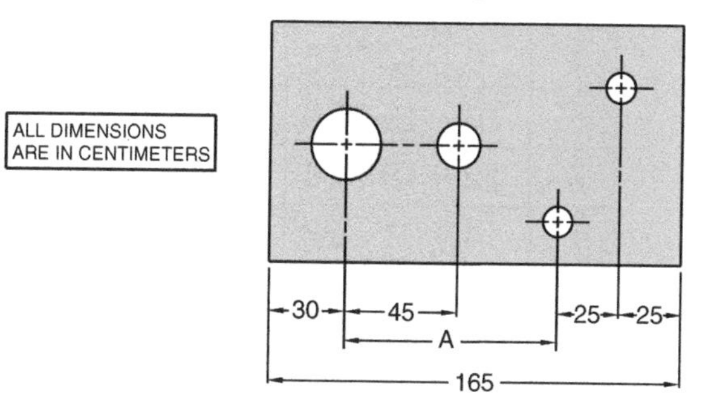

CENTERING TEMPLATE

______________ **10. HVAC:** A sheet metal worker installs 4″ × 12″ duct in various lengths for a forced air heating system. What is the length of C?

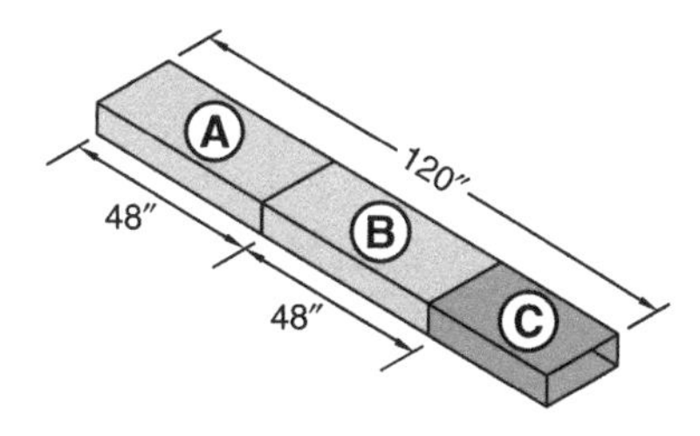

SECTION 1-4 MULTIPLYING WHOLE NUMBERS

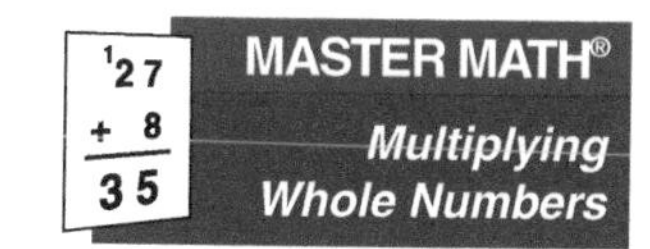

Multiplication is the process of adding one factor as many times as indicated by another factor. In other words, multiplication is a shortcut for addition. For example, 2 × 4 is the same as 2 + 2 + 2 + 2 = 8 and 4 + 4 = 8. A *factor* is a number used as a multiplier. The *product* is the number produced as a result of multiplication. The multiplication, or times, sign (×) indicates multiplication. **See Figure 1-8.**

A multiplication table is set up so that the factors run along the top row and down the left side column. The number that appears where the row and column of any two factors intersect is the product of those factors. **See Figure 1-9.**

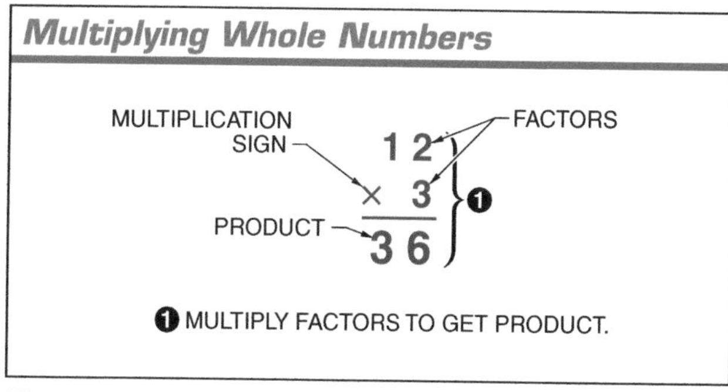

Figure 1-8. Multiplication is the process of adding one factor as many times as indicated by another factor.

Multiplication Table

1	2	3	4	5	6	7	8	9	10	11	12
2	4	6	8	10	12	14	16	18	20	22	24
3	6	9	12	15	18	21	24	27	30	33	36
4	8	12	16	20	24	28	32	36	40	44	48
5	10	15	20	25	30	35	40	45	50	55	60
6	12	18	24	30	36	42	48	54	60	66	72
7	14	21	28	35	42	49	56	63	70	77	84
8	16	24	32	40	48	56	64	72	80	88	96
9	18	27	36	45	54	63	72	81	90	99	108
10	20	30	40	50	60	70	80	90	100	110	120
11	22	33	44	55	66	77	88	99	110	121	132
12	24	36	48	60	72	84	96	108	120	132	144

Figure 1-9. Multiplication tables can be used to help learn and recall the products of multiplication.

In any multiplication problem, every digit in the first number must be multiplied by every digit in the second number. Numbers with only one digit each are easily multiplied through the memorization of the multiplication table. Also, numbers with two or more digits are easily multiplied by numbers with only one digit. In the last case, carrying is sometimes necessary. **See Figure 1-10.**

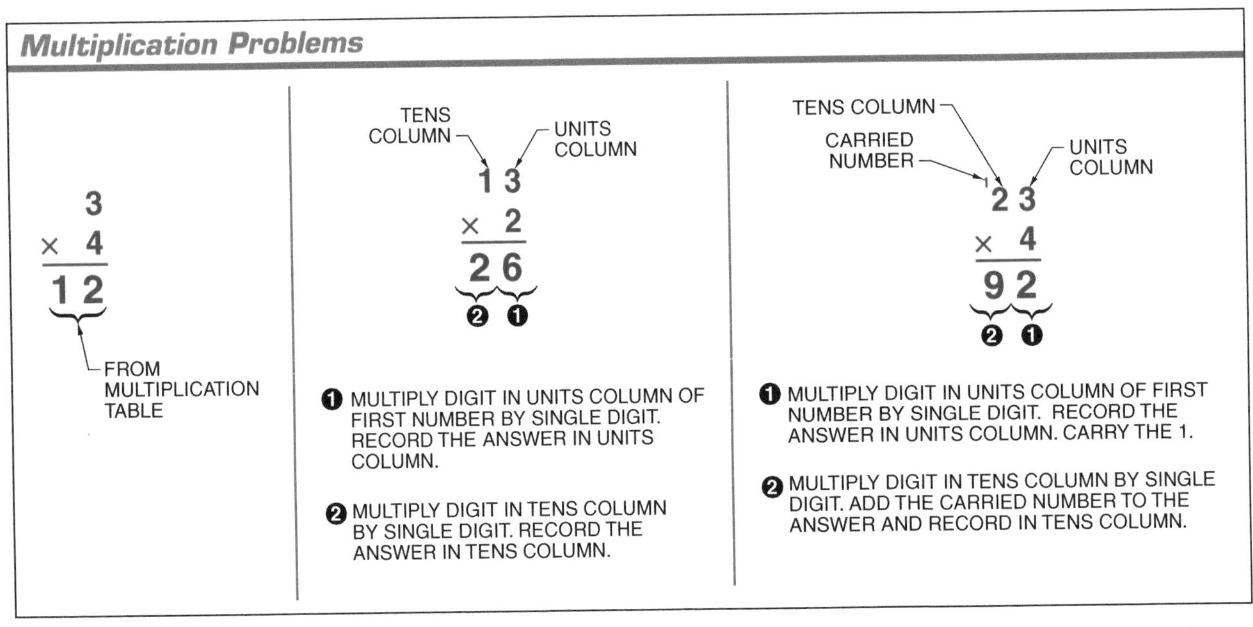

Figure 1-10. In multiplication, every digit of the first number must be multiplied by every digit of the second number.

To multiply numbers with two or more digits by numbers with only one digit, multiply the digit in the units column of the first number by the single digit (carry if necessary). Record the answer in the units column. Then multiply the digit in the tens column of the first number by the single digit. Record the answer in the tens column. Continue with this process until all the digits of the first number have been multiplied by the single digit.

When multiplying numbers with two or more digits each, every digit in the first number must be multiplied by every digit in the second number. **See Figure 1-11.** This results in multiple products that must be added together for the final product. Also, the place of each digit is critical to arriving at the correct answer.

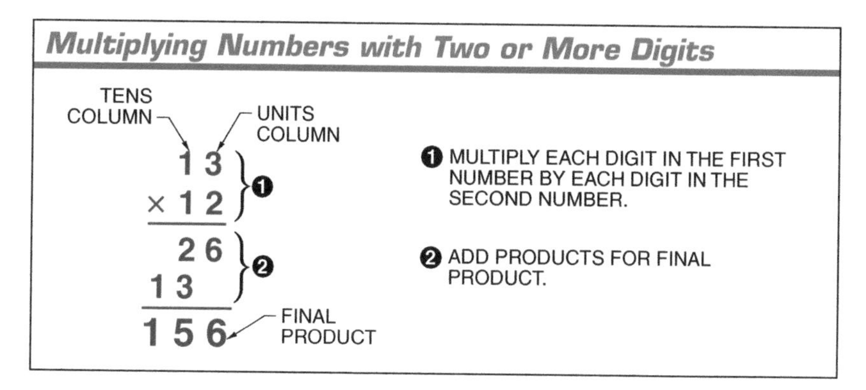

Figure 1-11. When multiplying numbers with two or more digits each, the products must be added together for the final product.

Only two numbers may be multiplied at any time. When multiplying more than two numbers, multiply the first two numbers. Then multiply the product by the next number, and so on. **See Figure 1-12.**

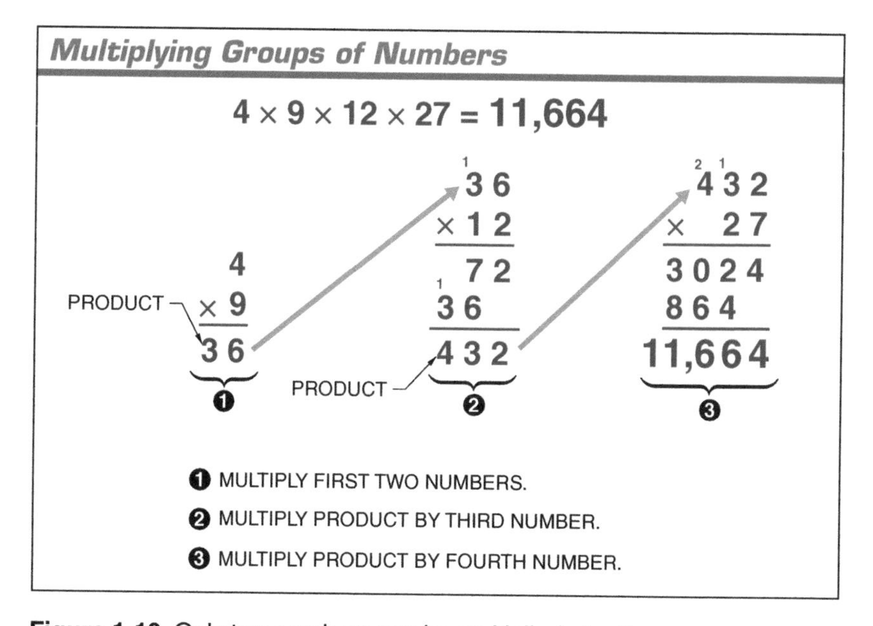

Figure 1-12. Only two numbers can be multiplied at a time.

If there are zeros in the numbers, they are handled in a special way. Any number multiplied by zero equals zero. For example, $0 \times 0 = 0$, $8 \times 0 = 0$, and $492 \times 0 = 0$. In contrast, when multiplying a number by 1, the product will always be that number. For example, $7 \times 1 = 7$ and $152 \times 1 = 152$.

When the factors end in zeros (for example, 10, 50, or 9000) the zeros can be dropped and the remaining numbers multiplied. Then, the number of zeros in both factors are counted and placed after the product. For example, when multiplying 270 by 2000, drop the four zeros and multiply 27 by 2, which equals 54. Then add the four zeros to 54 for a final product of 540,000.

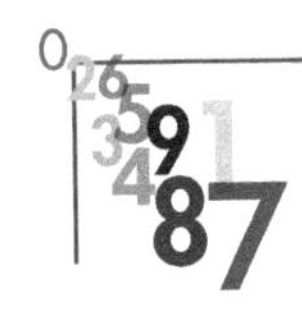

To check multiplication, reverse the order of the factors and perform the operation again. The same product will result if the operation is performed correctly.

Examples—Multiplying Whole Numbers

1. Multiply 24 by 3.

ANS: **72**

❶ Multiply 4 by 3 ($4 \times 3 = 12$). Record the 2. Carry the 1.

❷ Multiply 2 by 3 ($2 \times 3 = 6$). Add the 1. Record the 7.

$$\begin{array}{r} {}^{1}2\ 4 \\ \times\ \ 3 \\ \hline 7\ 2 \end{array} \Big\} ❶$$

2. Multiply 346 by 47.

ANS: **16,262**

❶ Multiply 346 by 7 ($346 \times 7 = 2422$). Record 2422, filling the units, tens, hundreds, and thousands places.

❷ Multiply 346 by 4 ($346 \times 4 = 1384$). Record 1384, filling the tens, hundreds, thousands, and ten thousands places. (Leave the units place blank.)

❸ Add 2422 and 13,840. (The zero is from the blank units place.)

$$\begin{array}{r} {}^{1}{}_{3}{}^{2}{}_{4}\ \ \\ 3\ 4\ 6 \\ \times\ \ 4\ 7 \\ \hline 2\ 4\ 2\ 2 \\ 1\ 3\ 8\ 4\ \ \ \\ \hline 1\ 6,2\ 6\ 2 \end{array}$$

QUICK REFERENCE

- *Multiply all digits of the first number by all digits in the second number. Carry as required.*
- *Add products together for final product.*

3. Multiply 13,456 by 2004.

ANS: **26,965,824**

❶ Multiply 13,456 by 4 and record the product (13,456 × 4 = 53,824).

❷ Bring down the two zeros. (Leave the units place blank.) Multiply the 13,456 by 2 (13,456 × 2 = 26,912) and record before the two zeros.

❸ Add 53,824 and 26,912,000. (The last zero is from the blank units place.)

```
   1 1 1 2 1 2
     1 3, 4 5 6
   ×   2 0 0 4
     5 3 8 2 4 } ❶
 2 6 9 1 2 0 0   } ❷
 2 6, 9 6 5, 8 2 4 } ❸
```

MATH EXERCISES – Multiplying Whole Numbers

Write each product in the blank provided.

________________ **1.** 144 × 12

________________ **2.** 71 × 35

________________ **3.** 1094 × 18

PRACTICAL APPLICATIONS – Multiplying Whole Numbers

________________ **4. Manufacturing:** A screw machine produces 97 units per hour. How many units are produced in 16 hr?

________________ **5. Welding:** A welder joins four pieces of channel iron. How long are the four pieces altogether?

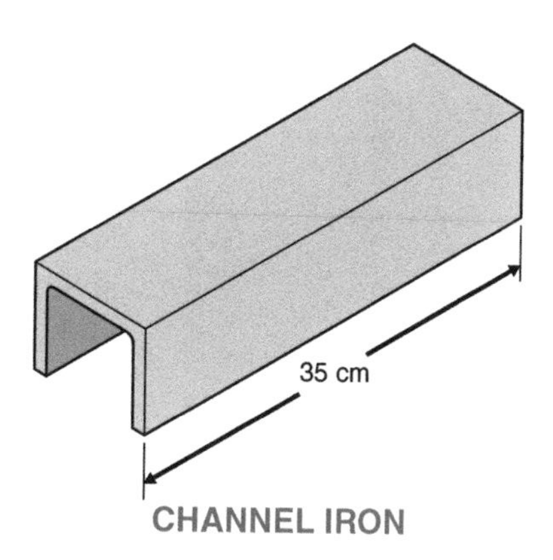

CHANNEL IRON

______________ 6. **Construction:** If one mile of railroad requires 116 t of steel rail at $290.00 per ton, what is the cost of the steel needed to construct a road 128 miles in length?

______________ 7. **Mechanics:** A mechanic receives $356.00 for five days work and spends $13.00 per day. How much money is left?

______________ 8. **Electrical:** An electrical supplier purchases 32 cases of duplex outlets. Each case contains 10 outlets. At a cost of $2.00 per outlet, how much do the 32 cases cost?

SECTION 1-5 DIVIDING WHOLE NUMBERS

MASTER MATH® *Dividing Whole Numbers*

Dividing whole numbers requires an understanding of division. *Division* is the process of determining the number of times one number is contained in another number. In a division equation, the *dividend* is the number being divided, and the *divisor* is the number the dividend is "divided by." The *quotient* is the number that is produced as a result.

Division is the reverse of multiplication. For example, $56 \div 8 = 7$ is the reverse of $7 \times 8 = 56$. Division is indicated using the division sign (÷), long division sign ($\overline{)\quad}$), horizontal fraction bar (–), or slash (/). **See Appendix.**

When dividing single digit numbers, an equation can be written horizontally. For example, $6 \div 3 = 2$ or $9 \div 3 = 3$. When dividing multiple digit numbers, the long division method is used. To carry out division using the long division method, the dividend is placed inside the division sign and the divisor to the left of the division sign. **See Figure 1-13.**

Before cutting a metal stud, the size of the piece is determined using subtraction.

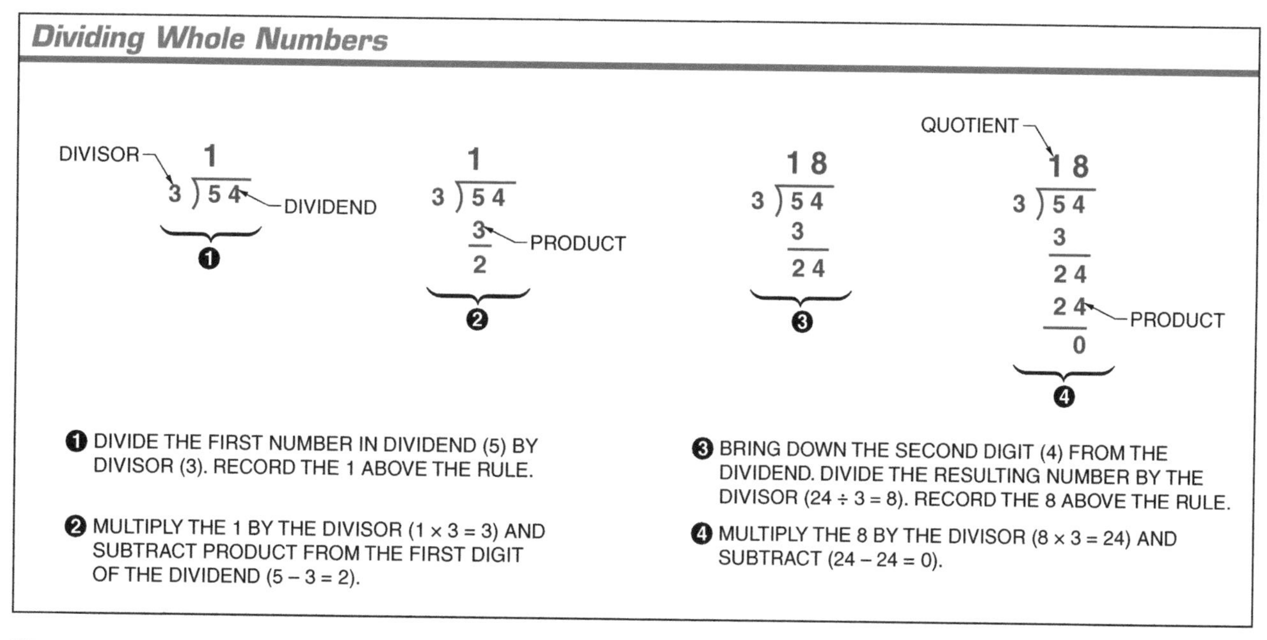

Figure 1-13. Division is the process of finding how many times a number contains another number.

When the divisor is too large to be divided into the first digit of the dividend, divide it into the number formed by the first and second digit (and so on as necessary). Also, if there is a number at the end of an operation that cannot be divided, this number is the remainder. A *remainder* is the undivided part of a quotient. Any remainder is placed over the divisor and expressed as a fraction. **See Figure 1-14.**

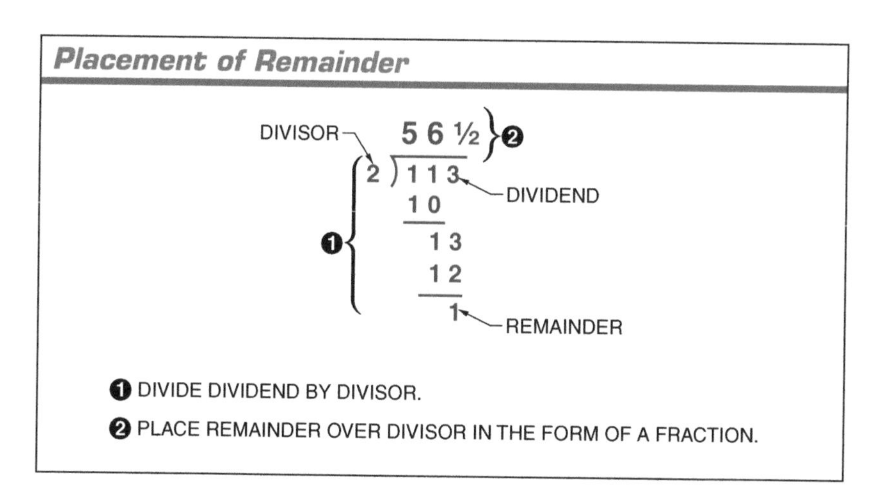

Figure 1-14. The remainder is the undivided part of a division operation.

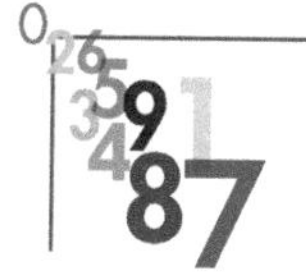

Division can be checked by multiplication. For example, to check 655 ÷ 5 = 131, multiply the quotient by the divisor (131 × 5 = 655). If the product equals the dividend, division was performed correctly.

Examples—Dividing Whole Numbers

1. Divide 820 by 5.

ANS: **164**

❶ Divide 8 by 5. Record the 1 above the 8.

❷ Multiply the 1 by the divisor 5 and subtract the product from 8 (8 – 5 = 3). Bring down the 2 from the dividend for a resulting number of 32.

❸ Divide 32 by 5. Record the 6 above the 2.

❹ Multiply the 6 by the divisor 5 and subtract the product from 32 (32 – 30 = 2). Bring down the 0 from the dividend for a resulting number of 20.

❺ Divide 20 by 5. Record the 4 above the 0.

❻ Multiply the 4 by the divisor 5 and subtract the product from 20 (20 – 20 = 0). There is no remainder.

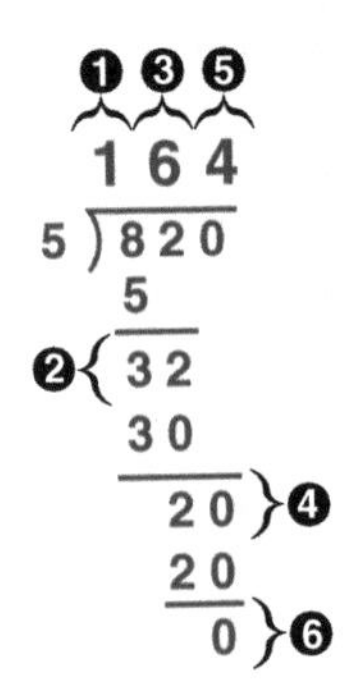

QUICK REFERENCE

- *Divide the dividend by the divisor.*
- *Record any remainders as fractions.*

2. Divide 270 by 3.

ANS: **90**

❶ Because 2 cannot be divided by 3, divide 27 by 3. Record the 9 above the 7.

❷ Multiply the 9 by the divisor 3 and subtract the product from 27 (27 – 27 = 0). Bring down the 0 from the dividend for a resulting number of 00.

❸ Divide 00 by 3. Record the resulting 0 above the 0 in the dividend. There is no remainder.

```
     ❶ ❸
     9 0
3 ) 2 7 0
    2 7
      0 0 } ❷
```

3. Divide 123 by 8.

ANS: **$15\frac{3}{8}$**

❶ Because 1 cannot be divided by 8, divide 12 by 8. Record the 1 above the 2.

❷ Multiply the 1 by the divisor 8 and subtract the product from 12 (12 – 8 = 4). Bring down the 3 from the dividend for the resulting number of 43.

❸ Divide 43 by 8. Record the 5 above the 3.

❹ Multiply the 5 by the divisor 8 and subtract the product from 43 (43 – 40 = 3). There are not digits left in the dividend so the 3 is placed over the numerator for a remainder of 3/8.

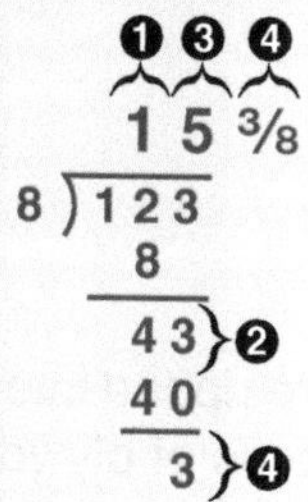

MATH EXERCISES – Dividing Whole Numbers

Write each quotient in the blank provided.

_______________ **1.** 1440 ÷ 20

_______________ **2.** 7200 ÷ 12

_______________ **3.** 414 ÷ 18

_______________ **4.** 1656 ÷ 23

PRACTICAL APPLICATIONS – Dividing Whole Numbers

_______________ **5. Pipefitting:** A pipefitter threads 28 pieces of pipe per hour. How many hours does it take to thread 1148 pieces of pipe?

_______________ **6. Construction:** A surveyor staked the boundaries in feet of six building lots. All of the lots had the same amount of frontage along Oak Street. What is the frontage length of each lot?

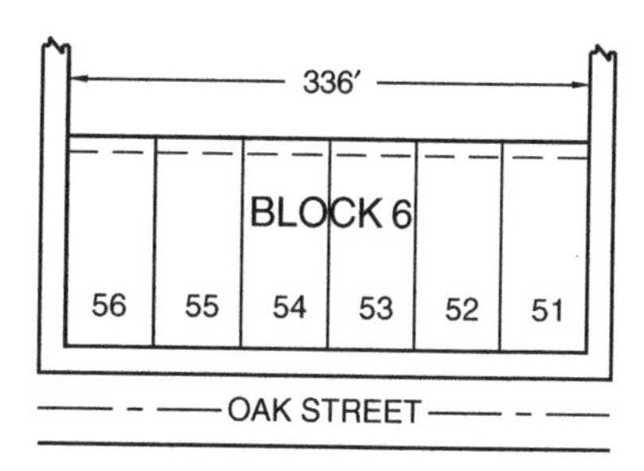

_______________ **7. Electrical:** If #28 copper conductor costs $11.00 per spool, how many spools can be purchased for $165.00?

_______________ **8. Construction:** Three pieces of 4′ × 8′ plywood are ripped into 6″ widths. How many 6″ pieces can be ripped from the three pieces of plywood? (*Disregard material removed by saw kerf.*)

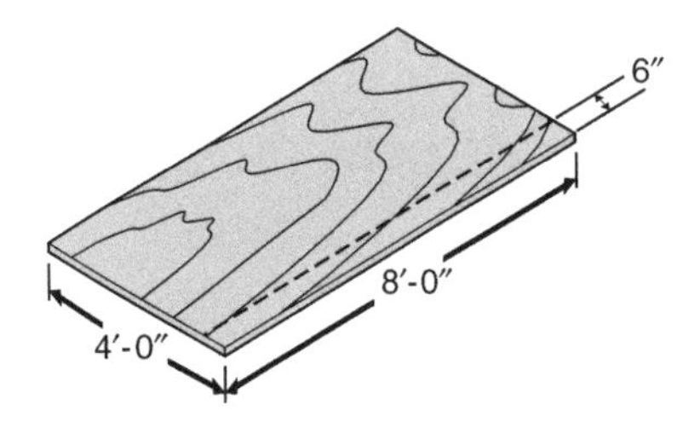

_______________ **9. Construction:** Floor tile for BR 2 costs $330.75. What is the cost per square foot?

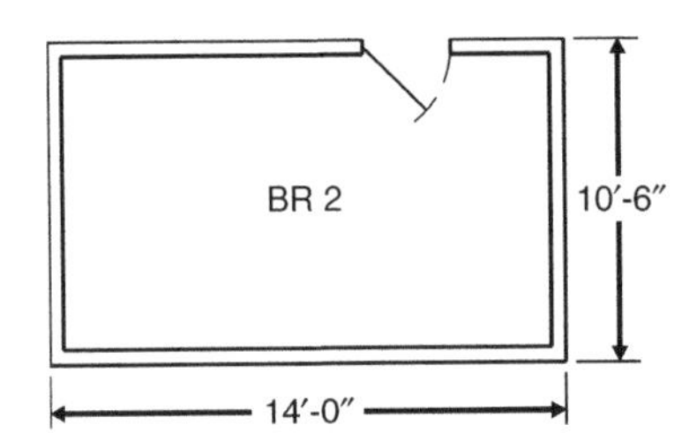

_______________ **10. Maintenance:** A front-end loader moves ¾ cu yd of soil per scoop. How many scoops are required to move 21 cu yd of soil?

For an interactive review of the concepts covered in Chapter 1, refer to the corresponding Quick Quiz® included on the Digital Resources.

QUICK QUIZ®
Working with Whole Numbers

Working with Whole Numbers 1

Review

Name ______________________ Date __________

Math Problems

Write out the following numbers as words.

1. 501 ______________________

2. 13,020 ______________________

Write out the following as Arabic or Roman numerals.

__________ **3.** One hundred forty-four

__________ **4.** Twenty-seven

__________ **5.** XIV

__________ **6.** VI

Calculate each sum, difference, product, or quotient.

__________ **7.** $12 + 25 + 33 + 47 + 54 + 66 + 78 + 99$

__________ **8.** $23{,}504 - 5946$

__________ **9.** 4375×86

__________ **10.** $25{,}600 \div 80$

1 Review (continued)

Practical Applications

_______________ 11. **Alternative Energy:** A utility produces electricity from multiple energy sources. Solar systems contribute 2% of the utility's generation, wind turbines contribute 9%, and hydroelectric generators contribute 7%. What is the total percentage?

_______________ 12. **Electrical:** An electrician cut off two 93′ pieces and four 12′ pieces of conductor from a 1240′ reel. How many feet are left on the reel?

_______________ 13. **Construction:** A truck delivers 8 pallets of common brick to a job site. Each pallet contains 484 bricks. How many total bricks were delivered?

_______________ 14. **Welding:** A welder fabricates 35 steel water tanks for $33,250.00. What is the welding cost per tank?

_______________ 15. **Construction:** For a particular job, a contractor has a budget of $375.00 for paint, $100.00 for paint rollers, and $120.00 for plastic drop cloths. If paint costs $15.00 per gallon, rollers cost $2.00 each, and plastic drop cloths cost $4.00 each, how many of each can the contractor purchase?

Working with Whole Numbers 1
Test

Name ______________________________ Date ______________

Math Problems

Write out the following numbers as words.

1. 7808 ______________________________

2. 734,507,000 ______________________________

Write the following as Arabic numerals.

______________ **3.** One hundred twenty-three

______________ **4.** Three thousand thirty-nine

______________ **5.** DCC

______________ **6.** XXXV

Calculate each sum, difference, product, or quotient.

______________ **7.** 3 + 12 + 84 + 14

______________ **8.** 328 – 172

______________ **9.** 1002 × 302

1 Test (continued)

______ **10.** 3798 ÷ 6

Practical Applications

______ **11. Alternative Energy:** A particular model of solar photovoltaic module delivers 185 W of electrical power. If an array consists of 32 modules, what is the total power output of the array?

______ **12. Boiler Operation:** The total heat in steam that is available for use is the sum of the sensible heat (heat required to heat water to boiling) and the latent heat (heat required to boil off the water as steam). If the sensible heat is 150 Btu and the latent heat is 970 Btu, what is the total heat in the steam?

______ **13. Welding:** A welding job requires a total of 165 in. of weld. If each welding electrode can be used to make 5 in. of weld, how many electrodes are required for the entire job?

______ **14. Boiler Operation:** The heat energy in 1 gal. of fuel oil is 141,000 British thermal units (Btu). What is the amount of heat energy in 8 gal. of fuel oil?

1 Test (continued)

_______________ **15. Construction:** A gallon of wall paint will cover 275 sq ft. Two coats of paint are needed on 3025 sq ft of wall area. How many total gallons are needed to paint the walls?

_______________ **16. HVAC:** An HVAC project requires 60,000 2″ × ¼″ self-tapping screws. The screws are in cartons of 10,000 that cost $209.00 each. What will the cost of the screws be for the project?

_______________ **17. Pipefitting:** A steam line is to be run to the cookers of 18 twin-screw oil expellers. Each cooker will have a ¾″ steam trap ($407.00), strainer ($118.00), 42′ of ¾″ pipe ($88.00), and assorted fittings and hangers ($61.00). What is the total material cost for installing steam to the 18 oil expellers?

_______________ **18. Welding:** Each I-beam connection in a structure requires 12 welds. If the structure contains 167 connections, what is the total number of required welds?

_______________ **19. Alternative Energy:** An ethanol refinery produces 17 gal. of ethanol for every bushel of corn. How many gallons of ethanol are produced from 56 bushels?

_______________ **20. Pipefitting:** A pipefitter must install a 4″ stainless steel pipe between two tanks in separate buildings through a tunnel system. The horizontal run of the tunnel is 756′. How many lengths of 21′ stainless steel pipe are needed?

1 Test (continued)

_______________ **21. Electrical:** A bundle of conduit contains fifty 10-ft sections, and 200 ft are used for a job. How many feet of conduit are left?

_______________ **22. Boiler Operation:** A blowdown tank contains 1200 gal. of water. The tank loses 6 gal. as flash steam evaporates, while 320 gal. are pumped into it. How many gallons are in the tank?

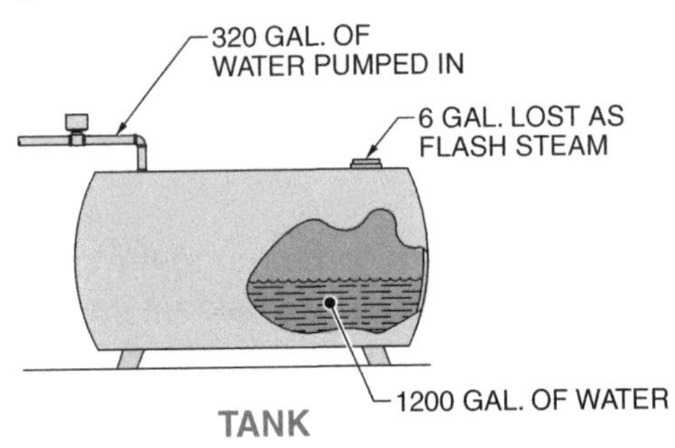

_______________ **23. Manufacturing:** Four bins contain 55 lb, 55 lb, 328 lb, and 173 lb of copper. What is the total weight?

_______________ **24. Mechanics:** In five months, a mechanic works on five projects. The mechanic spends 103 hr on project A, 60 hr on project B, 172 hr on project C, 148 hr on project D, and 89 hr on project E. How many total hours are spent on the five projects?

_______________ **25. HVAC:** If a 27″ length is cut from a piece of sheet metal that is 80″ long, how much of the sheet metal remains?

2 Working with Simple Fractions

A fraction is a part of a whole number. Understanding fractions and how fractions can be changed to equivalent fractions is essential in any trade. A carpenter may cut a board into several $3\frac{1}{2}$-foot pieces, or a plumber may need to use a $\frac{3}{8}$-inch diameter pipe.

OBJECTIVES

1. Explain how a fraction represents a part of a whole number.
2. Differentiate between proper and improper fractions.
3. Explain how an improper fraction and a mixed number are related.
4. Differentiate between prime factors and common factors.
5. Find prime factors and common factors of fractions.
6. Reduce fractions to lowest terms.
7. Reduce fractions to lowest terms with a given denominator.
8. Change fractions to higher terms with a given denominator.
9. Find the lowest common denominator for a group of fractions.
10. Change fractions to their lowest common denominator.

KEY TERMS

- fraction
- numerator
- denominator
- proper fraction
- improper fraction
- mixed number
- factor
- prime number
- prime factor
- common factor
- greatest common factor
- common denominator
- lowest common denominator (LCD)

Digital Resources
ATPeResources.com/QuickLinks
Access Code: 764460

SECTION 2-1 UNDERSTANDING FRACTIONS

A *fraction* is a part of a whole number. A whole number can be divided into any number of fractional parts. For example, 1 whole unit can be divided into 2 smaller parts, or 4 smaller parts, and so on.

Fractions are written by placing numbers above and below a fraction bar. For example, one-third is written $\frac{1}{3}$ and one-tenth is written $\frac{1}{10}$. Fraction bars can be horizontal or inclined. **See Figure 2-1.**

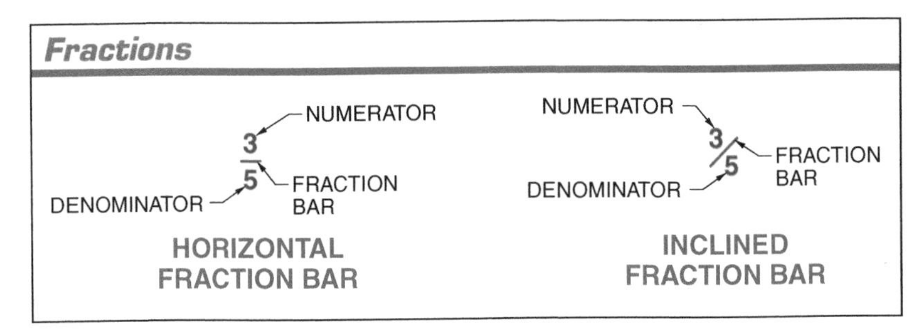

Figure 2-1. A fraction is represented by a numerator and a denominator separated by a fraction bar.

The *numerator* is the number above the fraction bar and shows the number of parts taken from the denominator. The *denominator* is the number below the fraction bar and shows how many parts the whole number has been divided into. For example, if a pipe is cut into four pieces, and three pieces are used, then three fourths ($\frac{3}{4}$) of the pipe are used. **See Figure 2-2.** The numerator 3 represents the number of parts taken from the denominator. The denominator 4 represents the number of parts into which the pipe was divided.

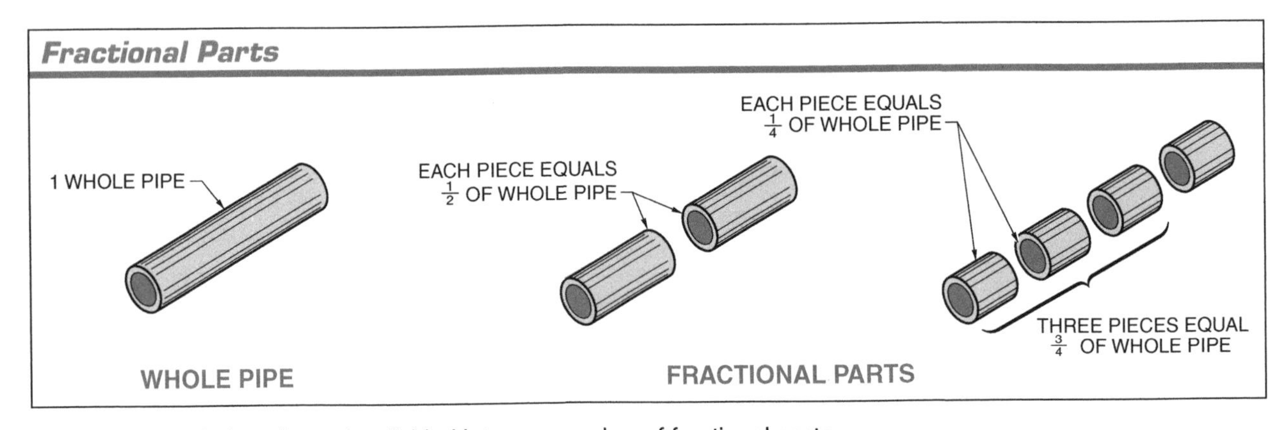

Figure 2-2. A whole unit can be divided into any number of fractional parts.

A ruler or tape measure is divided into inches and fractions of an inch. The large marks represent inches and are designated by whole numbers. The smaller marks represent fractions of an inch. The largest mark between inch marks represents $\frac{1}{2}''$. The largest mark between the inch and the $\frac{1}{2}''$ marks represents $\frac{1}{4}''$ and so on. **See Figure 2-3.** Some rulers and tape measures may be divided as low as $\frac{1}{32}''$ or $\frac{1}{64}''$.

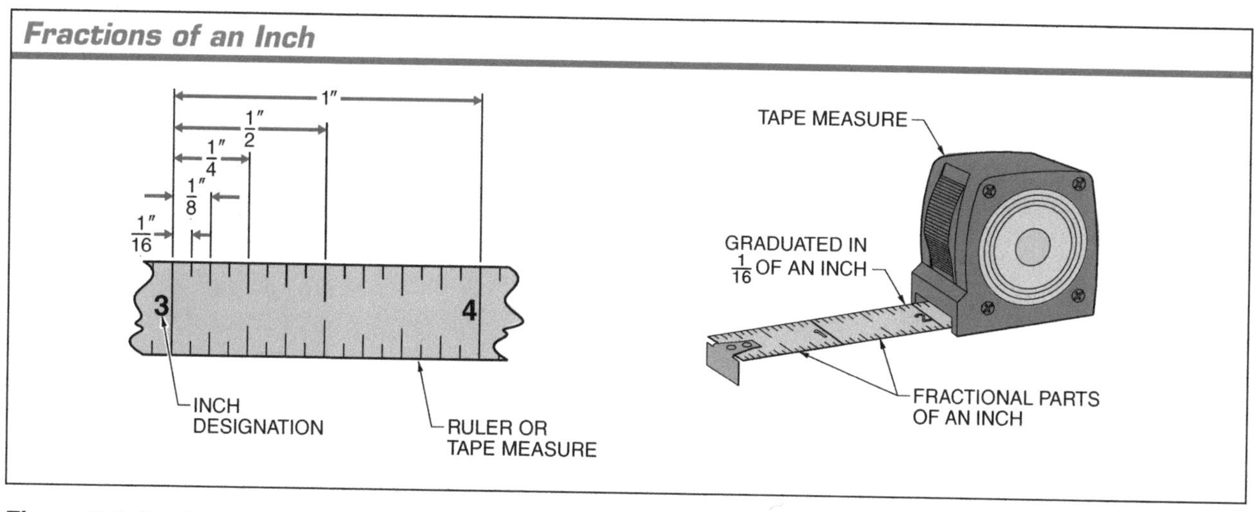

Figure 2-3. A ruler or tape measure is divided into inches and fractions of an inch.

A circle can be divided into fractional parts such as halves. **See Figure 2-4.** If ½ is divided into two equal parts, that part is now in fourths. If a fourth in turn is divided into two equal parts, that part is in eighths and so on.

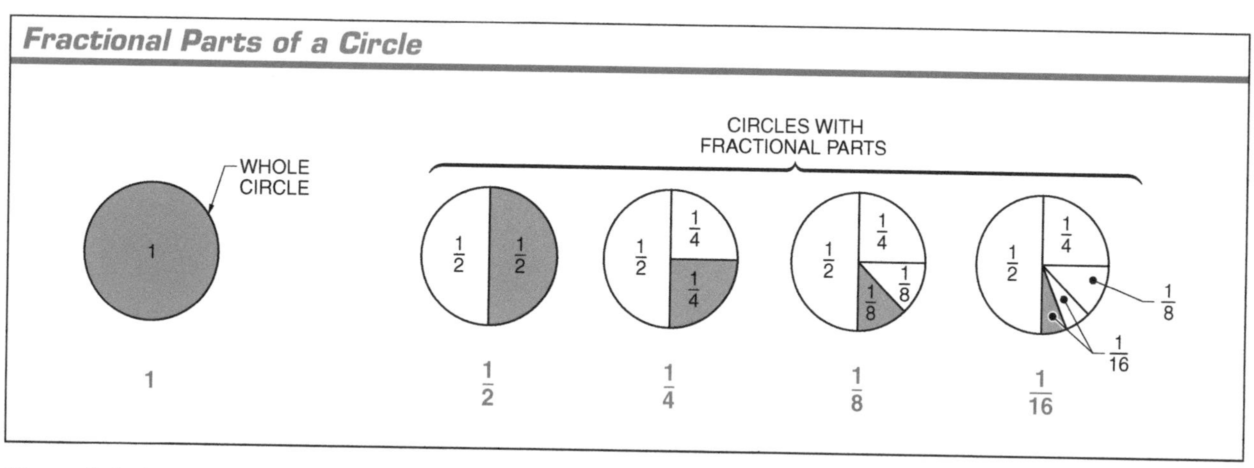

Figure 2-4. As a whole unit, a circle can be divided into fractional parts.

Proper Fractions, Improper Fractions, and Mixed Numbers

Fractions may be proper fractions, improper fractions, or mixed numbers. **See Figure 2-5.** A *proper fraction* is a fraction with the numerator smaller than the denominator, for example, ⅛, ¼, ½, 9⁄16, and 25⁄32. Proper fractions are always less than 1. An *improper fraction* is a fraction with the numerator larger than its denominator, for example, 7⁄5, 8⁄3, 4⁄3, 10⁄4, and 21⁄7. A *mixed number* is a combination of a whole number and a proper fraction, for example, 1⅛, 2¼, 4½, 6 9⁄16, and 11 63⁄64. Improper fractions and mixed numbers are always more than 1.

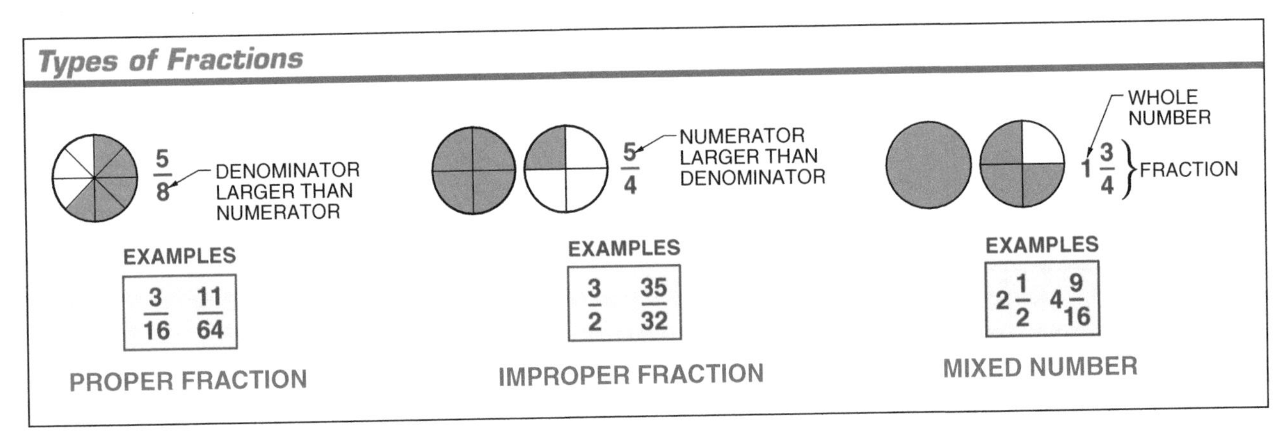

Figure 2-5. Fractions may be proper fractions, improper fractions, or mixed numbers.

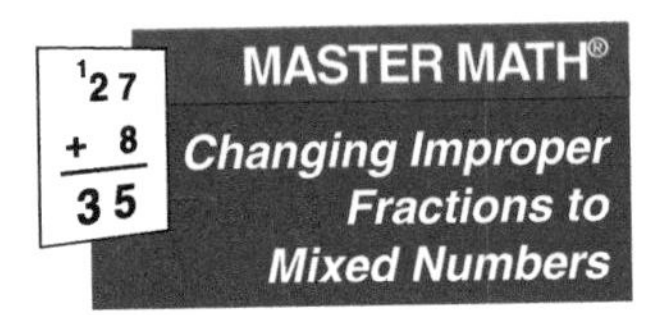

Improper fractions and mixed numbers are closely related. If one is known, the other can be found. **See Figure 2-6.** To change an improper fraction to a mixed number, divide the numerator by the denominator. For example, to change 4/3 to a mixed number, divide 4 by 3. The quotient is 1 with a remainder of 1, which is placed over the divisor.

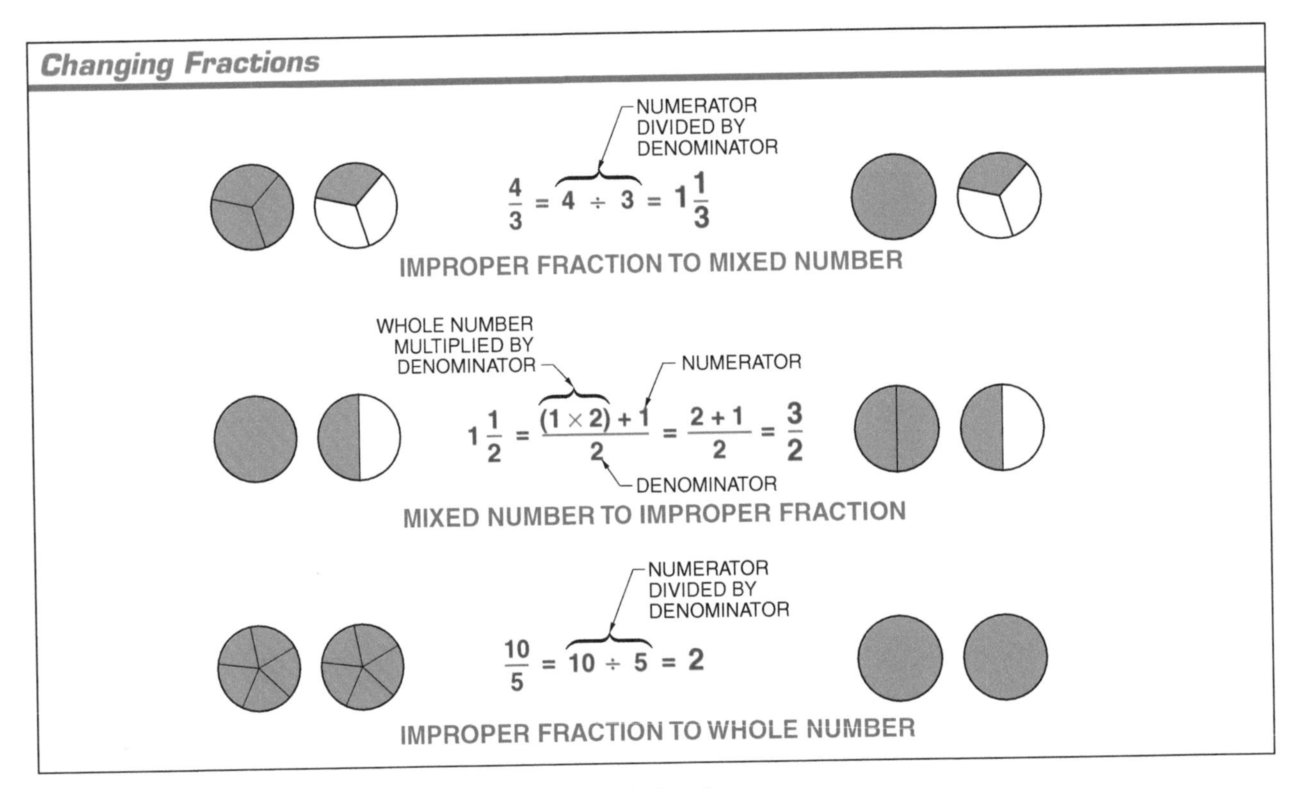

Figure 2-6. Fractions can be changed without altering their value.

To check the answer after changing an improper fraction to a mixed number, change the answer to an improper fraction. This should result in the original improper fraction.

To change a mixed number to an improper fraction, multiply the whole number by the denominator and then add the numerator. Place this new numerator above the denominator. For example, to change $3\frac{1}{2}$ to an improper fraction, multiply 1 by 2 ($1 \times 2 = 2$) and add 1. Place the resulting 3 above the divisor 2.

To check the answer after changing a mixed number to an improper fraction, change the answer to a mixed number. This should result in the original mixed number.

Changing improper fractions can also result in whole numbers. For example, to change $\frac{10}{5}$ to a mixed number, divide the numerator 10 by the denominator 5. The quotient is 2 with no remainder.

Examples — Proper Fractions, Improper Fractions, and Mixed Numbers

1. Change the improper fraction $\frac{5}{3}$ to a mixed number.

ANS: $1\frac{2}{3}$

❶ Divide 5 by 3 ($5 \div 3 = 1$ with a remainder of 2).

❷ Place the remainder (2) over the divisor (3).

$$\frac{5}{3} = \underbrace{3\overline{)5}}_{❶} = \underbrace{1\frac{2}{3}}_{❷} \qquad (5 - 3 = 2)$$

QUICK REFERENCE

- *Divide the numerator by the denominator.*
- *Place the remainder (if any) over the divisor (denominator).*

2. Change the mixed number $6\frac{8}{9}$ to an improper fraction.

ANS: $\frac{62}{9}$

❶ Multiply 6 by 9 ($6 \times 9 = 54$).

❷ Add 8 to 54 ($54 + 8 = 62$).

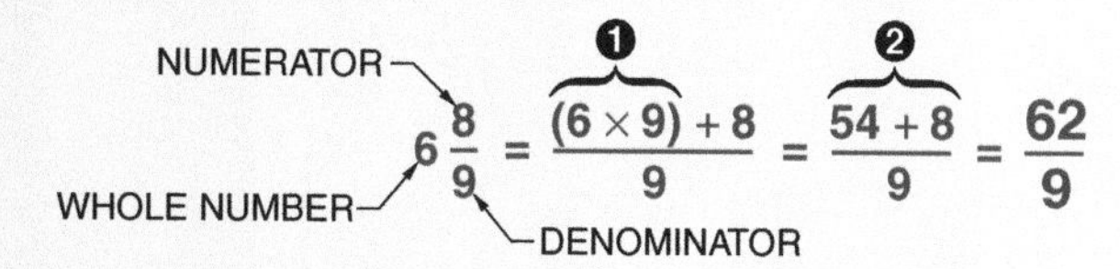

$$6\frac{8}{9} = \frac{(6 \times 9) + 8}{9} = \frac{54 + 8}{9} = \frac{62}{9}$$

QUICK REFERENCE

- *Multiply the whole number by the denominator.*
- *Add the product to the numerator and place over the denominator.*

3. Change the improper fraction $\frac{6}{3}$ to a whole number.

ANS: 2

❶ Divide 6 by 3.

$$\frac{6}{3} = 3\overline{)6} = 2$$

MATH EXERCISES—Proper Fractions, Improper Fractions, and Mixed Numbers

Change improper fractions to mixed numbers and mixed numbers to improper fractions.

_______________ **1.** $8\frac{3}{7}$

_______________ **2.** $12\frac{5}{9}$

_______________ **3.** $20\frac{1}{11}$

_______________ **4.** $\frac{8}{5}$

_______________ **5.** $83\frac{1}{3}$

_______________ **6.** $\frac{125}{2}$

_______________ **7.** $46\frac{7}{10}$

_______________ **8.** $\frac{94}{7}$

_______________ **9.** $\frac{763}{18}$

_______________ **10.** $23\frac{75}{76}$

SECTION 2-2 UNDERSTANDING FACTORS

In order to work with fractions, understanding the role of factors is helpful. A *factor* is a number used as a multiplier. The number 48 is a product of 12 and 4 (12 × 4 = 48), so 12 and 4 are factors of 48. Numbers may have more than one set of factors. The number 48 is also a product of 8 and 6, 2 and 24, and 1 and 48. Thus, 1, 2, 4, 6, 8, 12, 24, and 48 are all factors of 48. **See Figure 2-7.**

Some numbers have more than two factors in a set. For example, 3, 5, and 7 together are factors of 105 ($3 \times 5 \times 7 = 105$). Very large numbers can have many factors and many different sets of factors. However, only one set of factors can be used at one time to achieve the desired product.

To find the factor of a number, an upside-down division method can be used. **See Figure 2-8.** To factor a number, place the number inside the division bracket and select a divisor that will divide evenly into that number. Continue selecting numbers that will divide evenly into the resulting quotients until there is nothing left to divide. The divisors and the final quotient represent the factors of the original number.

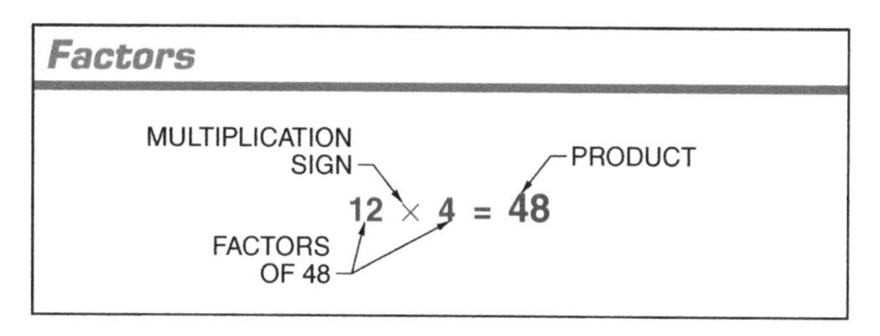

Figure 2-7. Factors are multipliers of a given product.

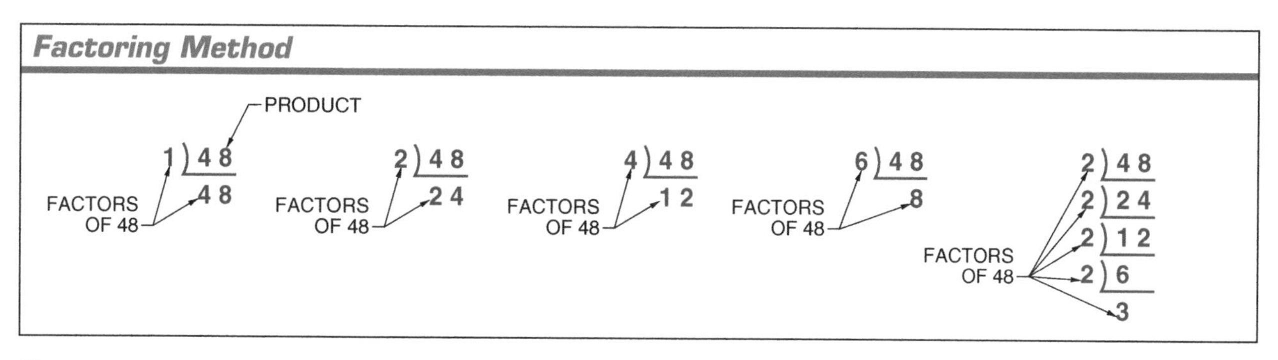

Figure 2-8. A product can have numerous factors as well as several sets of factors.

Finding Prime Factors

A *prime number* is a whole number that can only be divided an exact number of times by itself and the number 1. For example, the numbers 2, 3, 5, 7, 11, 13, 19, and 23 are prime numbers. A *prime factor* is a factor that is a prime number. **See Figure 2-9.**

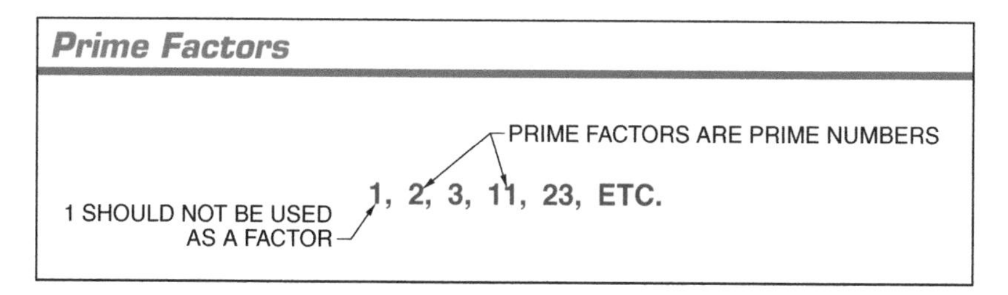

Figure 2-9. With the exception of the number 1, prime numbers can be used as factors.

To find the prime factors of any number, divide the given number by the smallest possible prime number. (The number 1 should not be used as a factor as the only result will be the number itself.)

A prime number is found through repeated effort, such as dividing by 2 first, then 3, 5, 7, 11, etc. Continue dividing by the smallest prime number until a quotient that is also a prime number is obtained. When a prime number is obtained, the factoring process is complete.

Example—Finding Prime Factors

1. Find the prime factors of 16.

ANS: **2, 2, 2, 2**

❶ Divide 16 by the smallest prime factor greater than 1 that will divide into it evenly (2).

❷ Divide 8 by the smallest prime factor greater than 1 that will divide into it evenly (2).

❸ Divide 4 by the smallest prime factor greater than 1 that will divide into it evenly (2).

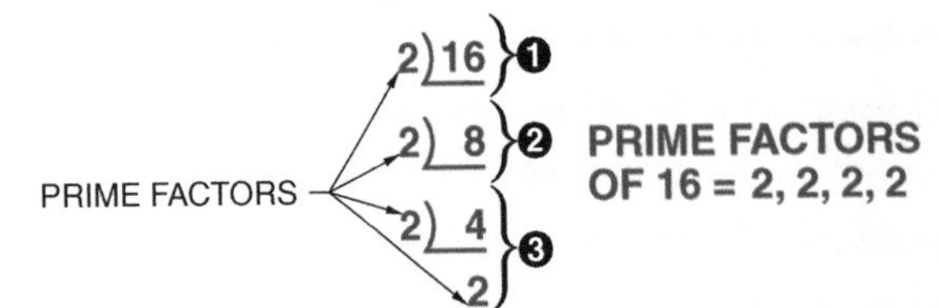

QUICK REFERENCE

- *Divide the number by the smallest prime factor greater than 1 that will divide evenly.*
- *Divide the resulting number by the smallest prime factor. Repeat until a quotient that is a prime number is obtained.*

MATH EXERCISES—Finding Prime Factors

Find the prime factors of the following.

________________ **1.** 78

________________ **2.** 2772

________________ **3.** 311

________________ **4.** 4235

________________ **5.** 2310

Finding Common Factors

A *common factor* is a factor that is common to two or more numbers. To find the common factors for two numbers, first determine the factors of each of the numbers. Then compare the list of factors for each of the given numbers. Any factors that are the same are considered common factors.

The common factors of 18 and 36 are 1, 2, 3, 6, 9, and 18. **See Figure 2-10.** The *greatest common factor* is the highest number in a group of factors. The number 18 is the highest of the common factors of 18 and 36.

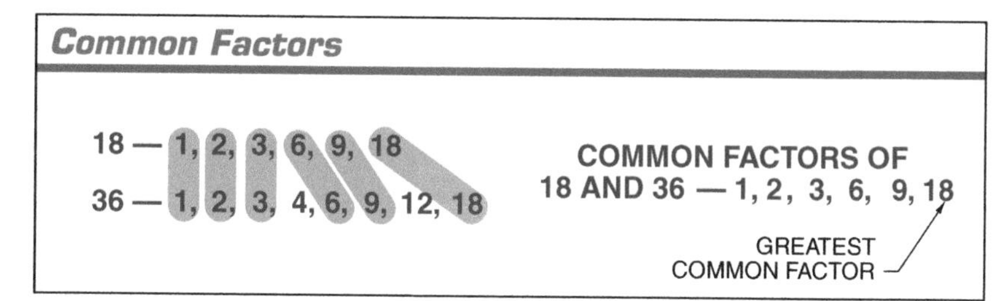

Figure 2-10. Common factors are factors that are common to two or more numbers.

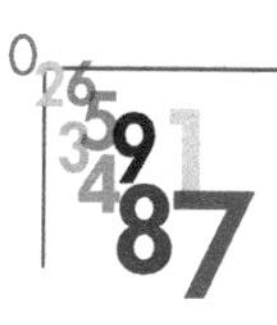

When asked to find the common factors between a set of given numbers, it is only necessary to find the factors that are less than or equal to the lowest of the given numbers.

Example—Finding Common Factors

1. Find the common factors of 28 and 84.

ANS: **1, 2, 4, 7, 14,** and **28**

❶ Find all the factors of 28.

❷ Find the factors of 84 that are lower than 28. It is only necessary to find the factors less than or equal to 28.

❸ Find the common factors.

❶ $28 \div 1 = 28$ $28 \div 2 = 14$ $28 \div 4 = 7$

❷ $84 \div 2 = 24$ $84 \div 3 = 28$ $84 \div 4 = 21$ $84 \div 6 = 14$ $84 \div 7 = 12$

❸ 28 – 1, 2, 4, 7, 14, 28
84 – 1, 2, 3, 4, 6, 7, 12, 14, 21, 24, 28

COMMON FACTORS = 1, 2, 4, 7, 14, 28

QUICK REFERENCE

- *Find all the factors of the given numbers.*
- *Find the common factors.*

MATH EXERCISES—Finding Common Factors

Find the common factors.

________________ **1.** 8, 38, and 44

________________ **2.** 16, 40, and 96

Find each greatest common factor.

________________ **3.** 16, 32, and 48

________________ **4.** 12, 64, and 90

________________ **5.** 36, 72, and 144

SECTION 2-3 FINDING EQUIVALENT FRACTIONS

Fractions can be changed to equivalent fractions without losing their value. For example, ½ is the equivalent of 2⁄4, and 4⁄8. The fraction ½ is in lowest terms. The terms of a fraction refer to both the numerator and the denominator. Fractions can be changed to lower or higher terms, and must be changed to common terms so that they can be added and subtracted.

Reducing Fractions to Lowest Terms

Fractions are easier to work with when reduced to the lowest terms possible. To reduce a fraction to its lowest terms, divide the numerator and denominator by their greatest common factor. **See Figure 2-11.** When only a 1 can be divided into both the numerator and denominator, the fraction is reduced to its lowest possible terms.

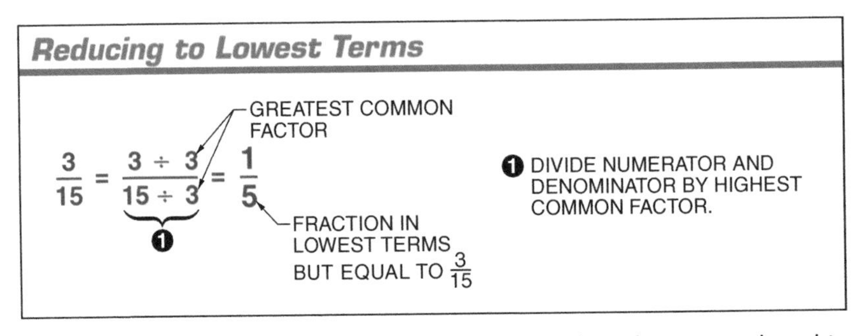

Figure 2-11. Fractions are easiest to work with when they are reduced to lowest terms.

Examples—Reducing Fractions to Lowest Terms

1. Reduce 24⁄36 to lowest terms.

***ANS:* ⅔**

❶ Find the greatest common factor of 24 and 36. The greatest common factor is 12.

❷ Divide the numerator and the denominator by 12.

$\frac{24}{36}$

24 — $1\overline{)24}$ = 24, $2\overline{)24}$ = 12, $3\overline{)24}$ = 8, $4\overline{)24}$ = 6

36 — $1\overline{)36}$ = 36, $2\overline{)36}$ = 18, $3\overline{)36}$ = 12, $4\overline{)36}$ = 9, $6\overline{)36}$ = 6

❶ { 24 — 1, 2, 3, 4, 6, 8, 12, 24
36 — 1, 2, 3, 4, 6, 9, 12, 18, 36 }

COMMON FACTORS = 1, 2, 3, 4, 6, 12

GREATEST COMMON FACTOR

❷ $\frac{24}{36} = \frac{24 \div 12}{36 \div 12} = \frac{2}{3}$

QUICK REFERENCE

- *Find the greatest common factor of the numerator and denominator.*
- *Divide the numerator and denominator by the greatest common factor.*

2. Reduce 5⁄15 to lowest terms.

***ANS:* ⅓**

❶ Find the greatest common factor of 5 and 15. The greatest common factor is 5.

❷ Divide the numerator and denominator by 5.

$\frac{5}{15}$

5 — $1\overline{)5}$ = 5

15 — $1\overline{)15}$ = 15, $3\overline{)15}$ = 5, $5\overline{)15}$ = 3

❶ { 5 — 1, 5
15 — 1, 3, 5 }

COMMON FACTORS = 1, 5

GREATEST COMMON FACTOR

❷ $\frac{5}{15} = \frac{5 \div 5}{15 \div 5} = \frac{2}{3}$

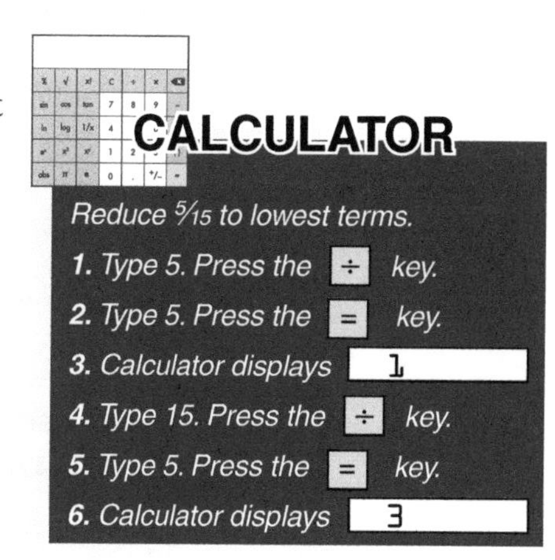

MATH EXERCISES—Reducing Fractions to Lowest Terms

Reduce to lowest terms.

________________ **1.** 5⁄10

________________ **2.** 8⁄36

______________ **3.** $\frac{12}{16}$

______________ **4.** $\frac{2}{3}$

PRACTICAL APPLICATIONS—Reducing Fractions to Lowest Terms

______________ **5. Construction:** Two pieces of plywood are glued together. One piece has a thickness of $\frac{1}{4}''$, and the other has a thickness of $\frac{5}{8}''$. What is the total thickness in lowest terms?

______________ **6. Construction:** A piece of corner bead is cut at $94\frac{10}{16}''$. Reduce $\frac{10}{16}$ to lowest terms.

______________ **7. Construction:** The total length of a steel lintel required above an $8'$-$4\frac{1}{2}''$ opening in a masonry wall is $9'$-$4\frac{14}{16}''$. Reduce $\frac{14}{16}$ to lowest terms.

______________ **8. Alternative Energy:** A rooftop solar energy system consists of 96 solar panels divided into groups. If one group contains 24 panels, what is its fraction of the total system in lowest terms?

Reducing Fractions to Lowest Terms with a Given Denominator

A fraction may need to be changed to an equivalent fraction where the denominator is already given. To reduce a fraction to lowest terms with a given denominator, divide the original denominator by the given denominator. Then divide the numerator and the original denominator by the quotient. For example, to reduce the fraction $\frac{8}{24}$ to thirds, divide the denominator by 3 and divide both terms of the fraction by the quotient. **See Figure 2-12.**

Reducing Fractions to Lowest Terms with Given Denominator

ORIGINAL DENOMINATOR $\frac{8}{24} = \frac{x}{3}$ GIVEN DENOMINATOR

❶ $24 \div 3 = 8$ QUOTIENT

❷ $\frac{8}{24} = \frac{8 \div 8}{24 \div 8} = \frac{1}{3}$

❶ DIVIDE ORIGINAL DENOMINATOR BY GIVEN DENOMINATOR.

❷ DIVIDE ORIGINAL TERMS BY QUOTIENT.

Figure 2-12. A fraction may need to be expressed in lowest terms with a given denominator.

Example—Reducing Fractions to Lowest Terms with a Given Denominator

1. Reduce $\frac{8}{32}$ to fourths.

ANS: $\frac{1}{4}$

❶ Divide the original denominator 32 by the given denominator 4 ($32 \div 4 = 8$).

❷ Divide the quotient 8 into both the numerator and denominator of the original fraction.

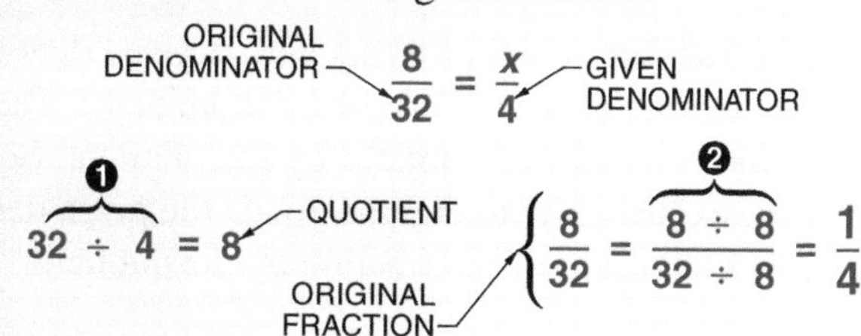

QUICK REFERENCE

- *Divide the original denominator by the given denominator.*
- *Divide the quotient into both terms of the original fraction.*

MATH EXERCISES—Reducing Fractions to Lowest Terms with a Given Denominator

__________ **1.** Reduce $\frac{15}{20}$ to fourths.

__________ **2.** Reduce $\frac{36}{40}$ to tenths.

__________ **3.** Reduce $\frac{24}{36}$ to sixths.

__________ **4.** Reduce $\frac{50}{75}$ to thirds.

__________ **5.** Reduce $\frac{12}{36}$ to ninths.

PRACTICAL APPLICATIONS—Reducing Fractions to Lowest Terms with a Given Denominator

________________ **6. Boiler Operation:** A length of boiler steam pipe is observed to expand by $^{12}/_{16}$″ when it is heated. What is this expansion in fourths of an inch?

________________ **7. Manufacturing:** A machinist needs to drill 4 holes $^{6}/_{16}$″ in diameter. What is the hole size in eighths of an inch?

Changing Fractions to Higher Terms with a Given Denominator

To change a fraction to higher terms with a given denominator, divide the given denominator by the original denominator. Then multiply the numerator and original denominator by the quotient. **See Figure 2-13.** For example, to change ¾ to eighths divide the given denominator by 4 and multiply both terms of the fraction by the quotient.

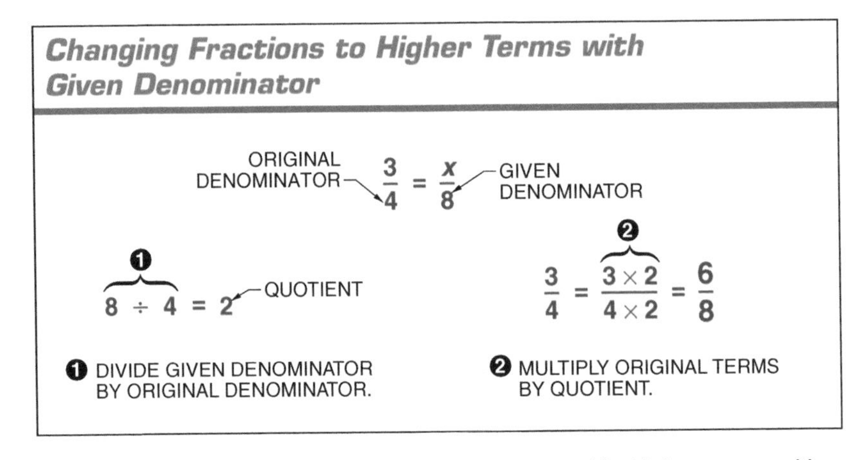

Figure 2-13. A fraction may need to be expressed in higher terms with a given denominator.

Addition and subtraction of fractions cannot be performed unless the fractions involved in the operation have common denominators. A fraction may need to be changed to higher or lower terms because the fraction it needs to be added to or subtracted from is already in those terms.

Example—Changing Fractions to Higher Terms with a Given Denominator

1. Change $\frac{3}{9}$ to twenty-sevenths.

ANS: $\frac{9}{27}$

❶ Divide given denominator $\frac{3}{9}$ by original numerator 9 ($27 \div 9 = 3$).

❷ Multiply both the numerator and denominator of the original fraction by the quotient 3.

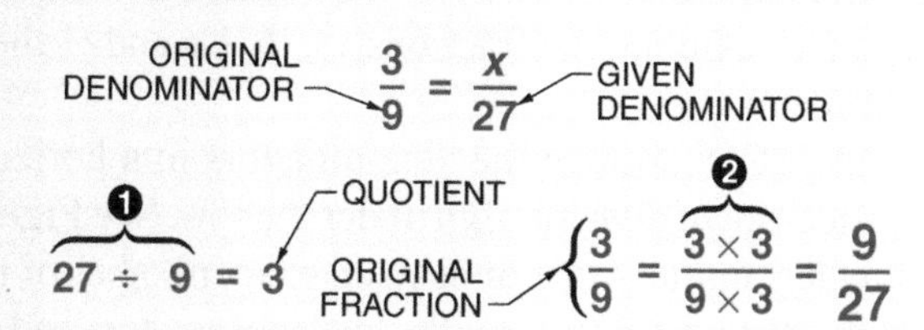

QUICK REFERENCE

- *Divide the given denominator by the original denominator.*
- *Multiply both terms of the fraction by the quotient.*

MATH EXERCISES—Changing Fractions to Higher Terms with a Given Denominator

____________ **1.** Change $\frac{5}{6}$ to twelfths.

____________ **2.** Change $\frac{15}{16}$ to sixty-fourths.

____________ **3.** Change $\frac{4}{5}$ to fortieths.

____________ **4.** Change $\frac{9}{25}$ to hundredths.

PRACTICAL APPLICATIONS—Changing Fractions to Higher Terms with a Given Denominator

____________ **5. Pipefitting:** A $\frac{3}{8}$″ pipe is being used to supply cooling water to a tank seal unit. What is the size of the pipe in sixty-fourths of an inch?

____________ **6. Mechanics:** An oil seal has an inside diameter of $\frac{7}{8}$″. What is the inside diameter of the seal in thirty-seconds of an inch?

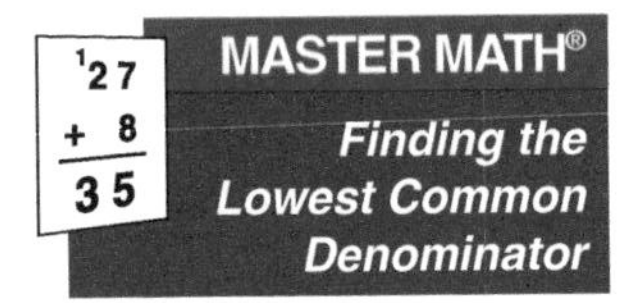

Finding the Lowest Common Denominator

A *common denominator* is a denominator that is the same among a group of fractions. For example, ⅛, 2⁄8, ⅜, 4⁄8, ⅝, 6⁄8, and ⅞ all have 8 as the common denominator. Fractions must have common denominators before addition and subtraction can be performed.

The *lowest common denominator (LCD)* is the smallest number into which the denominators of two or more fractions can divide an exact number of times. Among the fractions ½, ⅓, ⅙, and 1⁄9 there is no common denominator.

To find the LCD, arrange the denominators in a horizontal row, and draw an upside-down division sign under the row. **See Figure 2-14.** Then find the smallest number that divides an exact number of times into two or more of the denominators. Divide the denominators and place the quotients under the division sign. Bring down the denominators that are not divisible. Repeat these steps until division is no longer possible. Multiply all the divisors and the final quotients to arrive at the LCD.

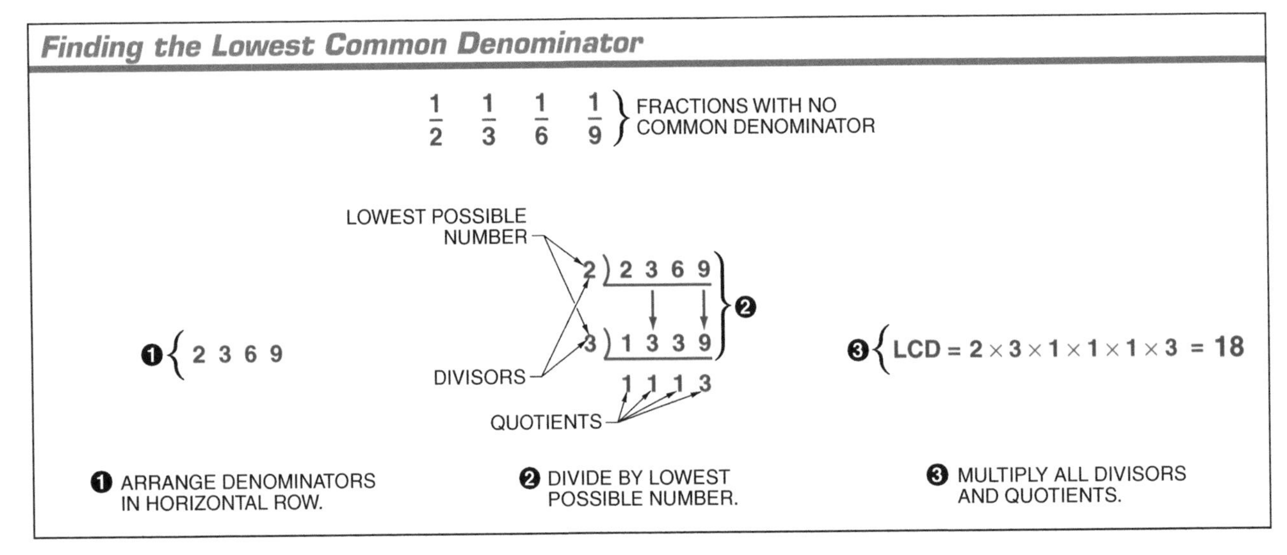

Figure 2-14. The lowest common denominator (LCD) is the smallest number into which the denominators of a group of two or more fractions divide an exact number of times.

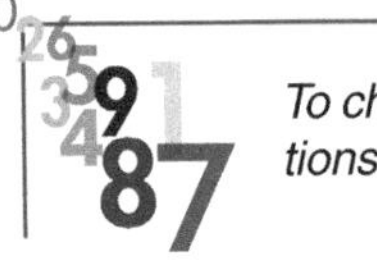

To check the LCD, divide it by all denominators of the original fractions. Each should divide into the LCD an even number of times.

Examples — Finding the Lowest Common Denominator

1. Find the lowest common denominator for ⅙, ⅜, 2⁄9, 5⁄12, 5⁄18, and 7⁄24.
   ANS: 72

   ❶ Arrange the denominators in a horizontal row and draw an upside-down division sign under them. Divide all numbers divisible by 2 (6, 8, 12, 18, and 24). Bring down the 9 because it cannot be divided by 2 an exact number of times.

❷ Divide all numbers divisible by 2 (4, 6, and 12). Bring down the 3 and the 9s.
❸ Divide all numbers divisible by 2 (2 and 6). Bring down the 3s and the 9s.
❹ Divide all numbers divisible by 3. Bring down the 1.
❺ Divide all numbers divisible by 3. Bring down the 1s.
❻ Multiply the divisors and the final quotients.

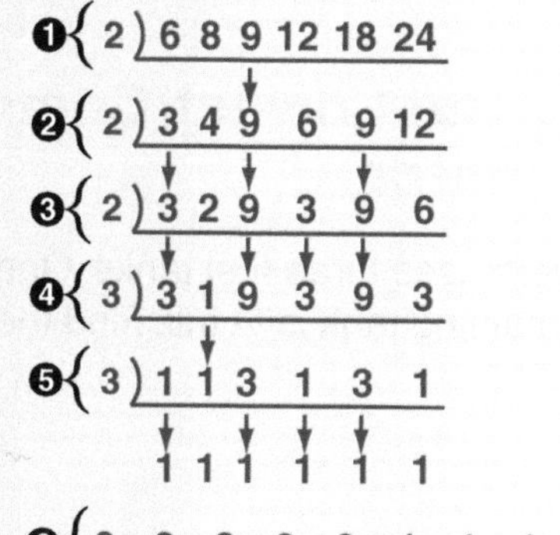

❻ $2 \times 2 \times 2 \times 3 \times 3 \times 1 \times 1 \times 1 \times 1 \times 1 = 72$

QUICK REFERENCE

- *Find the smallest number that divides an exact number of times into as many of the denominators as possible.*
- *Record the quotients and bring down the denominators that are not divisible.*
- *Repeat until division is no longer possible.*
- *Multiply the divisors and quotients to find the LCD.*

2. Find the LCD of ⅕, ⅝, and ⅗.
ANS: **40**

❶ Arrange the denominators in a horizontal row and draw an upside-down division sign under them. Divide all numbers divisible by 2. Bring down the 5s because 5 cannot be divided by 2 an exact number of times.
❷ Divide all numbers divisible by 2. Bring down the 5s.
❸ Divide all numbers divisible by 2. Bring down the 5s.
❹ Divide all numbers divisible by 5. Bring down the 1.
❺ Multiply the divisors and the final quotients.

❶ 2) 5 8 5
❷ 2) 5 4 5
❸ 2) 5 2 5
❹ 5) 5 1 5
1 1 1

❺ $2 \times 2 \times 2 \times 5 \times 1 \times 1 \times 1 = 40$

MATH EXERCISES—Finding the Lowest Common Denominator

Find each LCD.

________________ **1.** ⅙, ⅛, 1⁄12

________________ **2.** 1⁄12, 1⁄16, 1⁄24

______________ **3.** $\frac{3}{10}$, $\frac{4}{15}$, $\frac{7}{20}$

______________ **4.** $\frac{3}{7}$, $\frac{4}{11}$, $\frac{6}{7}$, $\frac{2}{13}$

PRACTICAL APPLICATIONS—Finding the Lowest Common Denominator

______________ **5. Boiler Operation:** A bank of boilers uses three fuel tanks. One tank is $\frac{7}{8}$ full, another tank is $\frac{1}{2}$ full, and the remaining tank is $\frac{1}{4}$ full. Find the LCD of these fractions.

______________ **6. Construction:** Drywall is available in $\frac{3}{8}''$, $\frac{1}{2}''$, and $\frac{5}{8}''$ thicknesses. What is the LCD of these fractions?

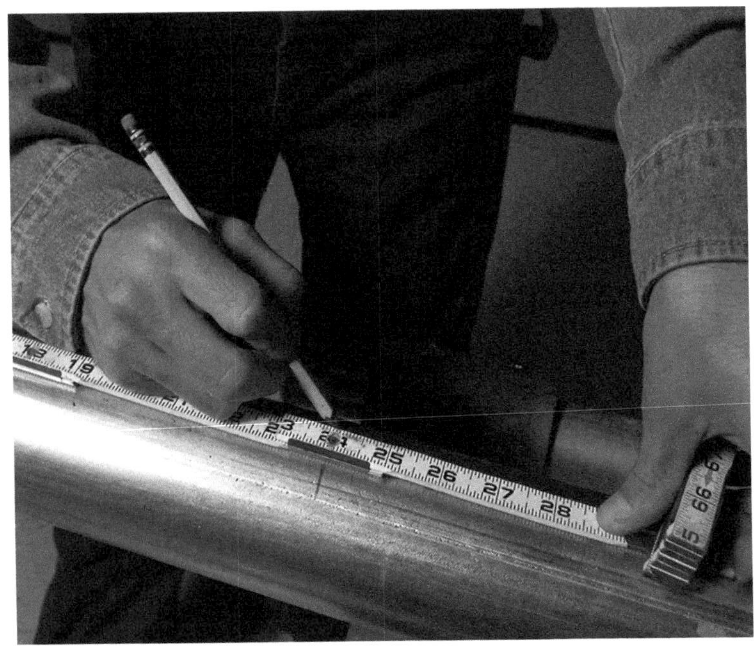

Folding rules and tape measures are generally marked in inches and fractions of an inch.

Changing Fractions to Their Lowest Common Denominator

The purpose of finding the LCD of a group of fractions is to give all of the fractions a common denominator. This is exactly the same as changing to higher terms with a given denominator. The given denominator is the LCD.

To give fractions a common denominator, divide the LCD by the denominator of the first fraction. Then multiply both terms of the fraction by the quotient. Do the same for the next fraction, and so on. For example, to change the fractions $\frac{3}{5}$ and $\frac{5}{6}$ to fractions with a common denominatior, divide the LCD (30) by each denominator ($30 \div 5 = 6$ and $30 \div 6 = 5$). Then multiply both terms of each fraction by the applicable quotient. **See Figure 2-15.**

Changing Fractions to Their LCD

THE LCD OF $\frac{3}{5}$ AND $\frac{5}{6}$ IS 30.

$30 \div 5 = 6 \qquad \frac{3 \times 6}{5 \times 6} = \frac{18}{30}$

$30 \div 6 = 5 \qquad \frac{5 \times 5}{6 \times 5} = \frac{25}{30}$

❶ DIVIDE THE LCD BY EACH DENOMINATOR.
❷ MULTIPLY BOTH TERMS OF BOTH FRACTIONS BY QUOTIENTS.

Figure 2-15. Changing fractions to their lowest common denominator (LCD) allows for addition and subtraction of common terms.

MATH EXERCISES—Changing Fractions to Their Lowest Common Denominator

Find each LCD.

______ **1.** 1/7, 3/8, 1/4, 2/3

______ **2.** 2/7, 1/3, 3/4, 1/2

______ **3.** 1/3, 2/9, 5/12, 3/8

______ **4.** 1/15, 3/10, 4/25, 1/30

______ **5.** 1/3, 2/7, 4/11

PRACTICAL APPLICATIONS—Changing Fractions to Their Lowest Common Denominator

______ **6. Plumbing:** A pipe is to be cut into sections 6½″, 11⅝″, and 24 11/16″ long. Change each fraction to its LCD so that the total length of the pipe can be determined.

____

7. Construction: A carpenter constructs a bookcase with different shelf heights. Each shelf is ¾″ thick. Change each fraction to its LCD so that the height of the bookcase can be determined.

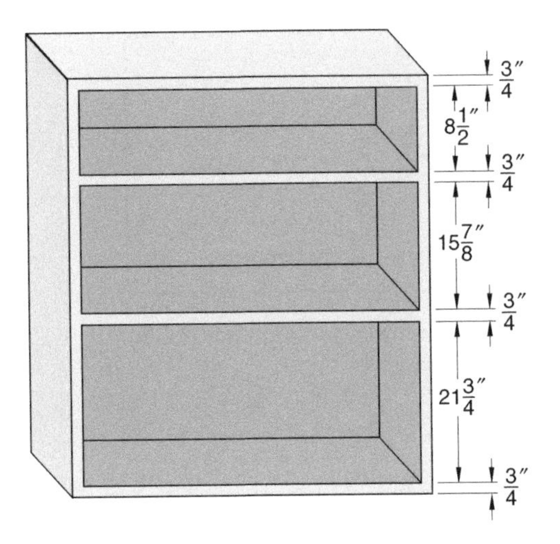

For an interactive review of the concepts covered in Chapter 2, refer to the corresponding Quick Quiz® included on the Digital Resources.

Working with Simple Fractions 2
Review

Name ______________________________ Date ______________

Math Problems

______________ **1.** Change $\frac{7}{5}$ to a mixed number.

______________ **2.** Change $2\frac{1}{8}$ to an improper fraction.

______________ **3.** Change $\frac{21}{7}$ to a whole number.

______________ **4.** Find the prime factors of 56.

______________ **5.** Find the common factors of 21 and 42.

Reduce each fraction to lowest possible terms or change to the given denominator.

______________ **6.** $\frac{24}{36}$

______________ **7.** $\frac{40}{64}$

______________ **8.** $\frac{20}{50}$ to fifths

______________ **9.** $\frac{100}{120}$

______________ **10.** $\frac{11}{12}$ to thirty-sixths

2 Review (continued)

_______________ **11.** $\frac{63}{81}$

Find each LCD and change the fractions.

_______________ **12.** $\frac{3}{4}$, $\frac{3}{8}$, $\frac{1}{2}$, $\frac{5}{16}$, and $\frac{1}{4}$

_______________ **13.** $\frac{7}{8}$, $\frac{3}{4}$, $\frac{5}{16}$, and $\frac{9}{32}$

Practical Applications

_______________ **14. Construction:** The thickness of a board measures $\frac{14}{16}''$. Reduce this to lowest terms.

_______________ **15. Pipefitting:** The diameter of one pipe is $\frac{3}{4}''$. The diameter of another pipe is $\frac{11}{16}''$. Find the LCD of the thicknesses of the pipes.

Working with Simple Fractions 2

Test

Name ______________________________ Date ______________

Math Problems

Change each improper fraction to a mixed number.

__________ 1. $\frac{102}{8}$

__________ 2. $\frac{46}{3}$

__________ 3. $\frac{60}{16}$

__________ 4. $\frac{75}{32}$

Change each mixed number to an improper fraction.

__________ 5. $35\frac{3}{5}$

__________ 6. $7\frac{1}{10}$

__________ 7. $6\frac{3}{4}$

__________ 8. $5\frac{7}{16}$

2 Test (continued)

Change each improper fraction to a whole number.

________________ **9.** $^{66}/_{11}$

________________ **10.** $^{90}/_{10}$

________________ **11.** $^{159}/_{53}$

Find the indicated factors.

________________ **12.** Find the prime factors of 76.

________________ **13.** Find the prime factors of 332.

________________ **14.** Find the common factors of 14 and 105.

________________ **15.** Find the common factors of 40, 168, and 296.

Reduce each fraction to lowest possible terms or change to the given denominator.

________________ **16.** $^{10}/_{75}$

________________ **17.** $^{20}/_{32}$

2 Test (continued)

________________ **18.** $43/123$

________________ **19.** $3/4$ to fortieths

________________ **20.** $14/28$

________________ **21.** $72/81$ to ninths

________________ **22.** $5/13$ to twenty-sixths

Find each LCD and change the fractions.

________________ **23.** $1/3$ and $6/7$

________________ **24.** $5/8$, $13/64$, and $1/16$

________________ **25.** $7/8$, $3/4$, $2/3$, and $5/16$

________________ **26.** $3/7$, $19/32$, $22/70$, and $7/8$

________________ **27.** $1/2$ and $2/9$

2 Test (continued)

Practical Applications

______________ **28. Pipefitting:** A pipefitter needs to cut a 45″ length of pipe in half. Therefore, each section is $^{45}/_{2}$″ long after cutting. Convert this improper fraction to a mixed number to determine the length of each pipe.

______________ **29. Welding:** A certain weld is specified to be $5\frac{3}{8}$″ long. What is the weld length as an improper fraction?

______________ **30. Agriculture:** A storage shed that is $18\frac{1}{2}$′ long is set back $40\frac{3}{4}$′ from the front property line. The distance from the back of the shed to the back property line is 150′. What is the length of the lot?

Working with Complex Fractions

Fractions cannot be added or subtracted if they have different denominators. It is necessary to find the lowest common denominator of two fractions or a group of fractions before they can be added or subtracted. Multiplication and division, however, do not require fractions to have a common denominator.

OBJECTIVES

1. Add fractions with common denominators.
2. Add fractions with uncommon denominators.
3. Add mixed numbers.
4. Subtract fractions with common denominators.
5. Subtract fractions with uncommon denominators.
6. Subtract mixed numbers.
7. Multiply fractions, whole numbers, and mixed numbers.
8. Divide fractions, whole numbers, mixed numbers, and complex fractions.

KEY TERMS

- cancellation
- complex fraction

Digital Resources
ATPeResources.com/QuickLinks
Access Code: 764460

SECTION 3-1 ADDING FRACTIONS

Fractions can be added together to make one sum. However, before fractions can be added, they must have the same denominator.

For example, there are two ½ gal. of paint. These halves added together yield a whole gallon because ½ + ½ = 1. However, if ¼ gal. of paint and ⅓ gal. of paint are combined, the fractions cannot be added. The lowest common denominator (LCD) must be found first. In any group of fractions being added, the denominators must indicate that all objects have been divided into the same number of parts.

$^{1}27 + 8 = 35$

MASTER MATH®

Adding Factions with Common Denominators

Adding Fractions with Common Denominators

To add any group of fractions with a common denominator, add the numerators and place the sum over the denominator. **See Figure 3-1.**

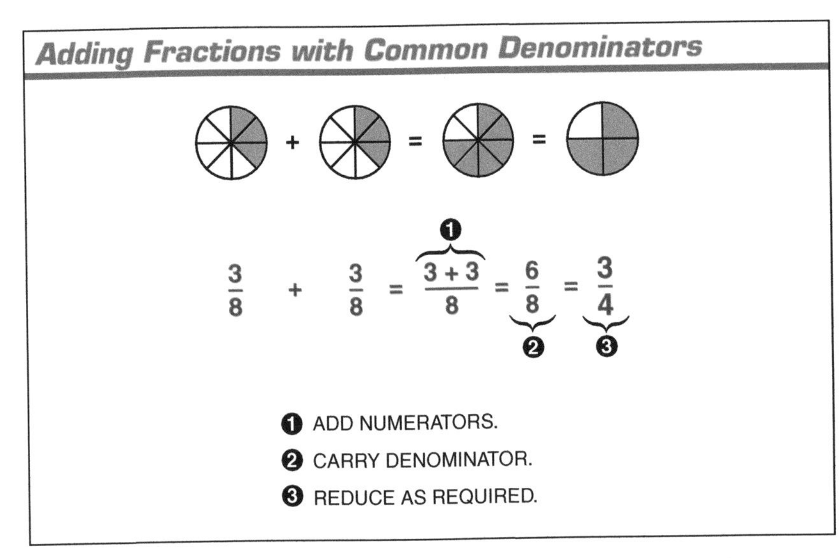

Figure 3-1. When adding fractions with common denominators, the numerators are added while the denominator remains the same.

Example—Adding Fractions with Common Denominators

1. Add ¼ + ²⁄₄ + ¾.

ANS: 1½

❶ Add the numerators (1 + 2 + 3 = 6).

❷ Place the 6 over the 4 (⁶⁄₄).

❸ Reduce as required.

$$\frac{1}{4} + \frac{2}{4} + \frac{3}{4} = \frac{1+2+3}{4} = \frac{6}{4} = 1\frac{2}{4} = 1\frac{1}{2}$$

QUICK REFERENCE

- *Add the numerators.*
- *Carry the denominator.*
- *Reduce as required.*

MATH EXERCISES—Adding Fractions with Common Denominators

_______________ **1.** $1/5 + 2/5 + 3/5$

_______________ **2.** $8/9 + 4/9 + 7/9 + 2/9$

_______________ **3.** $1/24 + 5/24 + 19/24 + 13/24$

PRACTICAL APPLICATIONS—Adding Fractions with Common Denominators

_______________ **4. Maintenance:** A custodian has four 1 gal. cans of paint. The cans are $1/8$, $3/8$, $5/8$, and $7/8$ full. How many gallons of paint does the custodian have?

_______________ **5. Manufacturing:** Find the center-to-center distance between the drilled holes of Bracket A.

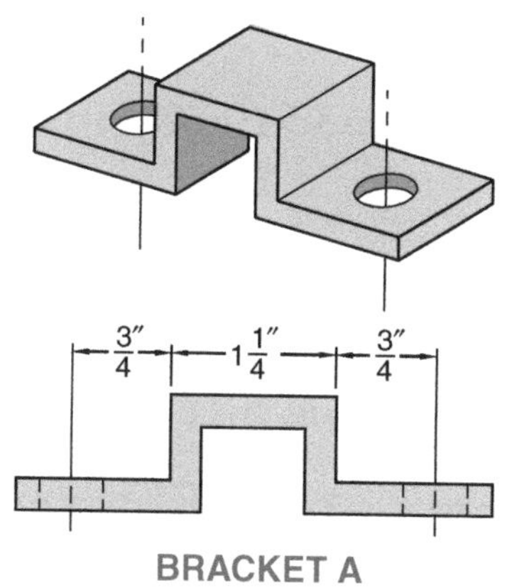

BRACKET A

_______________ **6. Pipefitting:** A pipefitter needs a pipe nipple that is $6\frac{3}{8}''$ long, not including the threads. The threads add $7/8''$ to each end of the pipe nipple. What is its total length?

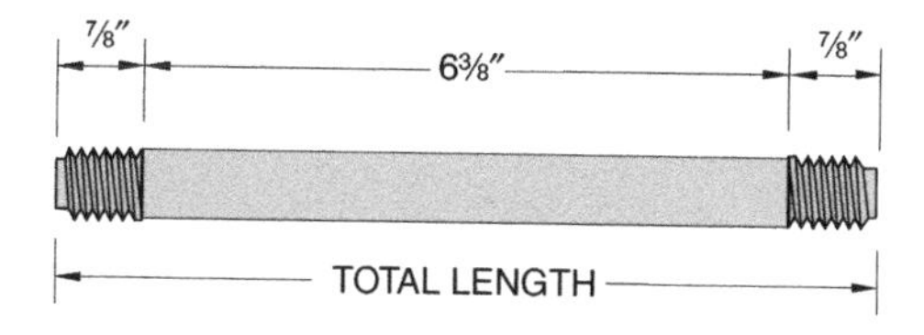

$^{1}27$
$+\ 8$
35

MASTER MATH®
Adding Factions with Uncommon Denominators

Adding Fractions with Uncommon Denominators

Fractions with uncommon denominators must be changed to fractions with a common denominator before they can be added. The lowest common denominator (LCD) is found in order to add fractions with uncommon denominators. **See Figure 3-2.** To add ½ and ¾, the fractions are changed to the LCD of 4. The changed fractions are ½ = 2/4 and ¾ = ¾. If the sum is an improper fraction, it should be reduced to a mixed number or whole number if possible.

Adding Fractions with Uncommon Denominators

$$\frac{1}{2} + \frac{3}{4} = \overbrace{\frac{2}{4} + \frac{3}{4}}^{❶} = \overbrace{\frac{2+3}{4}}^{❷} = \overbrace{\frac{5}{4}}^{❸} = \overbrace{1\frac{1}{4}}^{❹}$$

❶ CHANGE TO LCD.
❷ ADD NUMERATORS.
❸ CARRY DENOMINATOR.
❹ REDUCE AS REQUIRED.

Figure 3-2. The LCD must be found before adding fractions with uncommon denominators.

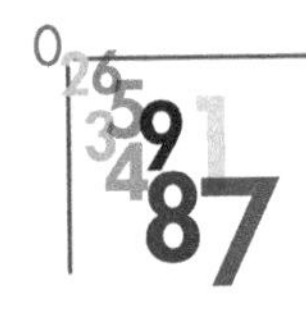

When adding fractions with uncommon denominators, the LCD of the denominators is actually the smallest whole number that is evenly divisible by each of the denominators.

Example—Adding Fractions with Uncommon Denominators

1. Add 7/8 + 2/5 + 3/10.
ANS: **1 23/40**

❶ Find the LCD (40) for the fractions.
❷ Divide the LCD by the denominator of each fraction (8, 5, and 10).
❸ Multiply both the numerators and denominators of the original fractions by the products (5, 8, and 4) of this division.
❹ Add the numerators of the changed fractions (35 + 16 + 12 = 63).
❺ Carry the 40 over (63/40).
❻ Reduce as required.

❶
5) 8 5 10
2) 8 1 2
2) 4 1 1
2) 2 1 1
1 1 1
$5 \times 2 \times 2 \times 2 \times 1 \times 1 \times 1 = 40$

❷
$40 \div 8 = 5$
$40 \div 5 = 8$
$40 \div 10 = 4$

❸
$\frac{7 \times 5}{8 \times 5} = \frac{35}{40}$
$\frac{2 \times 8}{5 \times 8} = \frac{16}{40}$
$\frac{3 \times 4}{10 \times 4} = \frac{12}{40}$

$$\overbrace{\frac{35+16+12}{40}}^{❹} = \overbrace{\frac{63}{40}}^{❺} = \overbrace{1\frac{23}{40}}^{❻}$$

QUICK REFERENCE

- *Change the fractions to their LCD.*
- *Add the numerators. Carry the denominator.*
- *Reduce as required.*

MATH EXERCISES—Adding Fractions with Uncommon Denominators

_______________ **1.** $\frac{1}{5} + \frac{1}{4} + \frac{1}{6}$

_______________ **2.** $\frac{1}{2} + \frac{5}{8} + \frac{9}{10}$

_______________ **3.** $\frac{3}{2} + \frac{7}{9} + \frac{2}{3}$

_______________ **4.** $\frac{31}{24} + \frac{11}{12} + \frac{3}{8}$

_______________ **5.** $\frac{25}{24} + \frac{5}{6} + \frac{7}{8}$

PRACTICAL APPLICATIONS—Adding Fractions with Uncommon Denominators

_______________ **6. Construction:** A small concrete wall requires ¾ cu yd of concrete for the footing and ½ cu yd of concrete for the wall. How many cubic yards of concrete are required for the footing and the wall combined?

_______________ **7. Culinary Arts:** A recipe calls for ½ cup of milk and ⅝ cup of water. What is the total amount of liquid required?

Adding Mixed Numbers

To add mixed numbers with common denominators, add the whole numbers, add the numerators, carry the denominator, and reduce as required. **See Figure 3-3.** To add mixed numbers with unlike denominators, add the whole numbers, find the LCD of the denominators, and add the fractions.

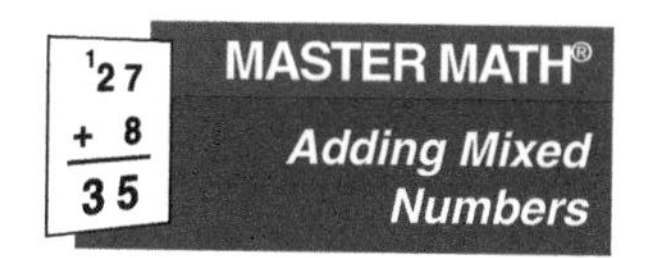

Adding Mixed Numbers

$1\frac{1}{3} + 1\frac{2}{3} = \overbrace{1+1}^{❶} = \overbrace{2\frac{1+2}{3}}^{❷} = \overbrace{2\frac{3}{3}}^{❸} = \overbrace{3}^{❹}$

❶ ADD WHOLE NUMBERS.
❷ ADD NUMERATORS.
❸ CARRY DENOMINATOR.
❹ REDUCE AS REQUIRED.

Figure 3-3. When adding mixed numbers, the whole numbers and fractions are added separately.

Example—Adding Mixed Numbers

1. Add $3\frac{1}{3} + 2\frac{1}{2}$

   ANS: $5\frac{5}{6}$

   ❶ Add the whole numbers (3 + 2 = 5).
   ❷ Find the LCD (6) for the fractions.
   ❸ Change the fractions by dividing the LCD by the original denominators, and multiply the fractions by the product.
   ❹ Add the numerators (2 + 3 = 5).
   ❺ Carry the 5 over ($\frac{5}{6}$).
   ❻ Add the result to the whole number.

❶ $3 + 2 = 5$

❷ $2\overline{)3\ 2}$; $3\ 1$; $2 \times 3 \times 1 = 6$

❸ $6 \div 3 = 2$; $6 \div 2 = 3$; $\frac{1 \times 2}{3 \times 2} = \frac{2}{6}$ $\frac{1 \times 3}{2 \times 3} = \frac{3}{6}$

❹ ❺ $\frac{2+3}{6} = \frac{5}{6}$ ❻ $5 + \frac{5}{6} = 5\frac{5}{6}$

QUICK REFERENCE

- *Add the whole numbers.*
- *Change the fractions to their LCD.*
- *Add the numerators. Carry the denominator.*
- *Add the whole number and the fraction.*
- *Reduce as required.*

MATH EXERCISES—Adding Mixed Numbers

_______________ **1.** $2\frac{5}{8} + 3\frac{7}{12} + 5\frac{11}{24}$

_______________ **2.** $1\frac{3}{20} + 2\frac{7}{12} + 3\frac{5}{15}$

_______________ **3.** $6\frac{2}{3} + 2\frac{2}{7} + 4\frac{4}{15}$

______________ **4.** $3\frac{5}{12} + \frac{7}{12} + 2\frac{9}{24}$

______________ **5.** $5\frac{1}{2} + \frac{3}{4} + 6\frac{2}{3}$

PRACTICAL APPLICATIONS—Adding Mixed Numbers

______________ **6. Plumbing:** Find the total length of the five pieces of pipe on the plumbing riser.

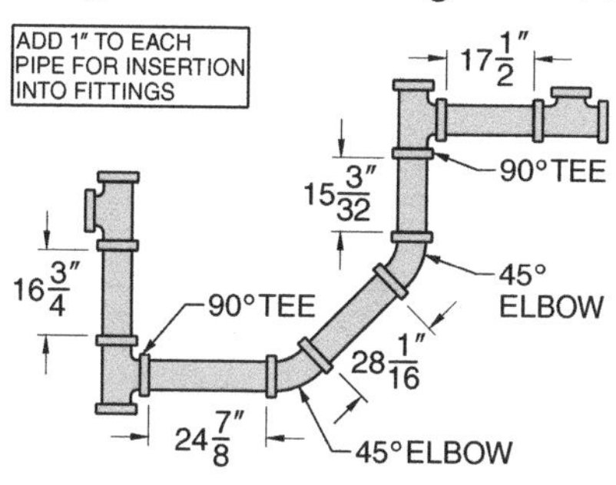

PLUMBING RISER

______________ **7. Construction:** Find the total length of four table legs from the table leg shown.

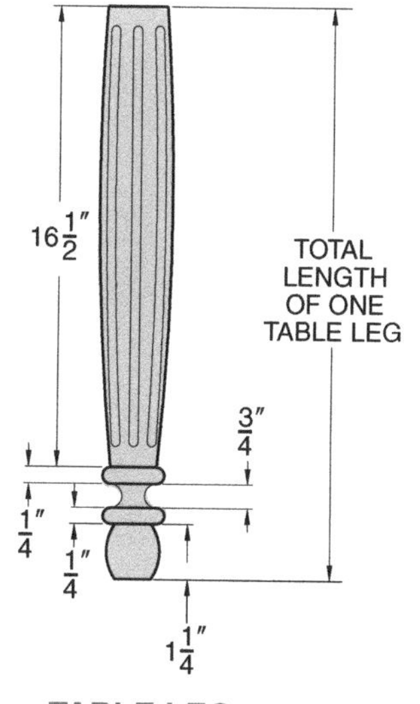

TABLE LEG

______________ **8. HVAC:** An HVAC technician at a sheet metal fabrication shop must determine the total length of three different workpieces. The workpieces have lengths of 17¾″, 21½″, and 8⅜″. What is the total length of the workpieces?

SECTION 3-2 SUBTRACTING FRACTIONS

As when adding fractions, fractions must have common denominators before they can be subtracted. When mixed numbers are subtracted, the whole numbers and the fractions are subtracted separately, and the two differences are added together.

MASTER MATH® *Subtracting Fractions with Common Denominators*

Subtracting Fractions with Common Denominators

To subtract fractions that have common denominators, subtract the numerators and place the difference over the denominator. **See Figure 3-4.**

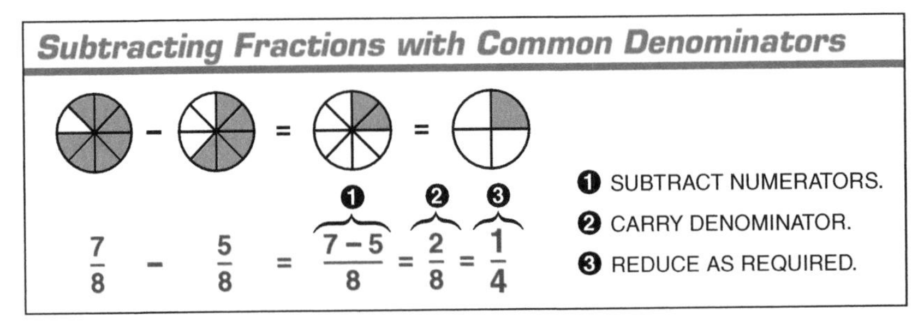

Figure 3-4. When subtracting fractions with common denominators, the numerators are subtracted while the denominator remains the same.

Example—Subtracting Fractions with Common Denominators

1. Subtract $\frac{2}{8}$ from $\frac{7}{8}$.

ANS: $\frac{5}{8}$

❶ Subtract 2 from 7 (7 – 2 = 5).

❷ Carry the 8 over.

$$\frac{7}{8} - \frac{2}{8} = \frac{7-2}{8} = \frac{5}{8}$$

QUICK REFERENCE

- *Subtract the numerators.*
- *Carry the denominator.*
- *Reduce as required.*

MATH EXERCISES—Subtracting Fractions with Common Denominators

__________ **1.** $\frac{8}{9} - \frac{5}{9}$

__________ **2.** $\frac{14}{12} - \frac{11}{12}$

__________ **3.** $\frac{20}{27} - \frac{6}{27}$

__________ **4.** $\frac{49}{70} - \frac{36}{70}$

PRACTICAL APPLICATIONS—Subtracting Fractions with Common Denominators

________________ **5. Manufacturing:** Find the missing dimension (X) on the shaft gauge.

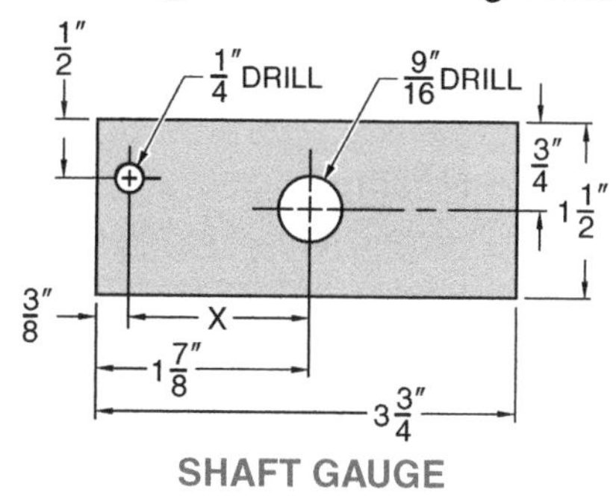

________________ **6. Plumbing:** A piece of copper pipe is $4\frac{13}{16}''$ long. Only $4\frac{3}{16}''$ are needed for the job. How much must be cut from the pipe?

Subtracting Fractions with Uncommon Denominators

To subtract fractions having uncommon denominators, find the LCD and change the fractions. Subtract one numerator from the other and carry the denominator. **See Figure 3-5.**

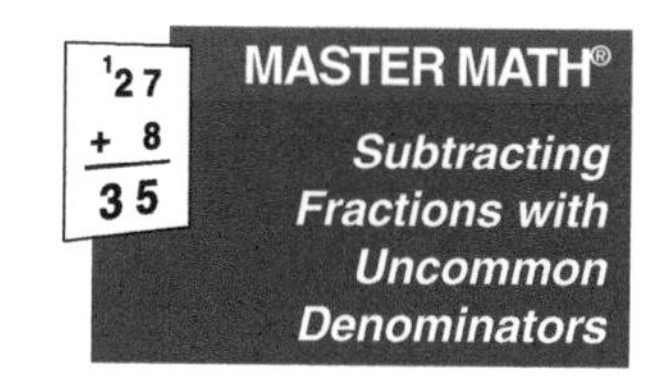

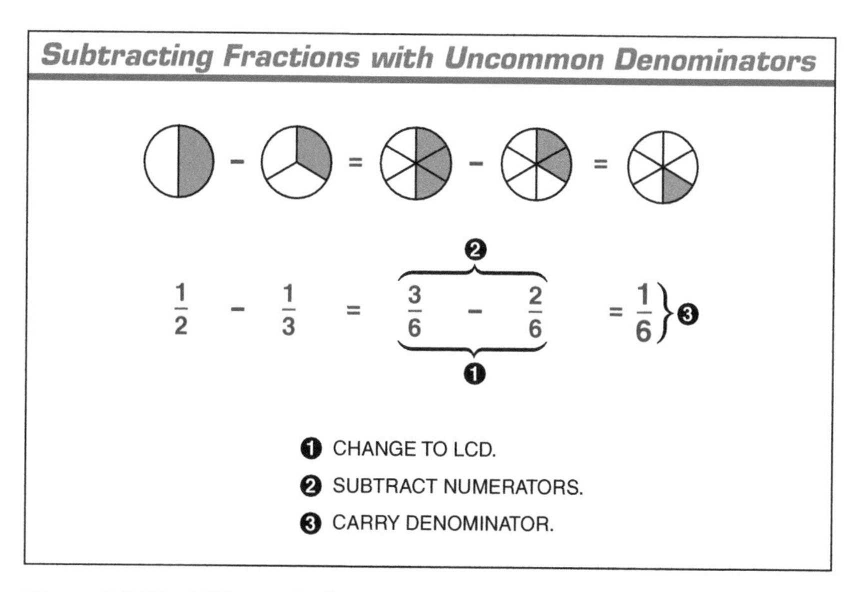

Figure 3-5. The LCD must be found before subtracting fractions with uncommon denominators.

Example—Subtracting Fractions with Uncommon Denominators

1. Subtract $\frac{7}{12}$ from $\frac{23}{36}$.

   ANS: $\frac{1}{18}$

   ❶ Find the LCD (36) for the fractions.
   ❷ Change the fractions by dividing the LCD by the original denominator and multiplying the fractions by the product.
   ❸ Subtract the numerators (23 – 21 = 2).
   ❹ Carry the denominator over ($\frac{2}{36}$), and reduce as required.

❶ $12\overline{)12\ \ 36}$ → 1 3; $12 \times 1 \times 3 = 36$

❷ $36 \div 12 = 3$; $36 \div 36 = 1$; $\frac{7 \times 3}{12 \times 3} = \frac{21}{36}$ $\frac{23 \times 1}{36 \times 1} = \frac{23}{36}$

❸ ❹ $\frac{23 - 21}{36} = \frac{2}{36} = \frac{1}{18}$

QUICK REFERENCE

- *Change the fractions to their LCD.*
- *Subtract the numerators.*
- *Carry the denominator.*
- *Reduce as required.*

MATH EXERCISES—Subtracting Fractions with Uncommon Denominators

______ 1. $\frac{3}{8} - \frac{4}{25}$

______ 2. $\frac{84}{120} - \frac{4}{35}$

______ 3. $\frac{6}{7} - \frac{4}{5}$

______ 4. $\frac{17}{32} - \frac{3}{8}$

PRACTICAL APPLICATIONS—Subtracting Fractions with Uncommon Denominators

______ 5. **Boiler Operation:** A fuel oil tank is $\frac{3}{4}$ full at the beginning of the day. A boiler uses $\frac{1}{8}$ of the tank that day. How full is the fuel tank at the end of the day?

________________ **6. Construction:** Find the final thickness of board A.

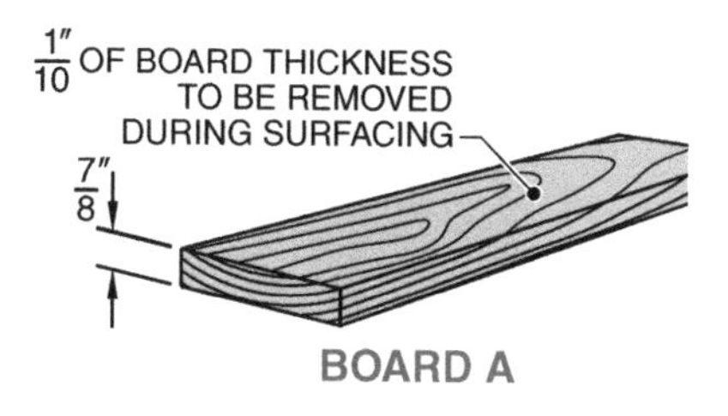

Subtracting Mixed Numbers

To subtract mixed numbers, subtract the whole numbers and fractions separately. Then add the differences. If the fractions do not have a common denominator, they must be changed to their LCD before subtraction. **See Figure 3-6.**

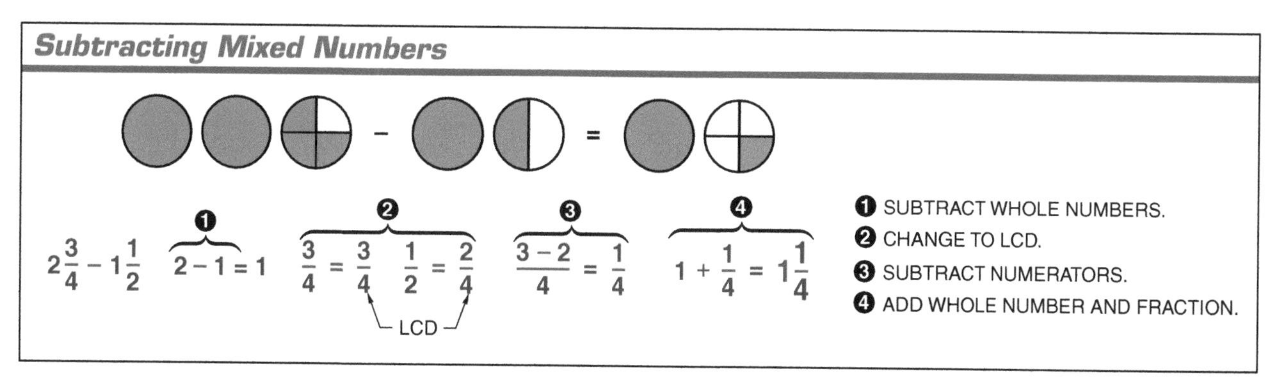

Figure 3-6. When subtracting mixed numbers, whole numbers and fractions are subtracted separately.

Example—Subtracting Mixed Numbers

1. Subtract 2½ from 10¾.

ANS: 8¼

❶ Subtract the whole numbers (10 – 2 = 8).
❷ Find the LCD (4) for the fractions.
❸ Change the fractions by dividing the LCD by the original denominators and multiply the fractions by the products.
❹ Subtract the numerators. Carry the denominator.
❺ Add the result to the whole number.

❶ $10 - 2 = 8$

❷ $2\overline{)2\ 4}$ → $1\ 2$; $2 \times 1 \times 2 = 4$

❸ $4 \div 2 = 2$; $4 \div 4 = 1$; $\frac{1 \times 2}{2 \times 2} = \frac{2}{4}$; $\frac{3 \times 1}{4 \times 1} = \frac{3}{4}$

❹ $\frac{3-2}{4} = \frac{1}{4}$

❺ $8 + \frac{1}{4} = 8\frac{1}{4}$

QUICK REFERENCE

- *Change the fractions to their LCD.*
- *Subtract the whole numbers.*
- *Subtract the numerators.*
- *Add the whole number and the fraction.*
- *Reduce as required.*

MATH EXERCISES—Subtracting Mixed Numbers

______ **1.** $5\frac{3}{4} - 2\frac{1}{2}$

______ **2.** $14\frac{5}{18} - 8\frac{2}{9}$

______ **3.** $6\frac{7}{8} - 3\frac{3}{5}$

______ **4.** $4\frac{2}{3} - 1\frac{3}{10}$

PRACTICAL APPLICATIONS—Subtracting Mixed Numbers

______ **5.** **Construction:** A carpenter measures and saws a piece of baseboard molding into three pieces. Disregarding the kerf (saw waste), find the length of the third piece (D).

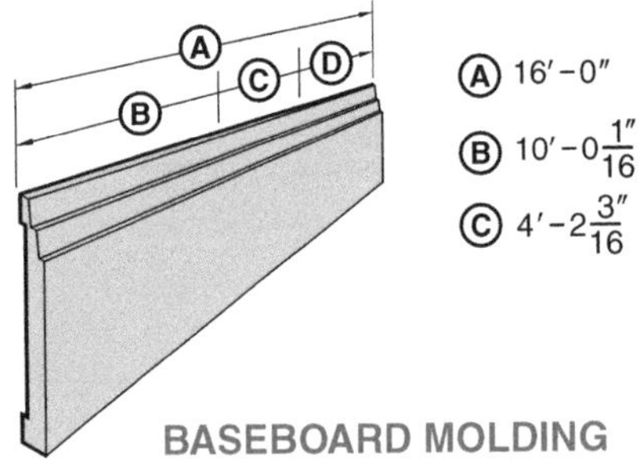

______ **6.** **Boiler Operation:** A plant has a 1000 gal. tank containing $949\frac{7}{8}$ gal. of #2 fuel oil. The boiler uses $170\frac{2}{3}$ gal. of the fuel oil. How many gallons are needed to refill the tank?

______ **7.** **Manufacturing:** A machinist has a steel bar $5\frac{1}{4}$″ long from which pieces $2\frac{1}{2}$″ long and $2\frac{1}{16}$″ long are cut. Disregarding the saw kerf, find the length of the remaining steel bar.

______ **8.** **Construction:** A contractor has $24\frac{1}{2}$ bags of cement mix for a job. Only $17\frac{1}{4}$ bags are needed. How many bags of cement mix are left?

SECTION 3-3 MULTIPLYING FRACTIONS

Fractions can be multiplied horizontally or vertically. Horizontal placement of fractions is the most common because identification of the numerators and denominators is more obvious.

Fraction combinations that may be multiplied include two or more fractions, improper fractions, a fraction by a whole number, a mixed number by a whole number, and two mixed numbers. Each fraction combination follows a slightly different rule.

Pipe may need to have several pieces and fittings joined together to achieve a specific length.

Multiplying Two or More Fractions

To multiply two or more fractions, first multiply the numerators of the fractions. Then, multiply the denominators of the fractions. Reduce if required. **See Figure 3-7.**

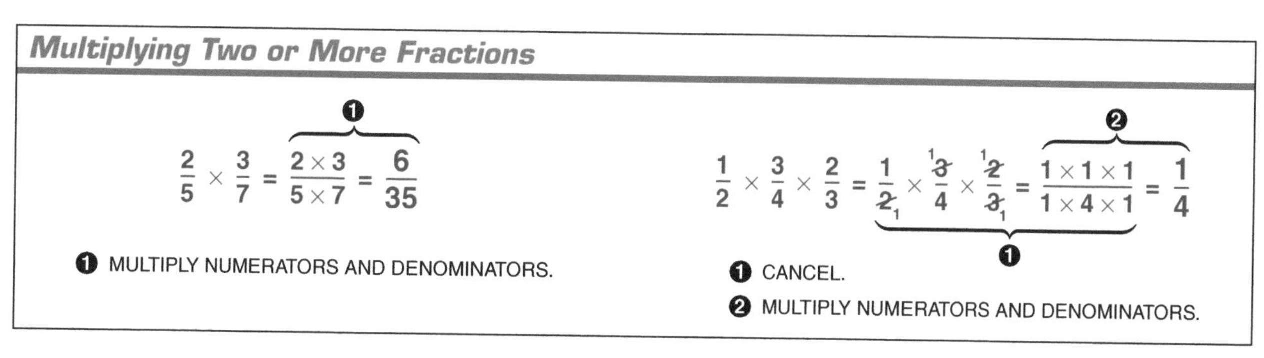

Figure 3-7. When two or more fractions are multiplied, the numerators are multiplied and the denominators multiplied.

Cancellation can be used to simplify the multiplication process. *Cancellation* is a method of removing common factors from both the numerator and denominator. A numerator and then a denominator are divided by a number that divides evenly into both, and this process is repeated until all the fractions are in lowest terms. Then the fractions can be multiplied as necessary. **See Figure 3-8.**

Cancellation

CANCELLATION IS USED TO SIMPLIFY NUMERATOR AND DENOMINATOR

$$\frac{18 \times 5 \times 6}{10 \times 3 \times 12} = \frac{18 \times \cancel{5}^1 \times 6}{\cancel{10}_2 \times 3 \times 12} = \frac{18 \times 1 \times \cancel{6}^2}{2 \times \cancel{3}_1 \times 12} = \frac{18 \times 1 \times \cancel{2}^1}{\cancel{2}_1 \times 1 \times 12} = \frac{\cancel{18}^3 \times 1 \times 1}{1 \times 1 \times \cancel{12}_2} = \frac{3}{2} = 1\frac{1}{2}$$

Figure 3-8. Cancellation is used to simplify mathematical operations.

Examples—Multiplying Two or More Fractions

1. Multiply $\frac{3}{8}$ by $\frac{1}{8}$.

ANS: $\frac{3}{64}$

❶ Multiply the numerators and denominators.

$$\frac{3}{8} \times \frac{1}{8} = \underbrace{\frac{3 \times 1}{8 \times 8}}_{❶} = \frac{3}{64}$$

2. Multiply $\frac{2}{3}$ by $\frac{2}{6}$.

ANS: $\frac{2}{9}$

❶ Cancel the 2 (of the $\frac{2}{3}$) and the 6 (of the $\frac{2}{6}$) by dividing by 2. No more cancellation is possible.

❷ Multiply the numerators and denominators.

$$\frac{2}{3} \times \frac{2}{6} = \underbrace{\frac{\cancel{2}^{1}}{3} \times \frac{2}{\cancel{6}_{3}}}_{❶} = \underbrace{\frac{1 \times 2}{3 \times 3}}_{❷} = \frac{2}{9}$$

QUICK REFERENCE

- *Multiply the numerators and denominators. (Cancel as required.)*
- *Reduce as required.*

MATH EXERCISES—Multiplying Two or More Fractions

__________ **1.** $\frac{1}{2} \times \frac{1}{5}$

__________ **2.** $\frac{1}{6} \times \frac{3}{7}$

__________ **3.** $\frac{2}{3} \times \frac{4}{5}$

__________ **4.** $\frac{5}{10} \times \frac{5}{6}$

__________ **5.** $\frac{3}{32} \times \frac{8}{10}$

__________ **6.** $\frac{7}{16} \times \frac{4}{5}$

__________ **7.** $\frac{3}{10} \times \frac{5}{9} \times \frac{9}{4} \times \frac{12}{10}$

__________ **8.** $\frac{5}{10} \times \frac{24}{3} \times \frac{9}{5} \times \frac{11}{4}$

PRACTICAL APPLICATIONS – Multiplying Two or More Fractions

________________ **9. Electrical:** An electrician bending conduit must calculate the shrink for an offset bend by multiplying the shrink constant by the offset rise. The shrink constant is $\frac{3}{15}$ and the offset rise is $\frac{3}{4}''$. What is the shrink?

________________ **10. Welding:** An oxygen tank for oxyfuel welding is $\frac{4}{5}$ full. During the fabrication of a component, $\frac{1}{2}$ of the oxygen is used. How full is the tank after the job is complete?

________________ **11. Construction:** A carpenter is cutting custom dentil molding. The width of a tooth needs to be $\frac{5}{8}$ the height of the tooth. If the height of a tooth is $\frac{3}{4}''$, what is its width?

Multiplying Improper Fractions

To multiply one improper fraction by another improper fraction, multiply the numerators and multiply the denominators. Use cancellation if possible. Reduce if required. Change all answers from improper fractions to mixed numbers or whole numbers. **See Figure 3-9.**

Multiplying Improper Fractions

$$\frac{6}{4} \times \frac{10}{3} = \overbrace{\frac{\overset{2}{\cancel{6}}}{\underset{2}{\cancel{4}}} \times \frac{\overset{5}{\cancel{10}}}{\underset{1}{\cancel{3}}}}^{❶} = \underbrace{\frac{2 \times 5}{2 \times 1} = \overbrace{\frac{10}{2}}}_{❸}\overbrace{= \frac{5}{1}}^{❷} = 5$$

❶ CANCEL.

❷ MULTIPLY NUMERATORS AND DENOMINATORS.

❸ REDUCE AS REQUIRED.

Figure 3-9. Improper fractions are multiplied like proper fractions, reducing as required.

Examples — Multiplying Improper Fractions

1. Multiply $\frac{5}{3}$ by $\frac{2}{1}$.

   ANS: **$3\frac{1}{3}$**

   ❶ Multiply the numerators and denominators.

   ❷ Reduce as required.

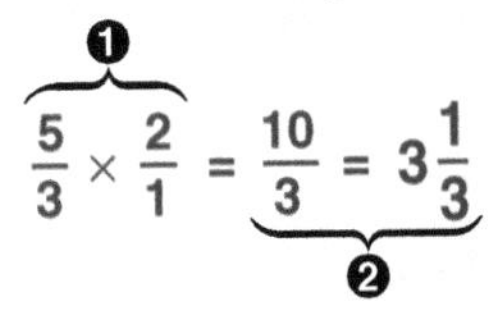

$$\underbrace{\frac{5}{3}\times\frac{2}{1}}_{❶} = \underbrace{\frac{10}{3} = 3\frac{1}{3}}_{❷}$$

> **QUICK REFERENCE**
>
> - *Multiply the numerators and denominators. (Cancel as required.)*
> - *Reduce as required.*

2. Multiply $\frac{18}{3}$ by $\frac{24}{9}$.

   ANS: **16**

   ❶ Cancel the 18 (of $\frac{18}{3}$) and the 9 (of $\frac{24}{9}$) by dividing by 9. Cancel the 3 (of $\frac{18}{3}$) and the 24 (of $\frac{24}{9}$) by dividing by 3. No more cancellation is possible.

   ❷ Multiply the numerators and denominators.

   ❸ Reduce as required.

$$\frac{18}{3}\times\frac{24}{9} = \overbrace{\frac{\cancel{18}^{\,2}}{\cancel{3}_{\,1}}\times\frac{\cancel{24}^{\,8}}{\cancel{9}_{\,1}}}^{❶} = \underbrace{\frac{2}{1}\times\frac{8}{1} = \frac{16}{1}}_{❷} = \overbrace{16}^{❸}$$

MATH EXERCISES — Multiplying Improper Fractions

_______________ **1.** $\frac{21}{3} \times \frac{27}{7}$

_______________ **2.** $\frac{23}{16} \times \frac{24}{9}$

_______________ **3.** $\frac{11}{10} \times \frac{5}{4}$

_______________ **4.** $\frac{4}{3} \times \frac{17}{12}$

_______________ **5.** $\frac{84}{21} \times \frac{7}{4}$

_______________ **6.** $^{18}/_{10} \times ^{5}/_{3}$

_______________ **7.** $^{17}/_{2} \times ^{17}/_{2}$

_______________ **8.** $^{3}/_{2} \times ^{20}/_{15}$

Multiplying a Fraction by a Whole Number

To multiply a fraction and any whole number, place the whole number over a denominator of 1 and proceed as when multiplying two fractions. Perform cancellation as required. Reduce as required. **See Figure 3-10.**

Multiplying a Fraction by a Whole Number

$$\frac{4}{9} \times 3 = \overbrace{\frac{4}{9} \times \frac{3}{1}}^{❶} = \underbrace{\frac{4}{\cancel{9}_{3}} \times \frac{\cancel{3}^{1}}{1}}_{❷} = \overbrace{\frac{4 \times 1}{3 \times 1}}^{❸} = \underbrace{\frac{4}{3} = 1\frac{1}{3}}_{❹}$$

❶ PLACE WHOLE NUMBER OVER DENOMINATOR OF 1.
❷ CANCEL.
❸ MULTIPLY NUMERATORS AND DENOMINATORS.
❹ REDUCE AS REQUIRED.

Figure 3-10. When multiplying a fraction by a whole number, the whole number is made into a fraction and the fractions are multiplied.

Example—Multiplying a Fraction by a Whole Number

1. Multiply $^{12}/_{16}$ by 4.

ANS: 3

❶ Place 4 over a denominator of 1.
❷ Cancel as required.
❸ Change improper fraction to whole number.

$$\frac{12}{16} \times 4 = \frac{12}{16} \times \underbrace{\frac{4}{1}}_{❶} = \overbrace{\frac{12}{\cancel{16}_{4}} \times \frac{\cancel{4}^{1}}{1}}^{❷} = \overbrace{\frac{12}{4} = 3}^{❸}$$

QUICK REFERENCE

- *Place the whole number over a denominator of 1.*
- *Multiply the numerators and denominators. (Cancel as required.)*
- *Reduce as required.*

MATH EXERCISES—Multiplying a Fraction by a Whole Number

_______________ 1. $2/9 \times 2$

_______________ 2. $1/10 \times 3$

_______________ 3. $1/20 \times 7$

_______________ 4. $3/11 \times 3$

_______________ 5. $4/5 \times 12$

_______________ 6. $3/9 \times 33$

_______________ 7. $4/7 \times 11$

PRACTICAL APPLICATIONS—Multiplying a Fraction by a Whole Number

_______________ 8. **Electrical:** A motor delivers 8/9 of the power it receives. How much power does the motor deliver if it receives 25 HP?

_______________ 9. **Maintenance:** Find the total amount of paint in the three, one-gallon cans that are each 3/5 full.

$\frac{3}{5}$ FULL

_______________ 10. **Manufacturing:** A machinist takes 1/6 of an hour to machine a component. How long does the machinist take to machine 25 components?

Multiplying a Mixed Number by a Whole Number

To multiply a mixed number and a whole number, multiply the whole numbers. Then multiply the fraction of the mixed number by the whole number over 1. Add the two products. Reduce if required. **See Figure 3-11.**

Multiplying a Mixed Number by a Whole Number

$$2\frac{1}{3} \times 4 = \overbrace{(2 \times 4)}^{❶} = \overbrace{(\frac{1}{3} \times \frac{4}{1})}^{❷} = \overbrace{8 + \frac{4}{3}}^{❸} = \overbrace{8\frac{4}{3} = 9\frac{1}{3}}^{❹}$$

❶ MULTIPLY WHOLE NUMBER OF MIXED NUMBER BY WHOLE NUMBER.

❷ MULTIPLY FRACTION OF MIXED NUMBER BY WHOLE NUMBER OVER 1.

❸ ADD PRODUCTS.

❹ CANCEL OR REDUCE.

Figure 3-11. When multiplying a mixed number and a whole number, the whole numbers are multiplied first. Then the single whole number is made into a fraction and the fractions are multiplied.

Example—Multiplying a Mixed Number by a Whole Number

1. Multiply $4\frac{7}{8}$ by 3.

ANS: **$14\frac{5}{8}$**

❶ Multiply the whole numbers ($4 \times 3 = 12$).

❷ Multiply $\frac{7}{8}$ (fraction of the mixed number) by 3 over 1 ($\frac{7}{8} \times \frac{3}{1} = \frac{21}{8}$).

❸ Reduce the fraction ($21 \div 8 = 2\frac{5}{8}$).

❹ Add the result to the whole number.

$$4\frac{7}{8} \times 3 = \overbrace{12}^{❶} + \overbrace{\frac{7}{8} \times \frac{3}{1}}^{❷} = \overbrace{12 + \frac{21}{8} = 12 + 2\frac{5}{8}}^{❸} = \overbrace{14\frac{5}{8}}^{❹}$$

QUICK REFERENCE

- *Multiply the whole numbers.*
- *Place the whole number over 1 and multiply the fractions.*
- *Reduce as required.*
- *Add the whole number and the fraction.*

MATH EXERCISES—Multiplying a Mixed Number by a Whole Number

__________ **1.** $81\frac{6}{7} \times 49$

__________ **2.** $35\frac{7}{8} \times 36$

__________ **3.** $47\frac{7}{10} \times 65$

______ **4.** $13\frac{3}{5} \times 22$

______ **5.** $3\frac{14}{19} \times 35$

______ **6.** $35\frac{6}{71} \times 5$

PRACTICAL APPLICATIONS—Multiplying a Mixed Number by a Whole Number

______ **7. Construction:** A trim carpenter needs 4 pieces of crown molding $13\frac{1}{2}'$ long to complete a job. How many linear feet of crown molding are needed?

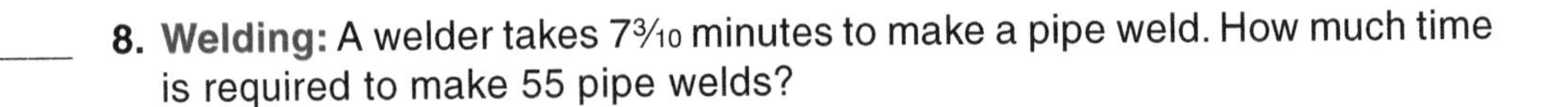

______ **8. Welding:** A welder takes $7\frac{3}{10}$ minutes to make a pipe weld. How much time is required to make 55 pipe welds?

______ **9. Construction:** What is the total rise of the stairs in the stair plan?

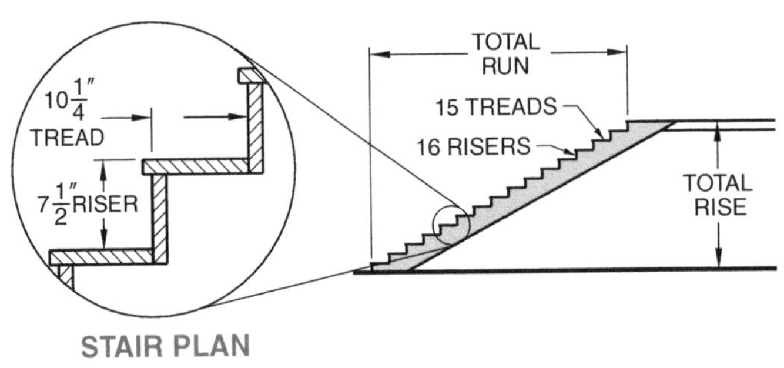

STAIR PLAN

______ **10. Construction:** What is the total run of the stairs in the stair plan?

Multiplying Mixed Numbers

To multiply one mixed number by another mixed number, change both mixed numbers to improper fractions and then multiply. Use cancellation if possible. Reduce as required. **See Figure 3-12.**

Multiplying Mixed Numbers

$$5\frac{2}{3} \times 1\frac{1}{6} = \overbrace{\frac{17}{3} \times \frac{7}{6}}^{❶} = \overbrace{\frac{17 \times 7}{3 \times 6}}^{❷} = \overbrace{\frac{119}{18} = 6\frac{11}{18}}^{❸}$$

❶ CHANGE MIXED NUMBERS TO IMPROPER FRACTIONS.

❷ MULTIPLY NUMERATORS AND DENOMINATORS.

❸ REDUCE AS REQUIRED.

Figure 3-12. When multiplying mixed numbers, the mixed numbers are made into improper fractions, multiplied, and reduced as required.

Example—Multiplying Mixed Numbers

1. $9\frac{1}{3} \times 7\frac{4}{5}$

ANS: $72\frac{4}{5}$

❶ Change $9\frac{1}{3}$ to the improper fraction $\frac{28}{3}$. Change $7\frac{4}{5}$ to the improper fraction $\frac{39}{5}$.

❷ Multiply the improper fractions ($\frac{28}{3} \times \frac{39}{5} = \frac{1092}{15}$).

❸ Reduce as required.

$$9\frac{1}{3} \times 7\frac{4}{5} = \overbrace{\frac{28}{3} \times \frac{39}{5}}^{❶} = \overbrace{\frac{28 \times 39}{3 \times 5}}^{❷} = \overbrace{\frac{1092}{15} = 72\frac{12}{15} = 72\frac{4}{5}}^{❸}$$

QUICK REFERENCE

- *Change the mixed numbers to improper fractions.*
- *Multiply the fractions. (Cancel as required.)*
- *Reduce as required.*

MATH EXERCISES—Multiplying Mixed Numbers

__________ **1.** $28\frac{4}{9} \times 16\frac{3}{4}$

__________ **2.** $38\frac{1}{2} \times 12\frac{1}{2}$

__________ **3.** $51\frac{4}{5} \times 72\frac{7}{9}$

__________ **4.** $8\frac{1}{2} \times 3\frac{2}{3}$

______ 5. $12\frac{2}{7} \times 28\frac{1}{3}$

______ 6. $3\frac{1}{5} \times 2\frac{2}{3}$

______ 7. $33\frac{1}{5} \times 20\frac{1}{2}$

______ 8. $17\frac{3}{10} \times 5\frac{3}{5}$

PRACTICAL APPLICATIONS—Multiplying Mixed Numbers

______ 9. **Boiler Operation:** One gallon of water weighs approximately $8\frac{1}{3}$ lb. How much does $25\frac{3}{4}$ gal. of water weigh?

______ 10. **Construction:** A trim carpenter applied paneling and molding to paneled walls. An area of $96\frac{1}{2}$ sq ft was completed in 1 hr. How many square feet could be completed in $7\frac{1}{4}$ hr?

SECTION 3-4 DIVIDING FRACTIONS

Fractions are divided horizontally. Fraction combinations that may be divided include two fractions, a fraction by a whole number, a mixed number by a whole number, a whole number by a fraction, and two mixed numbers. Each fraction combination follows a slightly different rule.

Dividing Two Fractions

To divide one fraction by another fraction, invert the divisor fraction. Then multiply the numerators and multiply the denominators. Reduce as required. **See Figure 3-13.**

Dividing Two Fractions

DIVISOR FRACTION

$$\frac{4}{9} \div \underbrace{\frac{3}{32}}_{\text{divisor fraction}} = \overbrace{\frac{4}{9} \times \frac{32}{3}}^{❶} = \overbrace{\frac{4 \times 32}{9 \times 3}}^{❷} = \overbrace{\frac{128}{27} = 4\frac{20}{27}}^{❸}$$

❶ INVERT DIVISOR FRACTION.
❷ MULTIPLY NUMERATORS AND DENOMINATORS.
❸ REDUCE AS REQUIRED.

Figure 3-13. When dividing one fraction by another fraction, the divisor fraction is inverted and numerators and denominators multiplied.

Example—Dividing Two Fractions

1. Divide $7/9$ by $2/3$.

ANS: **$1\frac{1}{6}$**

❶ Invert $2/3$.
❷ Cancel by dividing 3 into 3 and 3 into 9.
❸ Multiply $7/3$ by $1/20$.
❹ Reduce as required.

$$\frac{7}{9} \div \frac{2}{3} = \overbrace{\frac{7}{9} \times \frac{3}{2}}^{❶} = \overbrace{\frac{7}{\cancel{9}_3} \times \frac{\cancel{3}^1}{2}}^{❷} = \overbrace{\frac{7 \times 1}{3 \times 2}}^{❸} = \overbrace{\frac{7}{6} = 1\frac{1}{6}}^{❹}$$

QUICK REFERENCE

- *Invert the divisor fraction.*
- *Multiply the numerators and denominators. (Cancel as required.)*
- *Reduce as required.*

MATH EXERCISES—Dividing Two Fractions

____________________ **1.** $13/15 \div 2/3$

____________________ **2.** $25/35 \div 1/4$

____________________ **3.** $11/15 \div 1/2$

____________________ **4.** $3/8 \div 2/3$

____________________ **5.** $15/16 \div 1/4$

_______________ **6.** $\frac{4}{5} \div \frac{5}{8}$

_______________ **7.** $\frac{5}{8} \div \frac{1}{4}$

_______________ **8.** $\frac{9}{16} \div \frac{1}{4}$

PRACTICAL APPLICATIONS—Dividing Two Fractions

_______________ **9. Construction:** A painter has ⅞ gal. of stain. An interior door requires ¼ gal. of stain to finish. How many interior doors can be finished?

_______________ **10. Maintenance:** A scrap yard has $\frac{9}{10}$ t of scrap metal to be distributed into ¼ t bins. How many bins are needed for the metal? (*Round to the nearest whole number.*)

Dividing a Fraction by a Whole Number

To divide a fraction by a whole number, multiply the denominator of the fraction by the whole number. Carry the numerator, and reduce as required. Use cancellation if possible. **See Figure 3-14.**

Dividing a Fraction by a Whole Number

$$\frac{9}{10} \div 3 = \underbrace{\frac{9}{10 \times 3}}_{❶} = \underbrace{\frac{\overset{3}{\cancel{9}}}{10 \times \underset{1}{\cancel{3}}}}_{❷} = \underbrace{\frac{3}{10 \times 1} = \frac{3}{10}}_{❸}$$

❶ MULTIPLY DENOMINATOR BY WHOLE NUMBER.
❷ CANCEL.
❸ PLACE NUMERATOR OVER PRODUCT.

Figure 3-14. When dividing a fraction by a whole number, the denominator and whole number are multiplied and the numerator is carried over.

Examples — Dividing a Fraction by a Whole Number

1. Divide $\frac{3}{5}$ by 2.

ANS: $\frac{3}{10}$

❶ Multiply 5 by 2.

❷ Carry the 3.

$$\frac{3}{5} \div 2 = \underbrace{\frac{3}{5 \times 2}}_{❶} = \underbrace{\frac{3}{10}}_{❷}$$

QUICK REFERENCE

- *Multiply the denominator of the fraction by the whole number. (Cancel as required.)*
- *Carry the numerator.*
- *Reduce as required.*

2. Divide $\frac{4}{5}$ by 4.

ANS: $\frac{1}{5}$

❶ Multiply 5 by 4.

❷ Cancel the numerator 4 and the whole number 4 by dividing by 4. No more cancellation is possible.

❸ Multiply 5 by 1. Carry the 1.

$$\frac{4}{5} \div 4 = \underbrace{\frac{4}{5 \times 4}}_{❶} = \underbrace{\frac{\overset{1}{\cancel{4}}}{5 \times \underset{1}{\cancel{4}}}}_{❷} = \underbrace{\frac{1}{5 \times 1} = \frac{1}{5}}_{❸}$$

MATH EXERCISES — Dividing a Fraction by a Whole Number

__________ **1.** $\frac{30}{32} \div 5$

__________ **2.** $\frac{33}{55} \div 11$

__________ **3.** $\frac{31}{32} \div 3$

__________ **4.** $\frac{5}{8} \div 7$

__________ **5.** $\frac{7}{20} \div 4$

__________ **6.** $\frac{1}{2} \div 7$

__________ **7.** $\frac{3}{8} \div 16$

PRACTICAL APPLICATIONS—Dividing a Fraction by a Whole Number

_______________ **8. Maintenance:** A custodian paints three doors with ¾ gal. of paint. How much paint is needed to paint one door?

_______________ **9. Boiler Operation:** A bank of three identical boilers uses a total of ¾ of a tank of fuel each day. How much fuel does each boiler use each day?

_______________ **10. Alternative Energy:** A hydroelectric dam directs ⅗ of the river's water flow through its turbine generator units. If the flow is divided equally between the 12 units, how much of the river's total flow does each handle?

Dividing a Mixed Number by a Whole Number

To divide a mixed number by a whole number, change the mixed number to an improper fraction. Multiply the denominator of the fraction by the whole number, carry the numerator, and reduce as required. Use cancellation if possible. **See Figure 3-15.**

Dividing a Mixed Number by a Whole Number

$$8\frac{2}{3} \div 4 = \underbrace{\frac{26}{3}}_{❶} \div 4 = \underbrace{\frac{26}{3 \times 4}}_{❷} = \underbrace{\frac{\overset{13}{\cancel{26}}}{3 \times \underset{2}{\cancel{4}}}}_{❸} = \underbrace{\frac{13}{3 \times 2} = \frac{13}{6}}_{❹} = \underbrace{2\frac{1}{6}}_{❺}$$

❶ CHANGE MIXED NUMBER TO IMPROPER FRACTION.

❷ MULTIPLY DENOMINATOR BY WHOLE NUMBER.

❸ CANCEL.

❹ CARRY NUMERATOR.

❺ REDUCE AS REQUIRED.

Figure 3-15. When dividing a mixed number by a whole number, the mixed number is changed to an improper fraction, the denominator and whole number are multiplied, and the numerator is carried over.

Example – Dividing a Mixed Number by a Whole Number

1. $3\frac{3}{8} \div 3$

ANS: **$1\frac{1}{8}$**

❶ Change $3\frac{3}{8}$ to an improper fraction ($3\frac{3}{8} = \frac{27}{8}$).
❷ Multiply 8 by 3.
❸ Cancel the 27 and 3 by dividing by 3. No more cancellation is possible.
❹ Multiply 8 by 1. Carry the 9.
❺ Reduce as required.

$$3\frac{3}{8} \div 3 = \overbrace{\frac{27}{8}}^{❶} \div 3 = \overbrace{\frac{27}{8 \times 3}}^{❷} = \overbrace{\frac{^{9}\cancel{27}}{8 \times \cancel{3}_{1}}}^{❸} = \underbrace{\frac{9}{8 \times 1} = \frac{9}{8}}_{❹} = \overbrace{1\frac{1}{8}}^{❺}$$

QUICK REFERENCE

- *Change the mixed number to an improper fraction.*
- *Multiply the denominator of the fraction by the whole number.*
- *Carry the numerator. (Cancel as required.)*
- *Reduce as required.*

MATH EXERCISES – Dividing a Mixed Number by a Whole Number

_______________ **1.** $250\frac{1}{2} \div 5$

_______________ **2.** $333\frac{1}{3} \div 3$

_______________ **3.** $789\frac{2}{5} \div 4$

_______________ **4.** $877\frac{1}{7} \div 7$

_______________ **5.** $724\frac{2}{9} \div 8$

_______________ **6.** $5\frac{1}{2} \div 2$

_______________ **7.** $17\frac{7}{9} \div 8$

PRACTICAL APPLICATIONS—Dividing a Mixed Number by a Whole Number

_______________ **8. Construction:** A concrete mixer mixes 586⅔ cu ft of concrete in 5 hours. At the same rate, how much concrete is mixed in 1 hour?

_______________ **9. Manufacturing:** What is the width of each step of the step pulley?

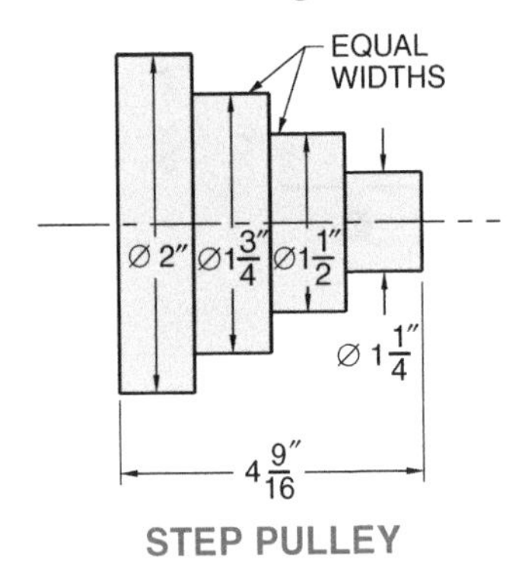

_______________ **10. Pipefitting:** A 43½″ piece of pipe is cut into three pieces. Disregarding the saw kerf, what is the length of each piece?

Dividing a Whole Number by a Fraction

To divide a whole number by a fraction, place the whole number over the denominator of 1 and invert the divisor fraction. Multiply the numerators and the denominators and reduce as required. Use cancellation if possible. **See Figure 3-16.**

Dividing a Whole Number by a Fraction

$$8 \div \frac{2}{3} \overset{❶}{=} \frac{8}{1} \div \frac{2}{3} \overset{❷}{=} \frac{8}{1} \times \frac{3}{2} \overset{❸}{=} \frac{\overset{4}{\cancel{8}}}{1} \times \frac{3}{\underset{1}{\cancel{2}}} \overset{❹}{=} \frac{4 \times 3}{1 \times 1} \overset{❺}{=} \frac{12}{1} = 12$$

❶ CHANGE WHOLE NUMBER TO FRACTION.
❷ INVERT DIVISOR FRACTION.
❸ CANCEL.
❹ MULTIPLY NUMERATORS AND DENOMINATORS.
❺ REDUCE AS REQUIRED.

Figure 3-16. When dividing a whole number by a fraction, the whole number is changed to a fraction, the divisor fraction is inverted, and the fractions are multiplied.

Example—Dividing a Whole Number by a Fraction

1. Divide 10 by $\frac{2}{5}$.

ANS: 25

❶ Change the 10 to $\frac{10}{1}$.
❷ Invert $\frac{2}{5}$.
❸ Cancel the 10 and the 2 by dividing by 2. No more cancellation is possible.
❹ Multiply the numerators and denominators ($5 \times 5 = 25$; $1 \times 1 = 1$).
❺ Reduce as required.

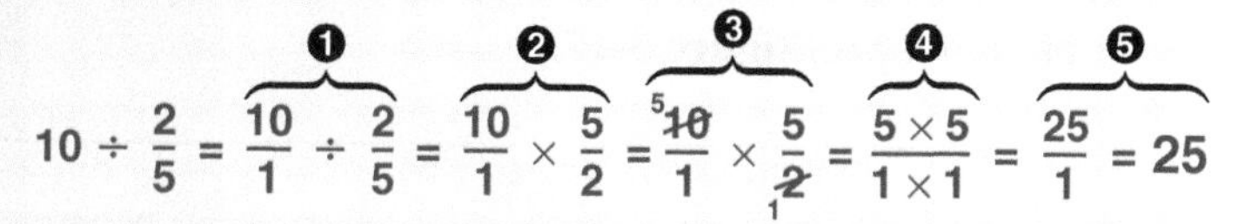

$$10 \div \frac{2}{5} = \frac{10}{1} \div \frac{2}{5} = \frac{10}{1} \times \frac{5}{2} = \frac{\overset{5}{\cancel{10}}}{1} \times \frac{5}{\underset{1}{\cancel{2}}} = \frac{5 \times 5}{1 \times 1} = \frac{25}{1} = 25$$

QUICK REFERENCE

- *Place the whole number over a denominator of 1.*
- *Invert the divisor fraction.*
- *Multiply the numerators and denominators. (Cancel as required.)*
- *Reduce as required.*

MATH EXERCISES—Dividing a Whole Number by a Fraction

________ **1.** $25 \div \frac{5}{7}$

________ **2.** $42 \div \frac{6}{7}$

________ **3.** $26 \div \frac{4}{9}$

________ **4.** $65 \div \frac{13}{15}$

________ **5.** $81 \div \frac{9}{11}$

PRACTICAL APPLICATIONS—Dividing a Whole Number by a Fraction

________ **6. HVAC:** Approximately how many chiller room refrigeration systems can be filled using a 125 lb cylinder? Each system contains $8\frac{1}{8}$ lb of R-134a refrigerant when filled.

________________ **7. Pipefitting:** How many ¾″ spacers can be cut from a 9″ length of PVC pipe?

Dividing Mixed Numbers

To divide a mixed number by a mixed number, change the mixed numbers to improper fractions and invert the divisor fraction. Multiply the numerators and denominators and reduce as required. Use cancellation if possible. **See Figure 3-17.**

Dividing Mixed Numbers

$$1\frac{1}{3} \div 1\frac{1}{9} \overset{❶}{=} \frac{4}{3} \div \frac{10}{9} \overset{❷}{=} \frac{4}{3} \times \frac{9}{10} \overset{❸}{=} \frac{\overset{2}{\cancel{4}}}{\underset{1}{\cancel{3}}} \times \frac{\overset{3}{\cancel{9}}}{\underset{5}{\cancel{10}}} \overset{❹}{=} \frac{2 \times 3}{1 \times 5} \overset{❺}{=} \frac{6}{5} = 1\frac{1}{5}$$

❶ CHANGE MIXED NUMBERS TO IMPROPER FRACTIONS.

❷ INVERT DIVISOR FRACTION.

❸ CANCEL.

❹ MULTIPLY NUMERATORS AND DENOMINATORS.

❺ REDUCE AS REQUIRED.

Figure 3-17. When dividing a mixed number by a mixed number, the mixed numbers are changed to improper fractions, and the fractions are multiplied and reduced.

Example—Dividing Mixed Numbers

1. 7²⁄₉ ÷ 4⅓

ANS: **1⅔**

1. Change 7²⁄₉ and 4⅓ to improper fractions (7²⁄₉ = ⁶⁵⁄₉ and 4⅓ = ¹³⁄₃).
2. Invert ¹³⁄₃.
3. Cancel the 65 and the 13 by dividing by 13. Cancel the 9 and the 3 by dividing by 3.
4. Multiply the numerators and denominators (5 × 1 = 5; 3 × 1 = 3).
5. Reduce as required.

$$7\frac{2}{9} \div 4\frac{1}{3} \overset{❶}{=} \frac{65}{9} \div \frac{13}{3} \overset{❷}{=} \frac{65}{9} \times \frac{3}{13} \overset{❸}{=} \frac{\overset{5}{\cancel{65}}}{\underset{3}{\cancel{9}}} \times \frac{\overset{1}{\cancel{3}}}{\underset{1}{\cancel{13}}} \overset{❹}{=} \frac{5 \times 1}{3 \times 1} \overset{❺}{=} \frac{5}{3} = 1\frac{2}{3}$$

QUICK REFERENCE

- *Change the mixed numbers to improper fractions.*
- *Invert the divisor fraction.*
- *Multiply the numerators and denominators. (Cancel as required.)*
- *Reduce as required.*

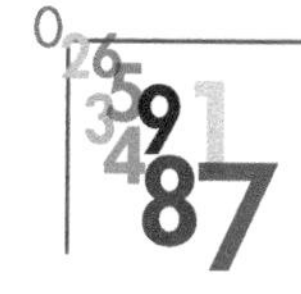

To divide mixed numbers, the mixed numbers must be changed to improper fractions. The reciprocal of the divisor is used to change the division problem into a multiplication problem.

MATH EXERCISES — Dividing Mixed Numbers

__________ **1.** $4\frac{2}{3} \div 3\frac{1}{2}$

__________ **2.** $8\frac{1}{9} \div 6\frac{2}{3}$

__________ **3.** $7\frac{1}{8} \div 5\frac{3}{4}$

__________ **4.** $3\frac{3}{8} \div 2\frac{1}{2}$

__________ **5.** $120\frac{2}{3} \div 2\frac{1}{2}$

__________ **6.** $3\frac{1}{3} \div 3\frac{1}{3}$

__________ **7.** $4\frac{1}{2} \div 2\frac{1}{4}$

PRACTICAL APPLICATIONS — Dividing Mixed Numbers

__________ **8. Manufacturing:** An automatic screw machine is set up to turn 1½″ lengths that are ½″ in diameter. Disregarding saw kerf, how many shafts can be turned from a 109½″ long piece of bar stock?

__________ **9. Pipefitting:** A pipe that is 40½″ long is cut into smaller pieces that are 6¾″ long. Disregarding saw kerf, how many small pieces can be cut from the longer piece?

______________ **10. Welding:** A welder deposits welds totaling 120¼″ in a 4½ hr shift. How many inches of welds are deposited per hour?

Dividing Complex Fractions

A *complex fraction* is a fraction that has a fraction, an improper fraction, a mixed number, or a mathematical process in its numerator, denominator, or both. For example, $\frac{8}{\frac{1}{3}}$, $\frac{3\frac{1}{3}}{\frac{1}{3}}$, and $\frac{2+\frac{1}{2}}{3}$ are all complex fractions.

Numerator Fractions. Dividing a complex fraction with a fraction in the numerator is the same as dividing a fraction by a whole number. To divide the numerator fraction, multiply its denominator by the denominator of the complex fraction. Carry the numerator and reduce as required. Use cancellation if possible. **See Figure 3-18.**

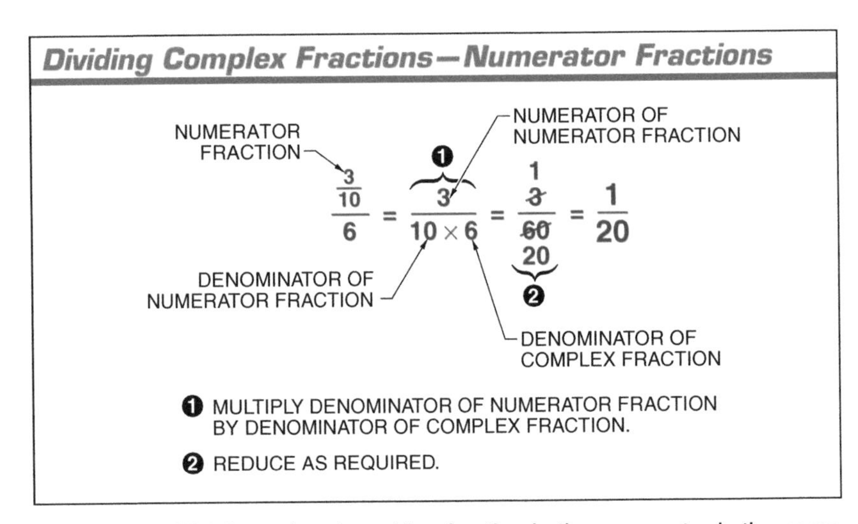

Figure 3-18. Dividing a fraction with a fraction in the numerator is the same as dividing a fraction by a whole number.

Example—Numerator Fractions

1. Divide ⅜ by 6.

ANS: 1⁄16

❶ Multiply 8 × 6 (8 × 6 = 48) and place under the numerator 3.

❷ Cancel to reduce.

$$\frac{\frac{3}{8}}{6} = \frac{3}{8 \times 6} = \frac{\cancel{3}^{1}}{\cancel{48}_{16}} = \frac{1}{16}$$

QUICK REFERENCE

- *Multiply the denominator of the numerator fraction by the denominator of the complex fraction.*
- *Carry the numerator and reduce as required.*

Denominator Fractions. Dividing a complex fraction with a fraction in the denominator is the same as dividing a whole number by a fraction. To divide a denominator fraction, change the numerator of the complex fraction into fraction form and invert the divisor fraction. Use cancellation if possible. Multiply the numerators and denominators and reduce as required. **See Figure 3-19.**

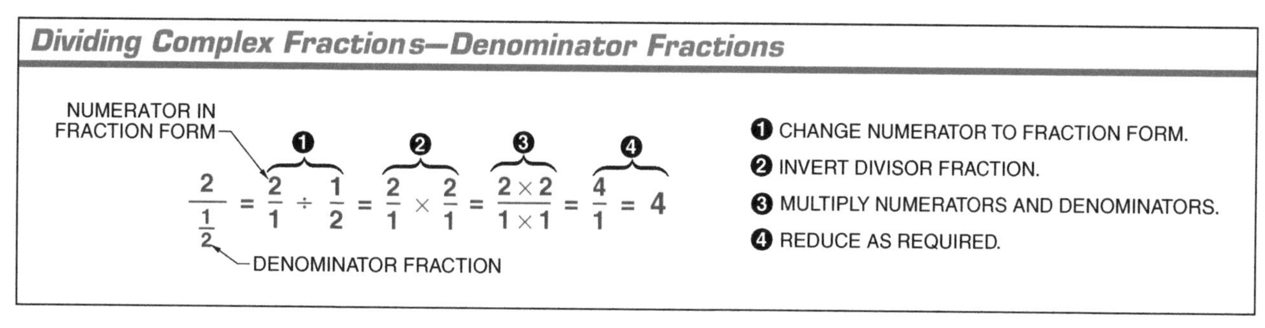

Figure 3-19. Dividing a fraction with a fraction in the denominator is the same as dividing a whole number by a fraction.

Example—Denominator Fractions

1. Divide 3 by ¼.
   ANS: **12**
   ❶ Change the 3 to 3⁄1.
   ❷ Invert the divisor fraction (4⁄1).
   ❸ Multiply the numerators and denominators.
   ❹ Reduce as required.

$$\frac{3}{\frac{1}{4}} = \overbrace{\frac{3}{1} \div \frac{1}{4}}^{❶} = \overbrace{\frac{3}{1} \times \frac{4}{1}}^{❷} = \overbrace{\frac{3 \times 4}{1 \times 1}}^{❸} = \overbrace{\frac{12}{1} = 12}^{❹}$$

QUICK REFERENCE

- *Change the numerator of the complex fraction into fraction form.*
- *Invert the divisor fraction.*
- *Multiply the numerators and denominators. (Cancel as required.)*
- *Reduce as required.*

Mixed Numbers. Dividing a complex fraction with a mixed number in both the numerator and denominator is the same as dividing two mixed numbers. To divide, change both mixed numbers to improper fractions and invert the divisor fraction. Multiply the numerators and denominators and reduce as required. Use cancellation if possible. **See Figure 3-20.**

Dividing Complex Fractions—Mixed Numbers

MIXED NUMBERS

$$\frac{2\frac{1}{2}}{1\frac{1}{3}} = \overbrace{\frac{5}{2} \div \frac{4}{3}}^{❶} = \overbrace{\frac{5}{2} \times \frac{3}{4}}^{❷} = \overbrace{\frac{5 \times 3}{2 \times 4}}^{❸} = \overbrace{\frac{15}{8} = 1\frac{7}{8}}^{❹}$$

❶ CHANGE MIXED NUMBERS TO IMPROPER FRACTIONS.
❷ INVERT DIVISOR FRACTION.
❸ MULTIPLY NUMERATORS AND DENOMINATORS.
❹ REDUCE AS REQUIRED.

Figure 3-20. Dividing a fraction with mixed numbers as both the numerator and denominator is the same as dividing two mixed numbers.

Example — Mixed Numbers

1. Divide $3\frac{1}{4}$ by $1\frac{1}{2}$.
   ANS: $2\frac{1}{6}$
   ❶ Change mixed numbers to improper fractions ($3\frac{1}{4} = \frac{13}{4}$ and $1\frac{1}{2} = \frac{3}{2}$).
   ❷ Invert the divisor fraction to $\frac{2}{3}$.
   ❸ Multiply the numerators and denominators.
   ❹ Reduce as required.

$$\frac{3\frac{1}{4}}{1\frac{1}{2}} = \overbrace{\frac{13}{4} \div \frac{3}{2}}^{❶} = \overbrace{\frac{13}{4} \times \frac{2}{3}}^{❷} = \overbrace{\frac{13 \times 2}{4 \times 3}}^{❸} = \overbrace{\frac{26}{12} = 2\frac{1}{6}}^{❹}$$

QUICK REFERENCE

- *Change the mixed numbers to improper fractions.*
- *Invert the divisor fraction.*
- *Multiply the numerators and denominators. (Carry as necessary).*
- *Reduce as required.*

Other Complex Fractions. To solve a complex fraction with a mathematical process in the numerator and/or denominator, solve the process first. Then divide the numerator by the denominator. **See Figure 3-21.**

Dividing Complex Fractions with Mathematical Processes

MATH PROCESS

$$\frac{(\frac{1}{3} \times \frac{3}{5}) + \frac{1}{3}}{\frac{1}{4} + \frac{3}{8}} = \frac{\overbrace{\frac{3}{15}}^{❶} + \frac{1}{3}}{\underbrace{\frac{2}{8} + \frac{3}{8}}_{❷}} = \frac{\overbrace{\frac{3}{15} + \frac{5}{15}}^{❸}}{\frac{5}{8}} = \frac{\frac{8}{15}}{\frac{5}{8}} = \overbrace{\frac{8}{15} \div \frac{5}{8}}^{❹} = \overbrace{\frac{8}{15} \times \frac{8}{5}}^{❺} = \overbrace{\frac{8 \times 8}{15 \times 5}}^{❻} = \frac{64}{75}$$

❶ SOLVE MATH PROCESS.
❷ FIND COMMON DENOMINATOR OF DENOMINATOR.
❸ FIND COMMON DENOMINATOR OF NUMERATOR.
❹ DIVIDE NUMERATOR BY DENOMINATOR.
❺ INVERT DIVISOR FRACTION.
❻ MULTIPLY NUMERATORS AND DENOMINATORS.

Figure 3-21. When a fraction problem includes a mathematical process, the mathematical process must be solved first.

Example — Mathematical Processes

1. Divide $(\frac{1}{2} \times \frac{1}{4}) + \frac{3}{8}$ by $\frac{3}{16} + \frac{3}{4}$.
   ANS: $\frac{8}{15}$
   ❶ Solve the math process ($\frac{1}{2} \times \frac{1}{4} = \frac{1}{8}$). Add the fractions.
   ❷ Divide the numerator by denominator ($\frac{4}{8} \div \frac{15}{16}$).
   ❸ Invert the divisor fraction ($\frac{4}{8} \times \frac{16}{15}$).
   ❹ Multiply numerators and denominators ($4 \times 16 = 64$; $8 \times 15 = 120$).
   ❺ Reduce as required.

$$\frac{(\frac{1}{2} \times \frac{1}{4}) + \frac{3}{8}}{\frac{3}{16} + \frac{3}{4}} = \overbrace{\frac{\frac{1}{8} + \frac{3}{8}}{\frac{3}{16} + \frac{3}{4}} = \frac{\frac{4}{8}}{\frac{15}{16}}}^{❶} = \overbrace{\frac{4}{8} \div \frac{15}{16}}^{❷} = \overbrace{\frac{4}{8} \times \frac{16}{15}}^{❸} = \overbrace{\frac{4 \times 16}{8 \times 15}}^{❹} = \overbrace{\frac{\cancel{64}^{8}}{\cancel{120}_{15}}}^{❺} = \frac{8}{15}$$

QUICK REFERENCE

- *Solve the math process.*
- *Invert the divisor fraction.*
- *Multiply the numerators and denominators. (Carry as necessary).*
- *Reduce as required.*

MATH EXERCISES—Dividing Complex Fractions

______________ 1. $\dfrac{\frac{1}{10}}{3}$

______________ 2. $\dfrac{3}{\frac{1}{2}}$

______________ 3. $\dfrac{\frac{17}{25}}{34}$

______________ 4. $\dfrac{16\frac{2}{3}}{33\frac{1}{3}}$

______________ 5. $\dfrac{30}{\frac{4}{30}}$

______________ 6. $\dfrac{16}{\frac{1}{4}}$

______________ 7. $\dfrac{\frac{13}{16}}{2}$

______________ 8. $\dfrac{2\frac{1}{4}}{\frac{5}{6}}$

PRACTICAL APPLICATIONS—Dividing Complex Fractions

______________ 9. **Construction:** Cripples (short braces) are to be cut from a 2×4 that is 8′ long. Disregarding the saw kerf, how many 8¼″ cripples can be cut from the 8′ piece?

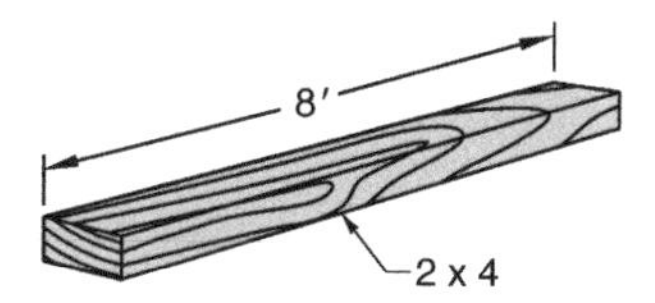

_______________ **10. Boiler Operation:** A fuel oil tank contains 4000 gal. This is ¼ of the fuel used in a week. How much fuel is used in a week?

_______________ **11. Agriculture:** A fence for a 1320′ pasture on one side has three strands of barbed wire with 6″ brace posts at 660′ spacing and 3½″ diameter line posts at 20′ spacing. Calculate the number of brace posts and line posts needed.

_______________ **12. Culinary Arts:** A recipe yields 3 gallons of barley soup. How many 6½ oz servings of barley soup will the recipe yield?

For an interactive review of the concepts covered in Chapter 3, refer to the corresponding Quick Quiz® included on the Digital Resources.

Working with Complex Fractions 3

Review

Name ______________________ Date ____________

Math Problems

______________ 1. $\frac{5}{32} + \frac{11}{32} + \frac{19}{32}$

______________ 2. $2\frac{1}{8} + 7\frac{5}{6}$

______________ 3. $\frac{7}{8} - \frac{3}{8}$

______________ 4. $\frac{1}{4} - \frac{1}{5}$

______________ 5. $15\frac{15}{32} - 7\frac{7}{16}$

______________ 6. $\frac{3}{8} \times \frac{2}{5}$

______________ 7. $\frac{2}{3} \times 5$

______________ 8. $4\frac{3}{8} \times 4$

______________ 9. $\frac{2}{5} \div 3$

______________ 10. $23 \div \frac{3}{8}$

3 Review (continued)

Practical Applications

_______________ 11. **Manufacturing:** Three pieces of steel rod measure 7¼″ each. Two other pieces of steel rod measure 5⅜″ each. What is the total length of the steel rod pieces?

_______________ 12. **Construction:** A landscape contractor has a stockpile containing 50 cu yd of topsoil. One job requires 11½ cu yd, and two small jobs require 4¼ cu yd each. How much topsoil is left in the stockpile?

_______________ 13. **Maintenance:** A barrel of SUS 40 oil that is ¾ full is used to fill the drive gearbox on a flash dryer. The operation uses ⅝ of the oil in the barrel. How much oil is left in the barrel?

_______________ 14. **Manufacturing:** An industrial coil-coating paint line has 730 gallons of paint that is to be used on a particular run. After an 8-hour shift, ⅔ of the paint has been used. How many gallons of paint remain?

_______________ 15. **Alternative Energy:** How many wind turbines can be constructed on 35¾ acres of land if each wind turbine requires 2¾ acres?

Working with Complex Fractions 3

Test

Name ______________________ Date ______________

Math Problems

______ 1. $\frac{3}{8} + \frac{3}{8} + \frac{3}{8}$

______ 2. $\frac{3}{8} + \frac{3}{4} + \frac{3}{16}$

______ 3. $3\frac{5}{32} + 11\frac{1}{2}$

______ 4. $12\frac{1}{4} + 3\frac{1}{2}$

______ 5. $\frac{15}{16} - \frac{5}{16}$

______ 6. $\frac{19}{32} - \frac{3}{16}$

______ 7. $\frac{7}{8} - \frac{3}{4}$

______ 8. $12\frac{1}{2} - 2\frac{1}{16} - 5\frac{7}{8}$

______ 9. $6\frac{3}{8} - 4\frac{15}{16}$

______ 10. $4\frac{3}{4} - 2\frac{3}{8}$

3 Test (continued)

_______________ **11.** $\frac{5}{8} \times \frac{3}{4}$

_______________ **12.** $\frac{7}{3} \times \frac{3}{2}$

_______________ **13.** $\frac{3}{4} \times \frac{16}{3} \times \frac{18}{24} \times \frac{12}{9}$

_______________ **14.** $\frac{3}{8} \times \frac{3}{4} \times \frac{1}{2}$

_______________ **15.** $\frac{3}{4} \times 2$

_______________ **16.** $\frac{1}{27} \times 135$

_______________ **17.** $5\frac{3}{8} \times 6$

_______________ **18.** $5\frac{1}{3} \times 3\frac{3}{4}$

_______________ **19.** $\frac{5}{9} \div \frac{1}{2}$

_______________ **20.** $\frac{1}{15} \times 3$

_______________ **21.** $\frac{3}{4} \div 12$

_______________ **22.** $6\frac{2}{3} \div 3$

3 Test (continued)

______ **23.** $7 \div \frac{1}{12}$

______ **24.** $108\frac{3}{4} \div 10\frac{7}{8}$

______ **25.** $1\frac{3}{10} \div 6\frac{1}{4}$

______ **26.** $\dfrac{\frac{1}{32}}{8}$

______ **27.** $\dfrac{\frac{5}{6}}{17}$

______ **28.** $\dfrac{4}{\frac{1}{10}}$

______ **29.** $\dfrac{4\frac{3}{4}}{3\frac{1}{8}}$

______ **30.** $\dfrac{\frac{3}{8}+\frac{3}{4}}{3}$

______ **31.** $\dfrac{\left(\frac{1}{4}\times\frac{1}{2}\right)+\frac{7}{8}}{\frac{3}{4}}$

3 Test (continued)

_______________ **32.** $\dfrac{\frac{2}{3}}{2\frac{1}{5}-\left(\frac{2}{5}\times\frac{1}{2}\right)}$

Practical Applications

_______________ **33. Construction:** A tractor operator works $4\frac{1}{4}$ hr, $4\frac{3}{4}$ hr, and $2\frac{1}{4}$ hr to grade the right-of-way on a road project. What is the total number of hours required to grade the right-of-way?

_______________ **34. HVAC:** An HVAC technician uses $71\frac{1}{2}$ boxes of $\frac{1}{4}$-20 UNC bolts and $18\frac{1}{4}$ boxes of $\frac{1}{4}$-28 UNF bolts to install an air conditioning unit. How many boxes of bolts did the technician use?

_______________ **35. Electrical:** An electrician needs to bend a 9′-6″ piece of conduit around an obstruction. The amount of shrink (reduced length because of the extra length required to bend around an obstacle) is $\frac{3}{8}$″. What is the length of the conduit after it is bent?

_______________ **36. Construction:** A carpenter cut $5\frac{3}{8}$″ off a $16\frac{1}{2}$″ board. Disregarding the saw kerf, what is the length of the remaining board?

3 Test (continued)

_______________ **37. Construction:** A mason uses $1\frac{1}{4}$ cu yd of sand from a pile containing $4\frac{1}{2}$ cu yd of sand. How much sand is left in the pile?

_______________ **38. Agriculture:** A farmer used $3\frac{5}{8}$ gal. of paint to paint a fence. A neighboring farmer uses $3\frac{1}{4}$ gal. to paint an identical fence. How much more paint did the first farmer use?

_______________ **39. Construction:** A cabinetmaker has 48 door pulls and uses $\frac{3}{4}$ of them on one job. How many door pulls were used on the job?

_______________ **40. Boiler Operation:** A boiler burns $1\frac{3}{4}$ gal. of fuel per minute. How many gal. of fuel are burned in 60 minutes?

_______________ **41. Construction:** A carpenter is laying out a stairway that has a unit run of $7\frac{5}{16}''$. There are 15 treads. What is the total run of the stairway?

3 Test (continued)

_______________ **42. Construction:** How many 7⅛″ lengths of #8 rebar can be cut from a 185¼″ long piece?

_______________ **43. Electrical:** An electrician needs a piece of conduit 42½″ in length to run between two junction boxes. This piece must be cut from a piece 47⅞″ in length. After the needed piece is cut, what is the length of the remaining conduit?

_______________ **44. Culinary Arts:** If a recipe requires 7½ qt of soy sauce but only 1 gal. of soy sauce is in storage, how much soy sauce needs to be purchased?

_______________ **45. Agriculture:** A gate for a perimeter fence requires the use of 1½″ × 5″ bolts. Calculate the thread length of the bolts.

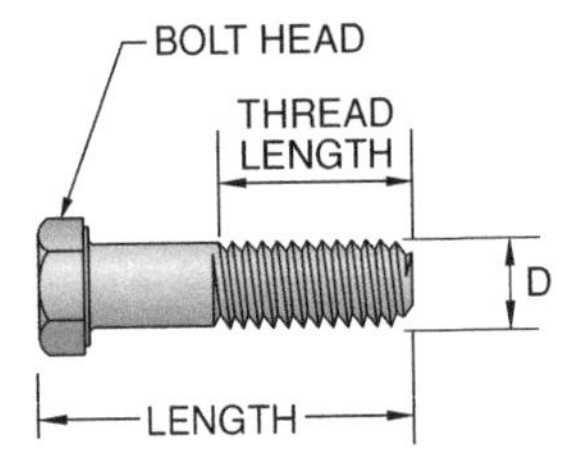

Thread Length	
Length	**Threads**
TO 6″	2D + ¼″
OVER 6″	2D + ½″

4 Working with Decimals

A decimal is a number expressed in base 10. Like whole numbers and fractions, decimals can be added, subtracted, multiplied, and divided. Fractions can be changed to decimals and decimals can be changed to fractions.

OBJECTIVES

1. Explain how a decimal represents part of a whole number.
2. Explain how to round decimals.
3. Add and subtract decimals.
4. Multiply decimals.
5. Divide decimals.
6. Change fractions to decimals.
7. Change decimals to fractions.
8. Change decimals to fractions with a given denominator.

KEY TERMS

- decimal
- decimal fraction
- repeating decimal
- rounding

Digital Resources
ATPeResources.com/QuickLinks
Access Code: 764460

SECTION 4-1 UNDERSTANDING DECIMALS

A *decimal* is a number expressed with 10 as its base. All numbers to the left of the decimal point are whole numbers. All numbers to the right of the decimal point are less than whole numbers. **See Figure 4-1.**

A *decimal fraction* is a fraction with the denominator of 10 or a multiple of 10. A decimal fraction will always have a 0 to the left of the decimal point, for example, 0.3 ($^3/_{10}$) and 0.35 ($^{35}/_{100}$).

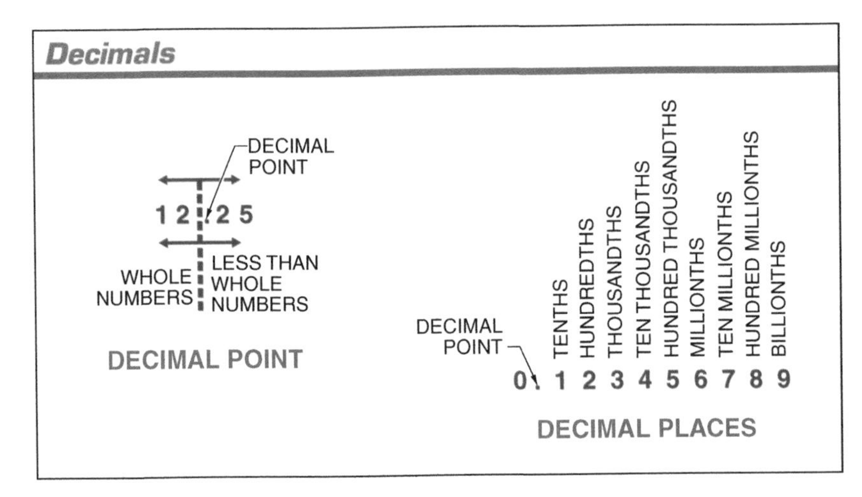

Figure 4-1. A decimal is a number expressed with 10 as its base.

Many monetary systems are based on decimals. **See Figure 4-2.** In both the United States and Canada, for example, the dollar ($1.00) is valued at 100 cents (100¢). Each penny is $^1/_{100}$ of a dollar ($0.01 or 1¢). Each nickel is $^5/_{100}$ of a dollar ($0.05 or 5¢) and so on.

Monetary Values

Currency	Value	Decimal	Fraction
DOLLAR	$1.00	1.00	$\frac{100}{100}$
HALF-DOLLAR	$0.50	0.50	$\frac{50}{100}$
QUARTER	$0.25	0.25	$\frac{25}{100}$
DIME	$0.10	0.10	$\frac{10}{100}$
NICKEL	$0.05	0.05	$\frac{5}{100}$
PENNY	$0.01	0.01	$\frac{1}{100}$

Figure 4-2. Many monetary systems are based on decimals.

When a decimal that represents a monetary amount needs to be written out the word "and" is used. For example, the amount $11.65 is written as "eleven dollars and sixty-five cents." The amount $3002.37 is written as "three thousand two dollars and thirty-seven cents."

Examples—Writing Out Monetary Amounts

1. Write out $0.68.
   ANS: **sixty-eight cents**
   The first decimal figure (6) is in the tenths place and represents 6⁄10 of a dollar, and the last decimal figure (8) is in the hundredths place, which represents 8⁄100 of a dollar.
2. Write out $62.29.
   ANS: **sixty-two dollars and twenty-nine cents**
   The number 62 represents dollars and are whole numbers. The first decimal figure (2) represents 2⁄10 of a dollar, and the last decimal figure (9) represents 9⁄100 of a dollar.
3. Write two thousand seven hundred eighty-three dollars and seventeen cents as a numerical value.
   ANS: **$2783.17**
   Write the whole number 2783, which represents the dollar amount. Then place the decimal point so that the 1 is in the tenths place, which represents 1⁄10 of a dollar. The 7 is in the hundredths place, which represents 7⁄100 of a dollar.

When reading decimals aloud, either "point" or "and" can be used for the decimal point. For example, 9.15 can be read as "nine point fifteen," or, using the name of the decimal place, it can be read as "nine and fifteen hundredths." For a decimal beginning with a zero, the zero may or may not be read. For example, 0.8 can be read as "zero point eight" or just "point eight."

Repeating Decimals

A *repeating decimal* is a decimal that has a repeating number or group of numbers that repeat infinitely (forever). The repeating part of a decimal is indicated with a rule above it. For example, the fraction 1⁄3 has a decimal equivalent of $0.333\overline{3}$. The rule above the last figure indicates that it repeats infinitely.

Rounding Decimals

Rounding is the process of reducing the number of places in a decimal. Decimals can be rounded up or down depending on the degree of accuracy required. Decimals that end in 4 or below are typically rounded down. Decimals that end in 5 or above are typically rounded up. For example, to round 35.768 to the hundredths place, add 1 to the 6, because "8" occupies the ten thousandths. Drop the rest of the decimal. The number 35.768 rounded to the hundredths is 35.77. **See Figure 4-3.**

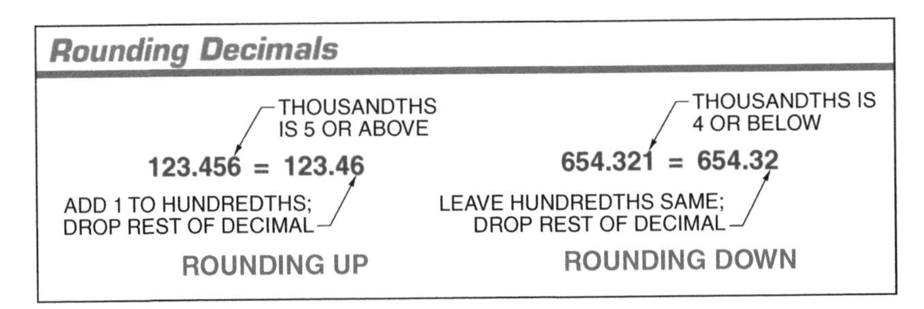

Figure 4-3. Decimals can be rounded to any decimal place depending on the degree of accuracy required.

In measurements, more places in a decimal indicate a higher degree of accuracy. When measuring tolerances on a machined part, the tolerance may be measured in hundredths or thousandths. For example, two parts may have measurements of 1.250″ ± 0.005″ and 1.25″ ± 0.01″. While the overall size of the parts is the same, the degree of accuracy is more critical for the part measured in thousandths.

Examples—Rounding Decimals

1. Round 2.75 to the tenths place.
ANS: **2.8**
Because 5 is in the hundredths place, the number is rounded up.

2. Round 45.8962 the thousandths place.
ANS: **45.896**
Because 2 is in the ten thousandths place, the number is rounded down.

3. Round 0.255559 to the hundredths place.
ANS: **0.26**
Because 5 is in the thousandths place, the number is rounded up.

MATH EXERCISES—Rounding Decimals

__________ **1.** 0.77 (*Round to the tenths place.*)

__________ **2.** 0.0091 (*Round to the hundredths place.*)

__________ **3.** 6.318299 (*Round to 5 places.*)

__________ **4.** 44.5233 (*Round to 3 places.*)

SECTION 4-2 ADDING AND SUBTRACTING DECIMALS

To add or subtract decimals, the numbers are stacked and aligned on their decimal points. Addition and subtraction are then performed as with whole numbers. **See Figure 4-4.**

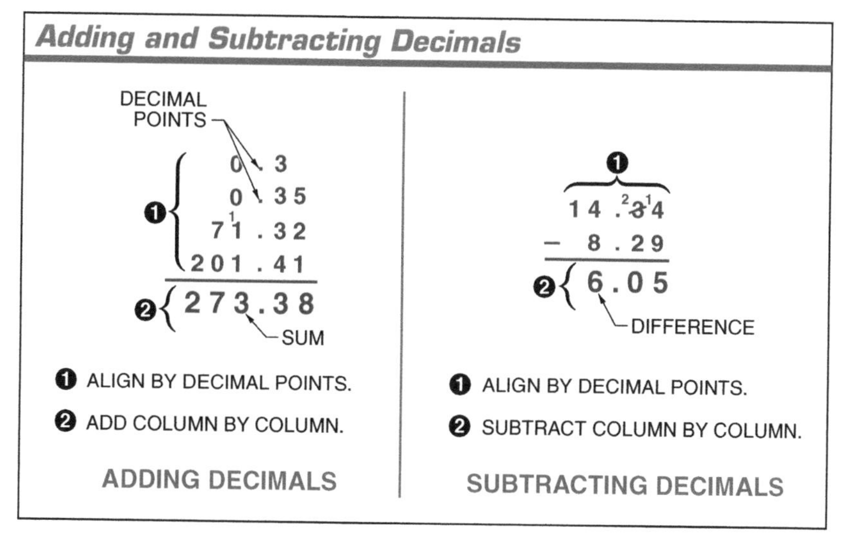

Figure 4-4. To add or subtract decimals, align by decimal points and perform the math process.

Units are added to or subtracted from units, tenths to or from tenths, hundredths to or from hundredths, etc. After the numbers are added or subtracted, the decimal point of the sum or difference is placed directly below the other decimal points.

Examples—Adding and Subtracting Decimals

1. Add \$27.07 and \$8.92.

ANS: **\$35.99**

❶ Align the numbers vertically by the decimal points.

❷ Add column by column, beginning with the right (7 + 2 = 9) and moving toward the left until the sum of all columns is found.

DECIMAL POINTS

❶ 27.07
+ 8.92

❷ 35.99

QUICK REFERENCE

- *Align by decimal points.*
- *Add column by column.*

2. Add \$0.04, \$24.34, \$740.01, and \$1.23.

ANS: **\$765.62**

❶ Align the numbers vertically by the decimal points.

❷ Add column by column, beginning with the right (4 + 4 + 1 + 3 = 12). Carry the 1 to the tens column. Continue moving toward the left until the sum of all columns is found.

```
        1
      0.04
     24.34
❶  740.01
      1.23
   -------
❷  765.62
```

3. Subtract \$275.50 from \$3000.24.

ANS: **\$2724.74**

❶ Align the numbers vertically by the decimal points.

❷ Subtract column by column, beginning with the right (4 – 0 = 4) and moving toward the left until all columns have been subtracted. Borrow as required.

```
    2 9 9 9  1
❶   3 0 0 0 . 2 4
   –  2 7 5 . 5 0
   --------------
❷   2 7 2 4 . 7 4
```

QUICK REFERENCE

- *Align by decimal points.*
- *Subtract column by column.*

4. Subtract \$540.10 from \$2346.32.

ANS: **\$1806.22**

❶ Align the numbers vertically by the decimal points.

❷ Subtract column by column, beginning with the right (2 – 0 = 2) and moving toward the left until all columns have been subtracted. Borrow as required.

```
    1 1
❶   2 3 4 6 . 3 2
   –  5 4 0 . 1 0
   --------------
❷   1 8 0 6 . 2 2
```

MATH EXERCISES—Adding and Subtracting Decimals

________________ **1.** 34.51 + 94.3545 + 2.09847

________________ **2.** 30,234.357 – 345.984

________________ **3.** 45.943 + 9.4329

________________ **4.** 474.84005 – 89.459323

_______________ **5.** 94.097 − 0.00875

PRACTICAL APPLICATIONS—Adding and Subtracting Decimals

_______________ **6. Construction:** A cabinet project requires 35.85 bd ft of cherry, 129.15 bd ft of pine, and 16.15 bd ft of fir. What is the total number of board feet required?

_______________ **7. Construction:** The total wall area of a room is 480 sq ft. The door opening is 16.675 sq ft. Two window openings total 20.375 sq ft. Disregarding the areas of the openings, what is the wall area of the room?

_______________ **8. Manufacturing:** How much does the wheel weigh after it has been machined?

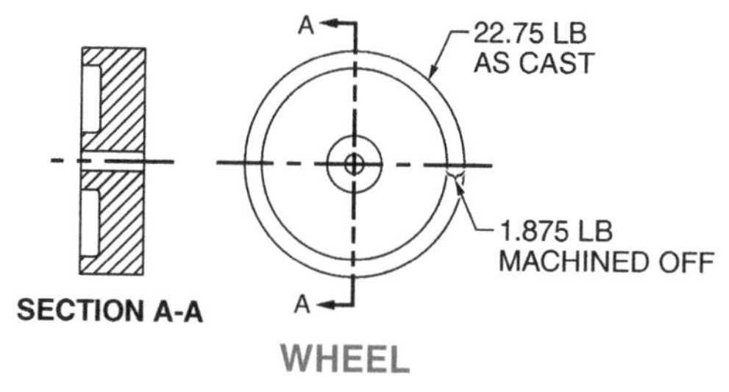

SECTION 4-3 MULTIPLYING DECIMALS

To multiply decimals, multiply in the same way as whole numbers. Add the number of decimal places in the multiplicand and multiplier, and move the decimal point toward the left using the sum. **See Figure 4-5.** For example, to multiply 2.36 by 0.17, multiply and place the decimal four places from the right (2.36 × 0.17 = **0.4012**).

To multiply a decimal by 10, 100, 1000, etc., move the decimal point of the multiplicand as many places to the right as there are zeros in the multiplier. If there are not enough figures, add zeros to the right. For example, to multiply 37.954 by 100, move the decimal point two places to the right (37.954 × 100 = **3795.4**).

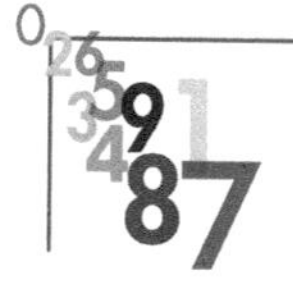

To multiply a decimal or a mixed decimal by 50, multiply the multiplicand by 100 and divide by 2. To multiply a decimal or a mixed decimal by 25, multiply the multiplicand by 100 and divide by 4.

Multiplying Decimals

MULTIPLICAND

$$\begin{array}{r} 2.36 \\ \times\ 0.17 \\ \hline 1652 \\ 2360 \\ \hline 0.4012 \end{array}$$

MULTIPLIER

❶ MULTIPLY SAME AS WHOLE NUMBER.

❷ POINT OFF EQUAL TO DECIMAL PLACES IN MULTIPLICAND AND MULTIPLIER.

MOVE DECIMAL POINT TO RIGHT AS MANY PLACES AS ZEROS IN MULTIPLIER

$37.954 \times 100 = 37.95.4$

MULTIPLYING DECIMALS BY 10, 100, 1000, ETC.

MULTIPLY MULTIPLICAND BY 100 AND DIVIDE BY 2

$0.4 \times 50 = \frac{0.4 \times 100}{2} = 20$

MULTIPLYING DECIMALS BY 50

MULTIPLY MULTIPLICAND BY 100 AND DIVIDE BY 4

$0.4 \times 25 = \frac{0.4 \times 100}{4} = 10$

MULTIPLYING DECIMALS BY 25

Figure 4-5. Multiply decimals the same way as whole numbers and place the decimal point the sum of the number of decimal places in the multiplicand and multiplier.

Examples—Multiplying Decimals

1. Multiply $7.24 by 1.5.

ANS: **$10.86**

❶ Multiply each numeral in the same way as whole numbers (724 × 15).

❷ Move the decimal the number of spaces equal to the sum of the decimal places in the multiplicand and multiplier—in this case, three decimal places from the right.

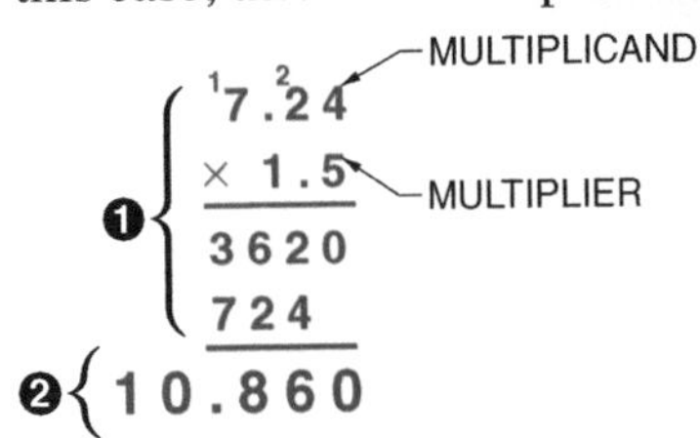

QUICK REFERENCE

- *Multiply each numeral in the same way as whole numbers.*
- *Move the decimal the number of spaces equal to the decimal places in multiplicand and multiplier.*

2. Multiply $0.35 by 25.

ANS: **$8.75**

❶ Multiply each numeral in the same way as whole numbers (35 × 25).

❷ Move the decimal the number of spaces equal to the sum of the decimal places in the multiplicand and multiplier—in this case, two decimal places from the right.

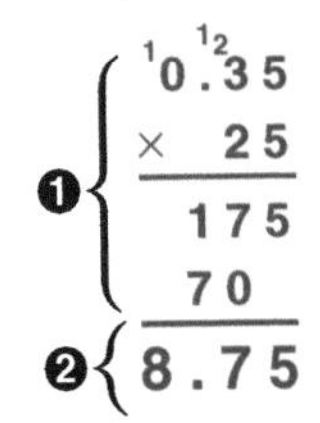

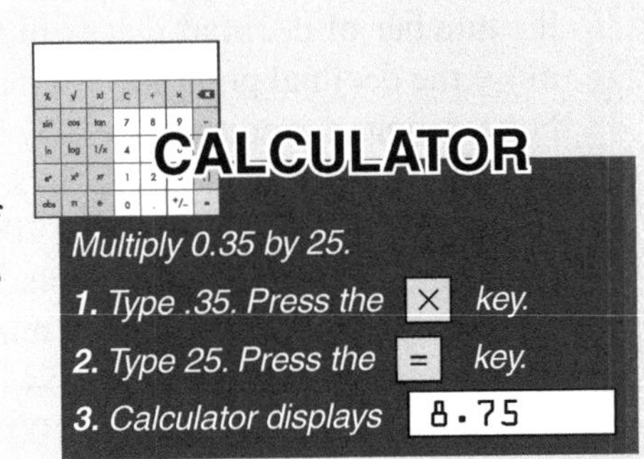

3. Multiply $11.62 by 6.0.

ANS: **$69.72**

❶ Multiply each numeral in the same way as whole numbers (1162 × 6).

❷ Move the decimal the number of spaces equal to the sum of the decimal places in the multiplicand and multiplier—in this case, three decimal places from the right.

```
      1³1.¹6 2
   ×      6.0
❶  ----------
       0 0 0 0
      6 9 7 2
   ----------
❷   6 9 . 7 2 0
```

4. Multiply 9.835 lbs by 100.

ANS: **983.5 lb**

❶ Multiply by moving the decimal point two places to the right.

9.835 × 100 = 9.83.5 = 983.5 ❶

5. Multiply $0.01 by 10,000.

ANS: **$100**

❶ Multiply by moving the decimal point four places to the right.

0.01 × 10,000 = 0.0100.00 = 100 ❶

MATH EXERCISES—Multiplying Decimals

__________ **1.** 0.35 × 4

__________ **2.** 0.785 × 25

__________ **3.** 0.287 × 0.356

__________ **4.** 0.002 × 0.014

__________ **5.** 1.0034 × 2.503

__________ **6.** 0.35 × 12.5

________ **7.** 56.98 × 1000

________ **8.** 24.575 × 50

________ **9.** 2.25 × 25

PRACTICAL APPLICATIONS—Multiplying Decimals

________ **10. Agriculture:** A wire fence consists of 16.5 sections each 9.75′ long. How long is the fence?

________ **11. Construction:** What is the total rise of the stairs?

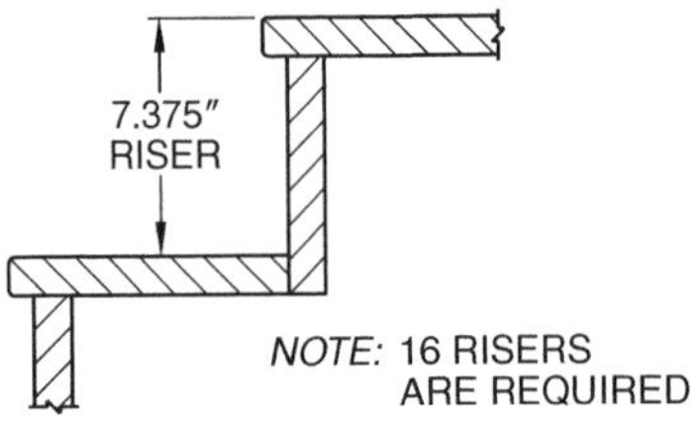

7.375″
RISER

NOTE: 16 RISERS
ARE REQUIRED

STAIR DETAIL

________ **12. Plumbing:** To complete a job, a plumber needs 10 sections of pipe that are each 24.125″ in length. How much pipe is needed?

SECTION 4-4 DIVIDING DECIMALS

To divide decimals, divide in the same way as whole numbers. Subtract the number of decimal places in the divisor from the number of decimal places in the dividend, and move the decimal point from the right using the difference. **See Figure 4-6.** For example, to divide 16.75 by 2.5, divide the same way as whole numbers (1675 ÷ 25 = 67). Point off one decimal place (difference in number of decimal places) from right to left (16.75 ÷ 2.5 = **6.7**).

Dividing Decimals

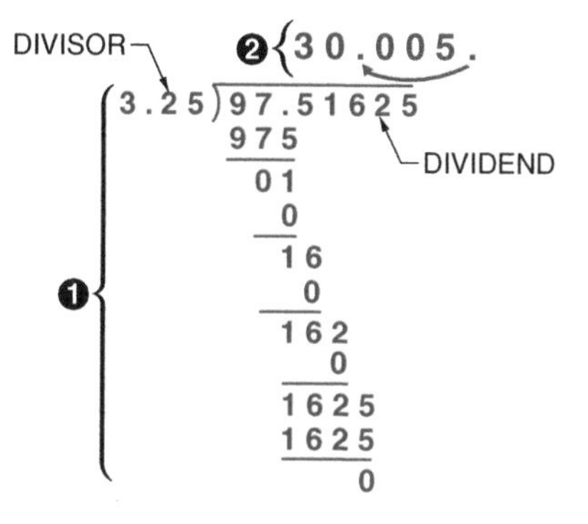

❶ DIVIDE SAME AS WHOLE NUMBERS.

❷ MOVE DECIMAL POINT TO LEFT THE NUMBER OF PLACES EQUAL TO DIFFERENCE IN DECIMAL PLACES BETWEEN DIVIDEND AND DIVISOR.

MOVE DECIMAL POINT TO LEFT AS MANY PLACES AS ZEROS IN DIVISOR

23.56 ÷ 1000 = 0.023.56

DIVIDING DECIMALS BY 10, 100, 1000, ETC.

MULTIPLY BY 2 AND MOVE DECIMAL POINT TWO PLACES TO LEFT

25.4 ÷ 50 = 25.4 × 2 = 0.50.8 = 0.508

DIVIDING DECIMALS BY 50

MULTIPLY BY 4 AND MOVE DECIMAL POINT TWO PLACES TO LEFT

25.4 ÷ 25 = 25.4 × 4 = 1.01.6 = 1.016

DIVIDING DECIMALS BY 25

Figure 4-6. Divide decimals the same way as whole numbers. Move the decimal point the number of decimal places equal to the difference between the decimal places in the dividend and the divisor.

There must be at least as many decimal places in the dividend as in the divisor. If the dividend has fewer decimal places than the divisor (the difference is negative), add zeros to the dividend. (Adding zeros to the right of a decimal does not alter its value.)

To divide a decimal or mixed decimal by 10, 100, 1000, etc., move the decimal point one place to the left for each zero in the divisor. Add zeros as needed.

Sprecher + Schuh

If 1200 cartons are produced in a 8.5 hr shift, then approximately 141 cartons are produced per hour.

To divide a decimal by 50, multiply by 2 and move the decimal point two places to the left. To divide a decimal by 25, multiply by 4 and move the decimal point two places to the left. To divide a decimal by 12.5, multiply by 8 and move the decimal point two places to the left. Add zeros as needed.

Examples—Dividing Decimals

1. Divide $203.74 by 5.

   ANS: **$40.75**

   ❶ Divide each number in the same way as whole numbers (20,374 ÷ 5 = 4074.8).

   ❷ Move the decimal point the number of decimal places equal to the difference between the dividend and the divisor (2 places). Since the dividend represents a monetary value, the quotient also represents a monetary value.

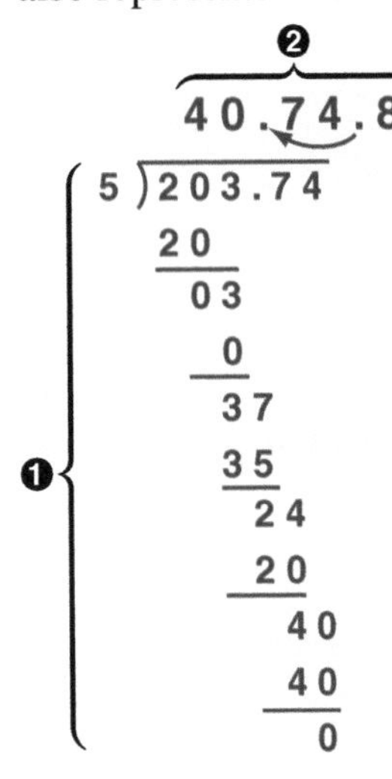

QUICK REFERENCE

- *Divide in the same way as whole numbers.*
- *Move the decimal point the number of decimal places equal to the difference between the dividend and divisor.*

2. Divide $326.52 by 12.

   ANS: **$27.21**

   ❶ Divide each number in the same way as whole numbers (32,652 ÷ 12 = 2721).

   ❷ Move the decimal point the number of decimal places equal to the difference between the dividend and the divisor (2 places). Since the dividend represents a monetary value, the quotient also represents a monetary value.

❷ 27.21.

12)326.52
24
86
84
25
24
12
12
0

❶

3. Divide $400.10 by 10.

   ANS: **$40.01**

   ❶ Divide by moving the decimal point one place to the left.

❶ 400.10 ÷ 10 = 40.010 = **40.01**

MATH EXERCISES—Dividing Decimals

____________ **1.** 3.036 ÷ 0.06

____________ **2.** 3.728 ÷ 0.16

____________ **3.** 0.864 ÷ 0.024

____________ **4.** 10.044 ÷ 0.36

____________ **5.** 0.125 ÷ 8000

____________ **6.** 3.16 ÷ 10

____________ **7.** 40.5 ÷ 25

____________ **8.** 34.75 ÷ 1000

____________ **9.** 245.25 ÷ 50

____________ **10.** 45.75 ÷ 25

PRACTICAL APPLICATIONS—Dividing Decimals

_______________ **11. Construction:** What is the length of C on the step gauge? All steps are equally spaced.

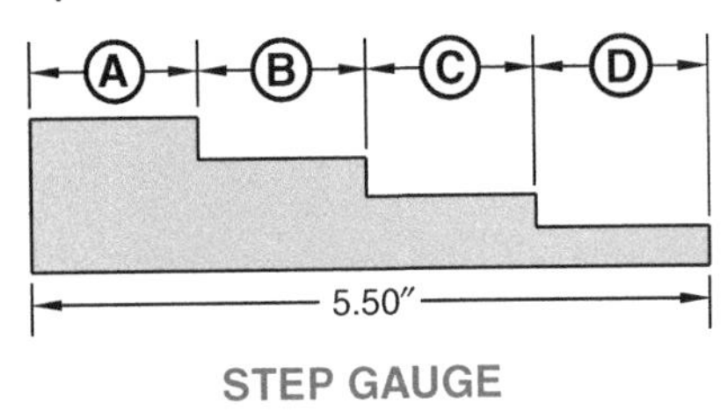

_______________ **12. Manufacturing:** Find the center-to-center distance between holes F and G of the hole gauge.

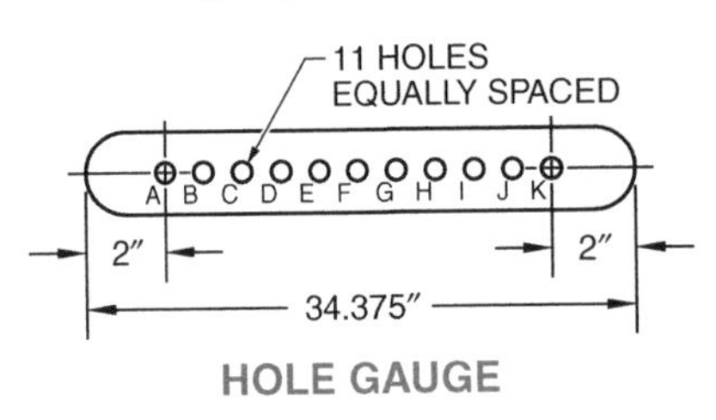

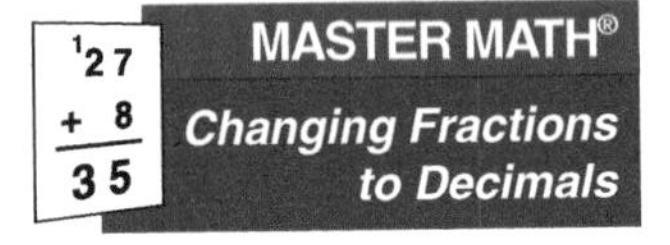

SECTION 4-5 CHANGING FRACTIONS TO DECIMALS

To change a fraction to a decimal, divide the numerator by the denominator. Add zeros if needed. **See Figure 4-7.** A decimal equivalents table can also be used.

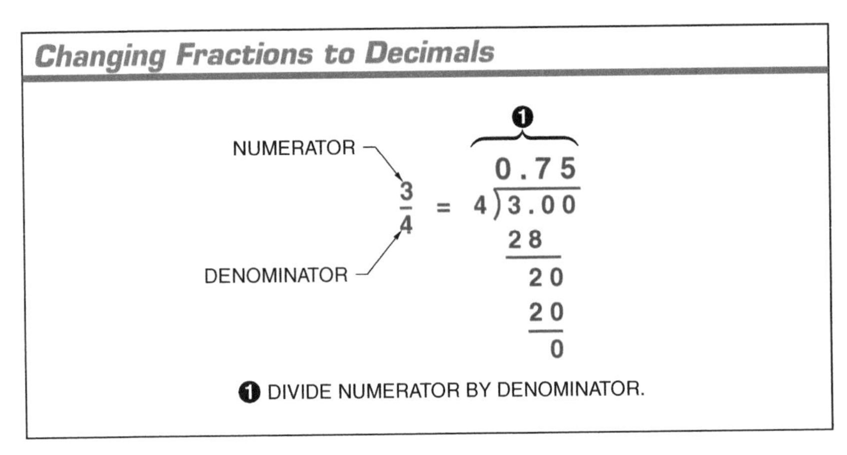

Figure 4-7. To change a fraction to a decimal, divide the numerator by the denominator.

Example—Changing Fractions to Decimals

1. Change ⅝ to a decimal.
ANS: **0.625**
❶ Divide the numerator 5 by the denominator 8.

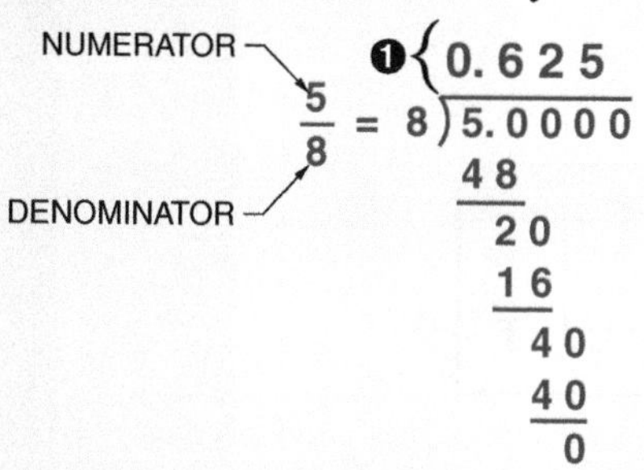

QUICK REFERENCE

- *Divide numerator by denominator.*

MATH EXERCISES—Changing Fractions to Decimals

Change the fractions to decimals.

__________________ **1.** 5⁄32

__________________ **2.** 35⁄100

__________________ **3.** 47⁄150 (*Round to 2 places.*)

__________________ **4.** 17⁄32

__________________ **5.** 3⁄8

__________________ **6.** 9⁄32

__________________ **7.** 3⁄4

__________________ **8.** 19⁄64

PRACTICAL APPLICATIONS—Changing Fractions to Decimals

_______________ **9. Plumbing:** What is the unknown dimension (X) in the diagram? Write the answer as a decimal.

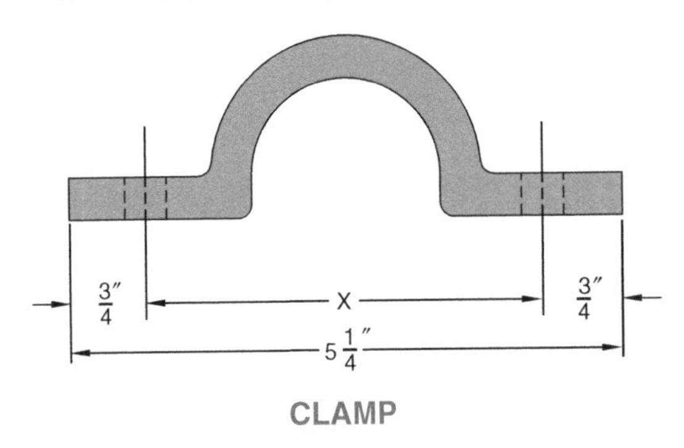

CLAMP

_______________ **10. Manufacturing:** The distance across flats on a bolt head is $\frac{11}{16}''$. What is the distance across flats on the bolt in decimal form?

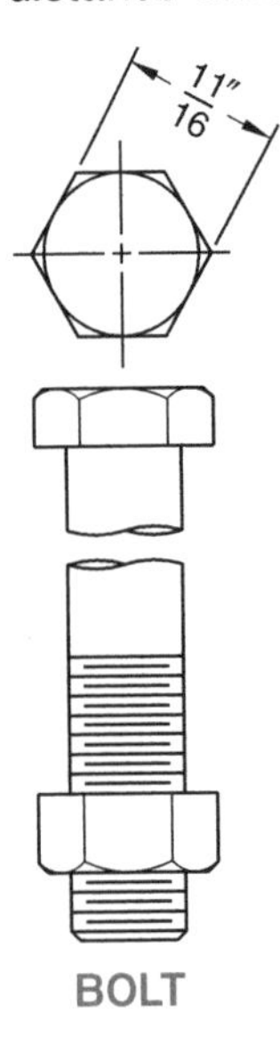

BOLT

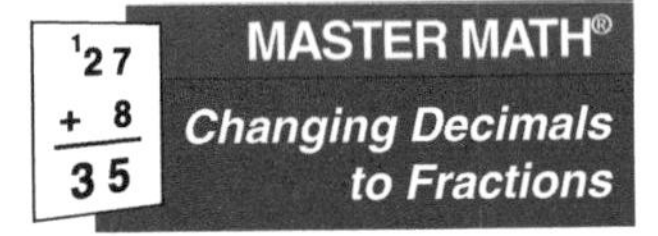

SECTION 4-6 CHANGING DECIMALS TO FRACTIONS

To change a decimal to a fraction, use the decimal as the numerator. For the denominator, place a 1 followed by as many zeros as there are figures in the decimal to the right of the decimal point. **See Figure 4-8.**

For example, to change the decimal 0.35 to a fraction, use the 35 as the numerator and place a 1 followed by two zeros (number of figures to right of the decimal point) as the denominator ($0.35 = \frac{35}{100}$).

Changing Decimals to Fractions

Figure 4-8. To change a decimal to a fraction, use the decimal as the numerator.

Examples—Changing Decimals to Fractions

1. Change 0.3 to a fraction.

ANS: 0.3 = 3⁄10

❶ Use the 3 as the numerator.

❷ Place 1 followed by a zero as the denominator.

DECIMAL

$0.3 = \frac{3}{10}$ ❶ ❷

2. Change 0.045 to a fraction.

ANS: 45⁄1000

❶ Use the 045 as the numerator.

❷ Place 1 followed by three zeros as the denominator (045⁄1000). In this example, the zero in the numerator has no value and can be dropped.

$0.045 = \frac{045}{1000} = \frac{45}{1000}$ ❶ ❷

3. Change the decimal 7.002 to a fraction.

ANS: 7 1⁄500

❶ Use 002 as the numerator.

❷ Place a 1 followed by three zeros as the denominator and drop the zeros in the numerator (2⁄1000). Reduce 2⁄1000 to 1⁄500.

$7.002 = 7\frac{002}{1000} = 7\frac{2}{1000} = 7\frac{1}{500} = 7\frac{1}{500}$ ❶ ❷

QUICK REFERENCE

- *Use the decimal as the numerator.*
- *Place 1 plus zeros equal to the number of decimal figures as the denominator. Reduce to lowest terms (if needed).*

MATH EXERCISES—Changing Decimals to Fractions

Change the decimals to fractions.

________________ **1.** 0.325

________________ **2.** 0.004

________________ **3.** 0.0205

________________ **4.** 9.3

________________ **5.** 0.1930

________________ **6.** 7.114

________________ **7.** 8.375

PRACTICAL APPLICATIONS—Changing Decimals to Fractions

________________ **8. Construction:** The rise of a stair is 3.75 inches. Give the measurement as a fraction.

________________ **9. Manufacturing:** A machined part has a dimensional tolerance of ±0.06. Give the tolerance as a fraction.

______________ **10. Manufacturing:** On the bracket, what are the overall length, width, and height as fractions?

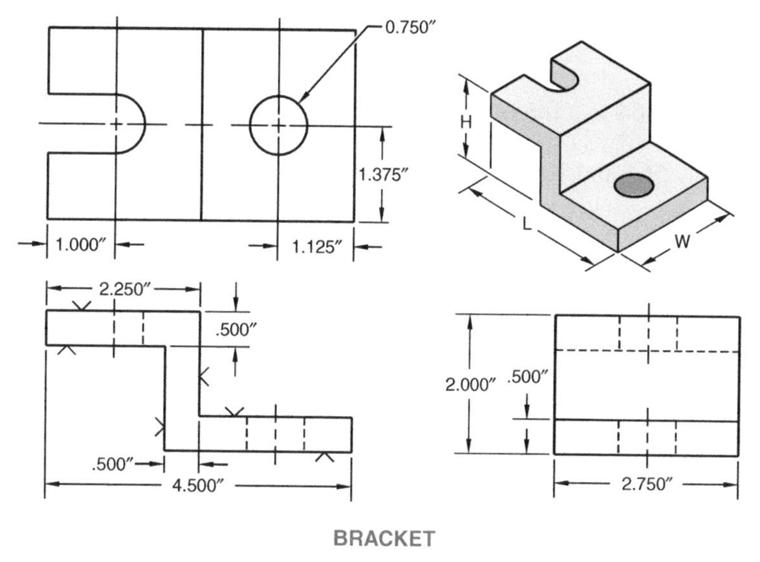

BRACKET

Decimals may need to be changed to fractions in order to make a measurement with a tape measure. Also, a decimal equivalents table may be used.

Changing Decimals to Fractions with a Given Denominator

To change a decimal to a fraction with a given denominator, change the decimal to a fraction and multiply the numerator and denominator by the given denominator. **See Figure 4-9.** Round if necessary.

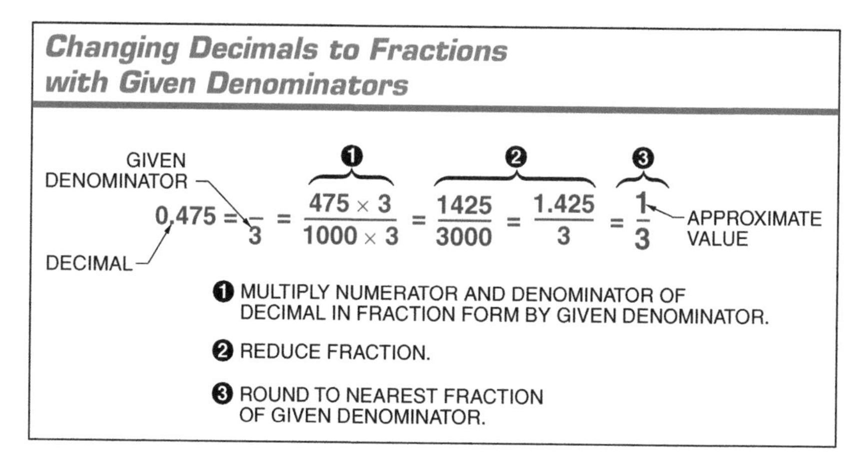

Figure 4-9. To change a decimal to a fraction, multiply the numerator and denominator by the given denominator.

Examples—Changing Decimals to Fractions with a Given Denominator

1. Change 0.564 to sixty-fourths.

ANS: **36/64**

❶ Change 0.564 to a fraction (0.564 = 564/1000).

❷ Multiply 564 and 1000 by 64 $\left(\frac{564 \times 64}{1000 \times 64} = \frac{36{,}096}{64{,}000}\right)$.

❸ Reduce the fraction $\left(\frac{36{,}096}{64{,}000} = \frac{36.096}{64}\right)$.

❹ Round to the nearest fraction of the given denominator. The approximate value is **36/64**.

$$0.564 = \overbrace{\frac{564}{1000}}^{❶} = \overbrace{\frac{564 \times 64}{1000 \times 64}}^{❷} = \overbrace{\frac{36{,}096}{64{,}000} = \frac{36.096}{64}}^{❸} = \overbrace{\frac{36}{64}}^{❹}$$

DECIMAL (0.564); APPROXIMATE VALUE (36/64)

2. Change 1.33 to ninths.

ANS: **1 3/9**

❶ Change 1.33 to a fraction (1.33 = 133/100).

❷ Multiply 133 and 100 by 9 $\left(\frac{133 \times 9}{100 \times 9} = \frac{1197}{900}\right)$.

❸ Reduce and round the fraction $\left(\frac{1197}{900} = \frac{11.97}{9} = \frac{12}{9}\right)$.

❹ Convert to a mixed number. (12/9 = 1 3/9).

$$1.33 = \overbrace{\frac{133}{100}}^{❶} = \overbrace{\frac{133 \times 9}{100 \times 9}}^{❷} = \overbrace{\frac{1197}{900} = \frac{11.97}{9} = \frac{12}{9}}^{❸} = \overbrace{1\frac{3}{9}}^{❹}$$

MATH EXERCISES—Changing Decimals to Fractions with a Given Denominator

________________ **1.** Change 0.756 to twelfths.

________________ **2.** Change 0.875 to sixty-fourths.

________________ **3.** Change 0.719 to thirty-seconds.

_______________ **4.** Change 0.45 to sixteenths.

_______________ **5.** Change 0.622 to eighths.

_______________ **6.** Change 0.93 to fourths.

PRACTICAL APPLICATIONS – Changing Decimals to Fractions with a Given Denominator

_______________ **7. Mechanics:** The diameter of a drive motor shaft is 0.9375. What is the diameter in sixteenths?

_______________ **8. Construction:** A reveal needs to be cut into a piece of maple that is 0.5625″ wide and 0.4375″ deep. What is the reveal dimensions in sixteenths?

_______________ **9. Manufacturing:** The recommended clearance hole for a 7⁄16–14 tap is 0.48875″. What is the clearance hole in thirty-seconds?

_______________ **10. Plumbing:** The outside diameter of ¼″ copper tube is 0.375″. What is the outside diameter in eighths?

_______________ 11. **Agriculture:** A 24.2′ × 30.5′ area is to be excavated to a depth of 8.5′. How many cubic feet of earth will need to be removed? Express the decimal number in eighths.

_______________ 12. **Alternative Energy:** Given that the standard flow rate per unit area of an unglazed solar collector is 51.5 lb/(hr ft²) and the standard flow rate for a glazed solar collector is 14.7 lb/(hr ft²), what is the difference in the standard flow rates? Express the decimal number in fourths.

For an interactive review of the concepts covered in Chapter 4, refer to the corresponding Quick Quiz® included on the Digital Resources.

Working with Decimals 4
Review

Name ______________________ Date ____________

Math Problems

______________ **1.** $34 \div 0.8$

______________ **2.** 34.9805×1.156

______________ **3.** $37.03 + 0.521 + 0.9 + 1000 + 4000.0014$

Change the following to fractions and reduce to lowest terms.

______________ **4.** 4.375

______________ **5.** 14.75

Change the following to decimals.

______________ **6.** $8\frac{5}{8}$

______________ **7.** $12\frac{9}{16}$

Practical Applications

______________ **8.** **Welding:** Manganese bronze contains the following amounts of metals per pound: copper 0.89 lb, tin 0.10 lb, and manganese 0.01 lb. How much copper is in a shaft that weighs 235 lb?

4 Review (continued)

______ **9. Mechanics:** The full-loaded current of a 1 HP, 230 V, 3ϕ, AC induction motor is 3.6 A. What is the full-loaded current as a fraction?

______ **10. Construction:** A type S, size 4 steel bar has a diameter of 0.250″. What is the diameter as a fraction?

______ **11. Manufacturing:** A machinist cuts five pieces from a bar of steel 25.5″ long. How much bar is left if the five pieces are 3.5″, 3.75″, 4.75″, 4.24″, and 4.0625″ long? Disregard the saw kerf.

______ **12. Construction:** An iron bar is 24″ long, 2⅝″ wide, and ⅛″ thick. A cubic inch of iron weighs 0.261 lb. (cu in. = $l \times w \times t$). What is the weight of the iron bar?

______ **13. Welding:** A pound of white brass is composed of 0.64 lb of tin, 0.02 lb of copper, and 0.34 lb of zinc. How many pounds of tin are in a 7.125 lb white brass casting?

______ **14. Pipefitting:** What length copper pipe is required so that two 3.75″ long pieces, four 2.875″ long pieces, and one 12.75″ long piece can be cut from it?

______ **15. Construction:** A contractor seals four driveways with areas of 84.8 sq yd, 124.1 sq yd, 98.9 sq yd, and 344.7 sq yd. How many drums of sealant will be needed if each drum covers 150 sq yd?

Working with Decimals 4
Test

Name ______________________ Date ____________

Math Problems

_______ 1. 34.895×0.56

_______ 2. $900 - 0.009$ (*Round to the hundredths place.*)

_______ 3. $0.05643 \div 0.00456$

_______ 4. $9.4565 - 3.094562$

_______ 5. 3.54×25

_______ 6. $12 \div 0.54$ (*Round to the hundredths place.*)

_______ 7. $0.30084 + 345.323 + 87.0038 + 0.080926$

_______ 8. 134.124×1000

_______ 9. $94.094 - 38.003493$

_______ 10. $0.0454 + 0.857 + 67.0989 + 0.00004487 + 5.97$

4 Test (continued)

______ 11. 0.058 + 0.2591 + 2.15

______ 12. $1\frac{7}{8} \times 0.32$

______ 13. $4\frac{3}{5} \div 0.8$.

______ 14. $0.453 \div 0.3$

______ 15. Subtract forty-four ten thousandths from 12.3816.

______ 16. 6.08×4.29

______ 17. $83.6 \div \frac{2}{5}$

______ 18. What is the product of $7.6 \times 2.81 \times 4.5$?

______ 19. $79.62 \div 0.4$

______ 20. Subtract twenty-two ten thousandths from 10.0302.

______ 21. What is the sum of twenty-six and twenty-six hundredths + seven tenths + six and eighty-three thousandths + four and seven thousandths?

4 Test (continued)

Change the following to fractions and reduce to lowest terms.

__________ **22.** 0.750

__________ **23.** Change 0.90 to tenths.

__________ **24.** 25.25

__________ **25.** 836.125

__________ **26.** 0.250

__________ **27.** Change 0.9375 to sixteenths.

__________ **28.** Change 0.625 to eighths.

__________ **29.** Change 4.33 to thirds.

Change the following to decimals.

__________ **30.** $\frac{2}{25}$

__________ **31.** $6\frac{7}{8}$ (*Round to 3 places.*)

4 Test (continued)

______ **32.** $4\frac{5}{32}$ (*Round to the hundred thousandths place.*)

______ **33.** Change three hundred twenty-two ten thousandths to a decimal.

______ **34.** Change seventeen and twenty-six hundredths to a decimal.

______ **35.** Change six thousand, two hundred forty-six dollars and thirty-seven cents to a decimal.

Practical Applications

______ **36. Construction:** In estimating an interior painting job, ceilings and walls of five rooms have areas (after subtracting all openings) of 190.8 sq ft, 162.6 sq ft, 128.5 sq ft, 202.4 sq ft, and 98.2 sq ft. What is the total area to be painted?

______ **37. Electrical:** What is the cost of 36.5′ of electrical wiring at $6.50 per foot?

______ **38. Electrical:** An electrician needs three sections of ¾″ electrical metallic tubing (EMT), 12.75′ long each. What is the total length of EMT needed?

4 Test (continued)

______ **39. Construction:** A contractor tiles a 15.5′ × 13.8′ family room and a 22.3′ × 5.0′ hallway. What is the total area tiled?

______ **40. Construction:** A laborer works 7.86 hr, 5.83 hr, and 6.75 hr installing windows. At $10.50 per hour, how much money did the laborer make?

______ **41. Electrical:** A 121.5″ cable is divided into nine pieces. How long is each segment?

______ **42. HVAC:** An HVAC technician works a total of 40.25 hours in a week. Calculate the technician's weekly pay at a rate of $21.50 per hour. *(Round to the nearest cent.)*

______ **43. Boiler Operation:** The weight of water is calculated by multiplying the density of the water by its volume. If the density of water is 8.34 lb/gal., calculate the weight of water in an 800 gal. boiler.

4 Test (continued)

______ **44. Construction:** A tradesworker uses a calculator to add the dimensions 7⅜″ and 24⅛″ to repair a hole in a section of drywall. Change these fractions to decimals so that they can be entered into the calculator.

______ **45. Electrical:** A parallel circuit contains a blender with a load of 0.754 A, a waffle iron with a load of 5.701 A, a lamp with a load of 1.07 A, and an oscillating fan with a load of 4.0 A. Add the individual loads to calculate the total load on the circuit.

5 Working with Percentages

A percentage expresses a part of a whole in terms of hundredths. Many common problems, including determining the cost of an item, involve percentages. Percentages are also used to calculate the amount of tax charged for an item.

OBJECTIVES

1. Explain how a percentage represents part of a whole.
2. Express percentages as fractions and decimals.
3. Change mixed number percentages to fractions.
4. Change percentages greater than 100% to decimals and then fractions.
5. Calculate the percentage of a base amount.
6. Calculate the percent rate of a base amount.
7. Calculate a base amount using a percentage and percentage rate.
8. Calculate a marked price based on a discount and percent discount.
9. Calculate a discount and percent discount.
10. Calculate a net price and a net price after multiple discounts.
11. Calculate tax.
12. Calculate total cost.

KEY TERMS

- percentage
- percent rate
- base amount
- marked price
- discount
- net price
- percent discount
- multiple discount
- tax

Digital Resources
ATPeResources.com/QuickLinks
Access Code: 764460

SECTION 5-1 UNDERSTANDING PERCENTAGES

A *percentage* is a number that represents part of a whole and is expressed as part of 100. The percent sign (%) indicates the number is a part of 100. **See Figure 5-1.** For example, a copper casting alloy (100%) consists of 88% copper (88 parts of 100), 10% tin (10 parts of 100), and 2% zinc (2 parts of 100).

Percentages are typically written as whole numbers (15%) but can also be written as fractions (¼%), mixed numbers (15-⅓%), or decimals (8.9%). Percentages that represent more than one whole are greater than 100%, such as 225%.

The use of percentages depends on the task being performed. For example, fractions such as ¼ and decimals such as 0.25 may be used to represent 25% of a whole, such as 1 inch, when using a measuring tape or a ruler. Percentages are often used to indicate the amount of work that needs to be completed on a project, to calculate costs that involve discounts, and to calculate tax rates.

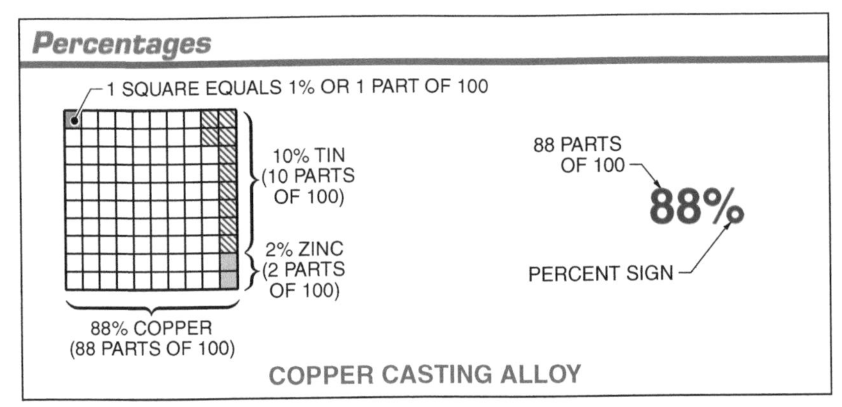

Figure 5-1. A percentage is part of a whole expressed in parts per hundred.

Fraction and Decimal Equivalents

All percentages have fraction and decimal equivalents. **See Appendix.** For example, the fraction ½ has a decimal equivalent of 0.5 and a percentage equivalent of 50% (½ = **0.5** or **50%**). Because percentages, fractions, and decimals all represent whole and partial numbers, each can be converted to another without changing their values.

To change a percentage to a fraction, set the numerator of the fraction equal to the percentage and set the denominator equal to 100. For example, to change 70% to a fraction, set the numerator equal to 70 and set the denominator equal to 100 (70% = **70⁄100**).

To change a fraction to a percentage, divide the numerator by the denominator and multiply by 100. For example, to change ⅛ to a percentage, divide 1 by 8 and multiply by 100 (1 ÷ 8 × 100 = **12.5%**)

To change a percentage to a decimal, divide by 100 (move the decimal point two places to the left). For example, to change 75% to a decimal, divide 75 by 100 (75% ÷ 100 = **0.75**).

To change a decimal to a percentage, multiply by 100 (move the decimal point two places to the right). For example, to change 0.25 to a percentage, multiply 0.25 by 100 (0.25 × 100 = **25%**).

Examples—Percentages and Their Fraction and Decimal Equivalents

1. Change 35% to a fraction.

ANS: 35/100

❶ Place percentage (35%) over 100.

$$35\% = \frac{35}{100} \Big\} ❶$$

2. Change 3/16 to a percentage.

ANS: **18.75%**

❶ Divide the numerator (3) by the denominator (16) (3 ÷ 16 = 0.1875).

❷ Multiply the quotient (0.1875) by 100.

$$\overbrace{\frac{3}{16} = 0.1875}^{❶} \times \overbrace{100}^{❷} = 18.75\%$$

3. Change 71% to a decimal.

ANS: **0.71**

❶ Divide the percentage (71%) by 100.

$$❶ \Big\{ \frac{71\%}{100} = 0.71$$

4. Change 0.54 to a percentage.

ANS: **54%**

❶ Multiply the decimal (0.54) by 100.

$$❶ \{ 0.54 \times 100 = 54\%$$

Changing Mixed Number Percentages

To change percentages that are mixed numbers to proper fractions, change the mixed number to an improper fraction, and place over 100. **See Figure 5-2.** Divide the improper fraction (numerator) by 100 (multiply by reciprocal). Reduce the fraction.

Changing Mixed Number Percentages to Proper Fractions

$$3\frac{3}{4}\% = \overbrace{\frac{15}{4}\%}^{❶} = \overbrace{\frac{\frac{15}{4}}{100} = \frac{15}{4} \times \frac{1}{100}}^{❷} = \overbrace{\frac{15}{400} = \frac{3}{80}}^{❸}$$

❶ CHANGE TO IMPROPER FRACTION.

❷ DIVIDE IMPROPER FRACTION BY 100 (MULTIPLY BY RECIPROCAL).

❸ REDUCE AS REQUIRED.

Figure 5-2. To change mixed number percentages to proper fractions, change the percentage to an improper fraction and divide by 100.

Examples—Changing Mixed Number Percentages to Proper Fractions

1. Change $43\frac{3}{4}\%$ to a proper fraction.

   ANS: $\frac{7}{16}$

   ❶ Change $43\frac{3}{4}$ to $\frac{175}{4}$.

   ❷ Divide improper fraction $\frac{175}{4}$ by 100 (multiply by reciprocal).

   ❸ Reduce as required.

$$43\frac{3}{4}\% = \underbrace{\frac{175}{4}}_{❶} = \underbrace{\frac{\frac{175}{4}}{100} = \frac{175}{4} \times \frac{1}{100}}_{❷} = \underbrace{\frac{175}{400} = \frac{7}{16}}_{❸}$$

2. Change $38\frac{3}{4}\%$ to a proper fraction.

   ANS: $\frac{31}{80}$

   ❶ Change $38\frac{3}{4}$ to $\frac{155}{4}$.

   ❷ Divide improper fraction $\frac{155}{4}$ by 100 (multiply by reciprocal).

   ❸ Reduce as required.

$$38\frac{3}{4}\% = \underbrace{\frac{155}{4}}_{❶} = \underbrace{\frac{\frac{155}{4}}{100} = \frac{155}{4} \times \frac{1}{100}}_{❷} = \underbrace{\frac{155}{400} = \frac{31}{80}}_{❸}$$

QUICK REFERENCE

- *Change to an improper fraction.*
- *Divide the fraction by 100.*
- *Reduce as required.*

Changing Percentages Greater Than 100%

Percentages greater than 100% are one whole unit and a part of another unit added together. To change a percentage greater than 100% to a decimal and then a fraction, divide the percentage by 100, divide the resulting decimal by 100, and reduce. **See Figure 5-3.**

Changing Percentages Greater Than 100% to Decimals and Fractions

$$120\% = \underbrace{\frac{120}{100}}_{❶} = 1.20 = \underbrace{1\frac{20}{100}}_{❷} = \underbrace{1\frac{\cancel{20}^{1}}{\cancel{100}_{5}} = 1\frac{1}{5}}_{❸}$$

❶ DIVIDE PERCENTAGE BY 100.

❷ DIVIDE DECIMAL NUMBER BY 100.

❸ REDUCE AS REQUIRED.

Figure 5-3. To change a percentage greater than 100% to a decimal, divide the percentage by 100. To change the decimal to a fraction, divide by 100 and reduce.

Examples — Changing Percentages Greater Than 100% to Decimals and Fractions

1. Change 125% to a decimal and then to a fraction.

ANS: **$1\frac{1}{4}$**

❶ Divide 125 by 100 ($125 \div 100 = 1.25$).

❷ Multiply 0.25 by 100 and place over 100 ($0.25 \times 100 = 25$; $\frac{25}{100}$).

❸ Reduce to $1\frac{1}{4}$.

$$125\% = \underbrace{\frac{125}{100}}_{❶} = 1.25 = \underbrace{1\frac{25}{100}}_{❷} = \underbrace{1\frac{\cancel{25}^{\,1}}{\cancel{100}_{\,4}}}_{❸} = 1\frac{1}{4}$$

QUICK REFERENCE

- *Divide by 100.*
- *Place the decimal over 100.*
- *Reduce as required.*

2. Change 330% to a decimal and then to a fraction.

ANS: **$3\frac{3}{10}$**

❶ Divide 330 by 100 ($330 \div 100 = 3.30$).

❷ Multiply 0.30 by 100 and place over 100 ($0.30 \times 100 = 30$; $\frac{30}{100}$).

❸ Reduce to $3\frac{3}{10}$.

$$330\% = \underbrace{\frac{330}{100}}_{❶} = 3.30 = \underbrace{3\frac{30}{100}}_{❷} = \underbrace{3\frac{\cancel{30}^{\,3}}{\cancel{100}_{\,10}}}_{❸} = 3\frac{3}{10}$$

MATH EXERCISES — Percentages and Their Fraction and Decimal Equivalents

________________ **1.** Change $\frac{1}{4}$ to a percent.

________________ **2.** Change 325% to a fraction.

________________ **3.** Change 14% to a fraction and reduce as required.

________________ **4.** Change 0.22 to a percent.

________________ **5.** Change 275% to a decimal.

______________ **6.** What is A in decimal form?

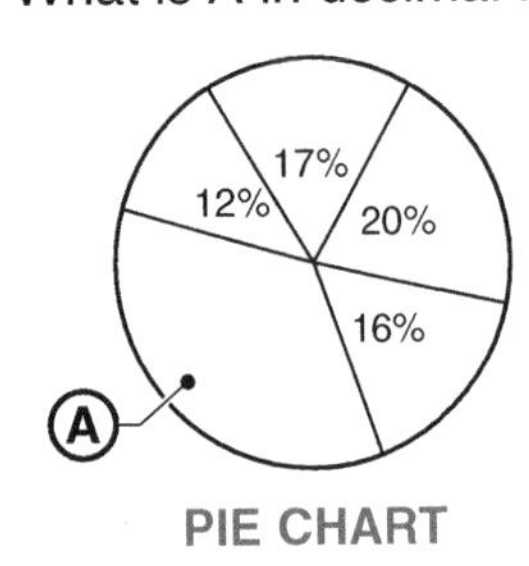

SECTION 5-2 SOLVING PERCENTAGE PROBLEMS

In percentage problems, there are three quantities that may need to be calculated: percentage, percent rate, and base amount. Each requires a specific calculation.

A percentage is a part of a whole amount. The *percent rate* is the percent value of the amount (how much) and is the quantity found before the percent sign. The *base amount* is the whole amount. For example, in the statement 3 is 25% of 12, the 3 is a percentage, 25% is the percent rate, and 12 is the whole amount.

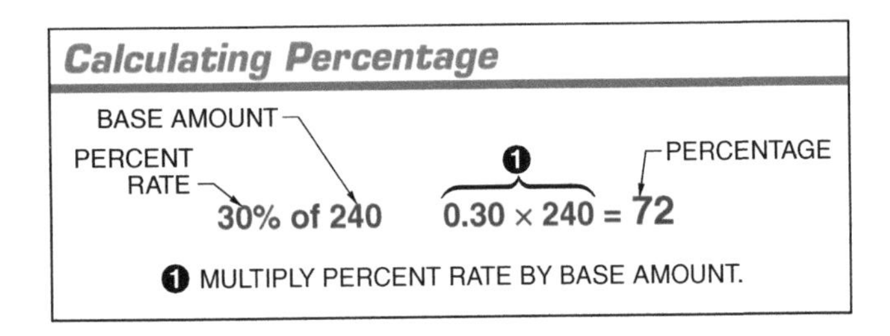

Figure 5-4. A percentage of a total can be found when both the percent rate and base amount are known.

Calculating Percentage

It is often necessary to calculate the percentage of a total, or base amount, such as the percentage of a job that has been completed. To find the percentage of a base amount, multiply the percent rate in decimal form by the base amount. **See Figure 5-4.** For example, to find 30% of 240, change 30% to a decimal (30 ÷ 100 = 0.30). Multiply 240 by 0.30 (240 × 0.30 = **72**).

Examples—Calculating Percentage

1. Calculate 8% of $245.50.

ANS: **$19.64**

➊ Change 8% to a decimal (8 ÷ 100 = 0.08). Multiply by 245.50.

$$\underbrace{0.08 \times 245.50}_{❶} = \$19.64$$

2. Calculate 2.7% of 54.

ANS: **1.458**

➊ Change 2.7% to a decimal (2.7 ÷ 100 = 0.027). Multiply by 54.

$$\underbrace{0.027 \times 54}_{❶} = 1.458$$

QUICK REFERENCE

- *Multiply the percent rate (decimal form) by the base amount.*

3. Calculate 135% of 750.

ANS: **1012.5**

❶ Change 135% to a decimal (135 ÷ 100 = 1.35). Multiply by 750.

$$\overbrace{1.35 \times 750}^{❶} = 1012.5$$

4. Calculate 21⅞% of 240.

ANS: **52.5**

❶ Change 21⅞% to a decimal (7 ÷ 8 = 0.875 + 21 = 21.987 ÷ 100 = 0.21875). Multiply by 240.

$$\overbrace{0.21875 \times 240}^{❶} = 52.5$$

MATH EXERCISES—Calculating Percentage

________________ **1.** Find 20% of 35.

________________ **2.** Find 22% of 40.

________________ **3.** Find 12% of 2125.

PRACTICAL APPLICATIONS—Calculating Percentage

________________ **4. Boiler Operation:** A boiler generates 20,000 lb of steam/hr, 75% of which is recovered as condensate. How much condensate is recovered?

________________ **5. Alternative Energy:** A solar photovoltaic system provided 21% of the electricity used by an office building in a certain month. If the total electricity used was 18,500 kWh, how much was provided by solar power?

______________ 6. **Manufacturing:** If 8.33% of a box of 840 bolts are defective, how many bolts are defective? *(Round answer.)*

Calculating Percent Rate

To find a percent rate, or "how much," divide the percentage by the base amount and multiply by 100. **See Figure 5-5.** Another method for finding the percent rate is to reduce the fraction (the percent rate divided by the base amount) to its lowest terms and then locate the fraction on the Decimal Equivalents table. **See Appendix.** Multiply the corresponding decimal by 100 to find the percent rate.

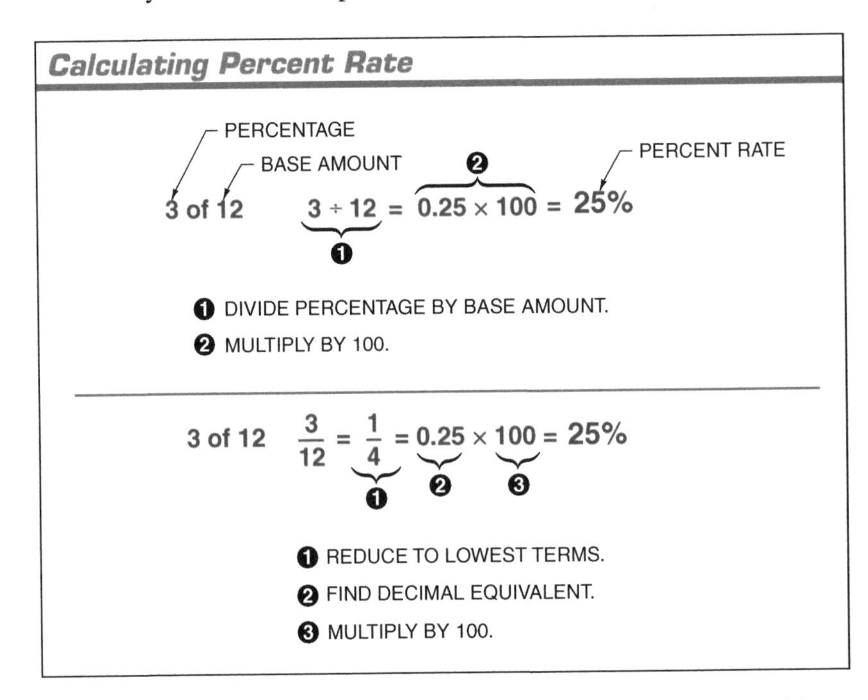

Figure 5-5. Percent rate can be found when both percentage and base amount are known.

Examples—Calculating Percent Rate

1. What percent of 50 is 4?

   ANS: 8%

   ❶ Divide 4 by 50 ($4 \div 50 = 0.08$).

   ❷ Multiply 0.08 by 100.

$$\underbrace{\frac{4}{50}}_{❶} = \underbrace{0.08 \times 100}_{❷} = 8\%$$

QUICK REFERENCE

- *Divide the percentage by the base amount.*
- *Reduce as required.*
- *Multiply by 100.*

2. What percent of 120 is 90?
ANS: 75%

❶ Divide 90 by 120 or $^{90}/_{120}$. Reduce $^{90}/_{120}$ to lowest terms (¾). Find ¾ on the Decimal Equivalents table, or divide 3 by 4 (3 ÷ 4 = 0.75).

❷ Multiply 0.75 by 100.

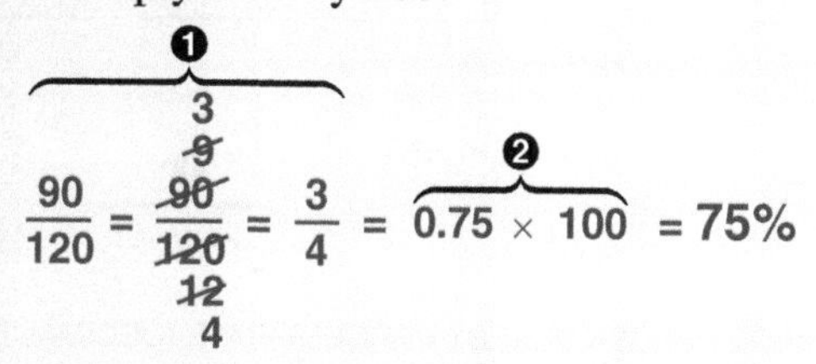

$$\frac{90}{120} = \frac{\overset{3}{\cancel{\overset{9}{\cancel{90}}}}}{\underset{4}{\cancel{\underset{12}{\cancel{120}}}}} = \frac{3}{4} = 0.75 \times 100 = 75\%$$

MATH EXERCISES—Calculating Percent Rate

_______________ **1.** What percent of 130 is 19.50?

_______________ **2.** What percent of 500 is 200?

_______________ **3.** What percent of 500 is 250?

_______________ **4.** What percent of 200 is 220?

PRACTICAL APPLICATIONS—Calculating Percent Rate

_______________ **5. Electrical:** A transformer has a 320 W rating. What percent of the rated power is lost by a 48 W loss?

_______________ **6. Construction:** A building is to be 40 stories high. What percent is complete after 16 stories are finished?

Calculating Base Amount

To find the base amount when the percentage and the percent rate are known, divide the percentage by the percent rate. **See Figure 5-6.**

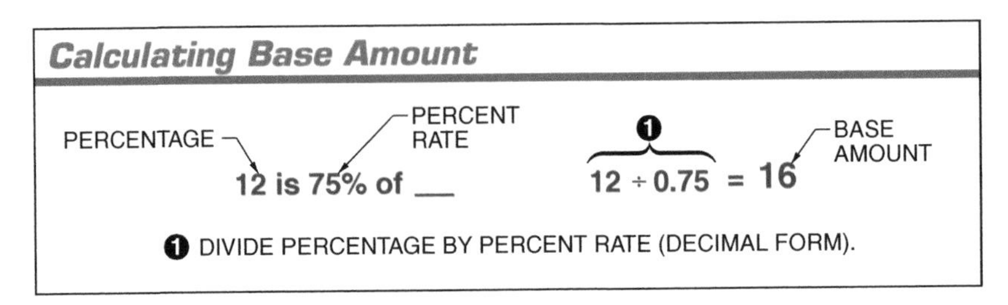

Figure 5-6. The base amount can be found when both percentage and percent rate are known.

Examples—Calculating Base Amount

1. Eight is 20% of what amount?

ANS: **40**

❶ Divide 8 by 0.20.

$$\frac{8}{0.20} = 40$$

2. Two is 50% of what amount?

ANS: **4**

❶ Divide 2 by 0.50.

$$\frac{2}{0.50} = 4$$

QUICK REFERENCE

- *Divide the percentage by the percent rate (decimal form).*

MATH EXERCISES—Calculating Base Amount

__________ **1.** One is 10% of what amount?

__________ **2.** Sixteen is 4% of what amount?

__________ **3.** Thirty-eight is 2% of what amount?

PRACTICAL APPLICATIONS—Calculating Base Amount

_______________ **4. Mechanics:** What is the horsepower of a 78% efficient motor that delivers 23.4 HP?

_______________ **5. Pipefitting:** An ironworker fabricates 120′ of large pipe, which is 30% of an order. How much pipe is needed to complete the order?

_______________ **6. Electrical:** An installer uses 44%, or 264′, of Cat 5e cable from a cable reel to complete a construction job. How many feet of cable were on the reel when it was purchased from the manufacturer?

SECTION 5-3 CALCULATING COSTS

Determining the cost of an item often requires percentage calculations. For example, the cost of 1400 bf of lumber, the discount on 100 ICFs, or the percent discount on 40 pieces of conduit may all need to be calculated.

It is important to understand how to calculate marked price, discount, and net price when determining costs. A *marked price* is the retail price of an item (the base amount). A *discount* is a reduction of a marked price (the percentage). *Net price* is the price of an item after a discount has been applied.

Items such as lumber may be discounted depending on the quantity purchased.

It is also necessary to understand how to determine percent discount and multiple discounts. The *percent discount* is the amount of a discount given as a percent (the percent rate). A *multiple discount* is the total discount that results by applying more than one discount to an item.

Calculating Marked Price

The marked price of an item is the base amount of the item. To calculate the marked price, divide the discount by the percent discount (decimal form). **See Figure 5-7.**

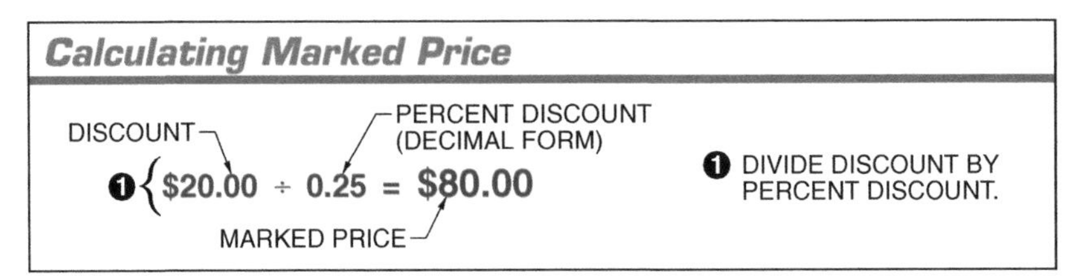

Figure 5-7. The marked price of an item is the base amount.

Example—Calculating Marked Price

1. Find the marked price for a product with a discount of $16.00 and a percent discount of 40%.
   ANS: **$40.00**
   ❶ Divide 16 by 0.40.
   ❶ {$16.00 ÷ 0.40 = **$40.00**

QUICK REFERENCE

- *Divide the discount by the percent discount (decimal form).*

Calculating Discount

The discount on the cost of an item is a percentage of the marked price. To find the discount, multiply the marked price by the percent discount (decimal form). **See Figure 5-8.**

When a discount has more than two decimal places, round the decimal to the nearest cent (hundredths place). For example, an item with a marked price of $10.25 and a percent discount of 15% costs 1.5375, or rounded, **$1.54**.

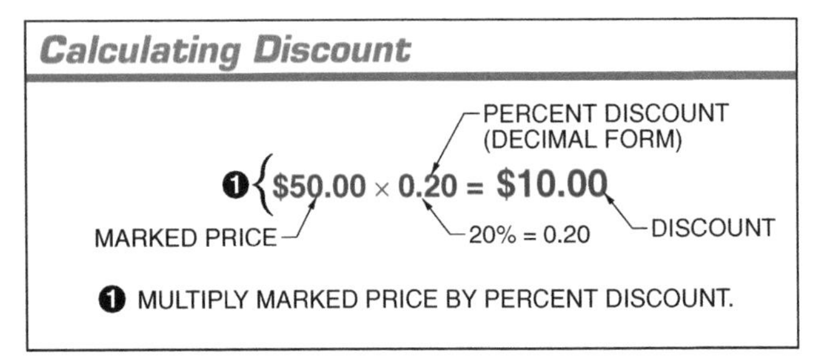

Figure 5-8. The discount on the cost of an item is a percentage of the marked price.

Example—Calculating the Discount

1. Calculate the discount on a product with a marked price of $122.00 and a discount of 60%.
   ANS: **$73.20**
   ❶ Multiply 122 by 0.60.
   ❶ {$122.00 × 0.60 = **$73.20**

QUICK REFERENCE

- *Multiply the marked price by the percent discount (decimal form).*

Calculating Percent Discount

The percent discount on the cost of an item is the percent rate. To find the percent discount if the discount and the marked price are known, divide the discount by the marked price, and multiply the quotient by 100 (move decimal two places to the right). **See Figure 5-9.**

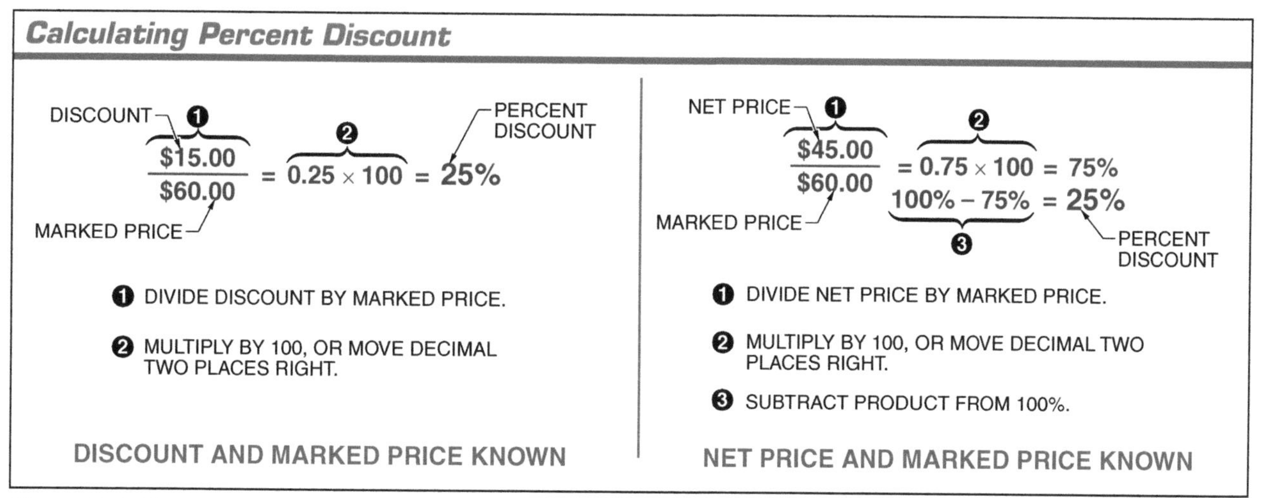

Figure 5-9. The percent discount on the cost of an item is the percent rate.

To find the percent discount if the net price and the marked price are known, divide the net price by the marked price and multiply by 100. Subtract the quotient from 100%.

Examples—Calculating Percent Discount

1. Find the percent discount of a product with a marked price of $40.00 and a discount of $10.00.

ANS: 25%

❶ Divide 10 by 40 (10 ÷ 40 = 0.25).

❷ Multiply 0.25 by 100.

❶ ❷
$10.00 ÷ $40.00 = 0.25 × 100 = 25%

QUICK REFERENCE

- *Divide the discount by the marked price.*
- *Multiply by 100.*

2. Find the percent discount of a product with a marked price of $220.00 and a net price of $165.00.

ANS: 25%

❶ Divide 165.00 from 220.00 (165 ÷ 220 = 0.75)

❷ Multiply 0.75 by 100 (0.75 × 100 = 75).

❸ Subtract 75 from 100%.

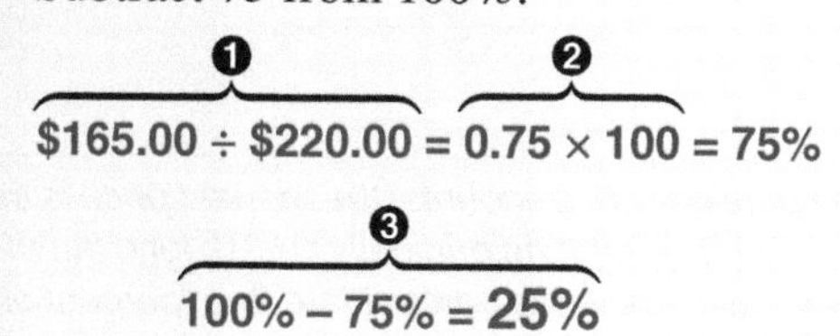

QUICK REFERENCE

- *Divide the net price by the marked price.*
- *Multiply by 100.*
- *Subtract from 100%.*

Calculating Net Price

The net price of an item is the price paid after the discount has been applied. To find the net price of an item, subtract the discount from the marked price. **See Figure 5-10.**

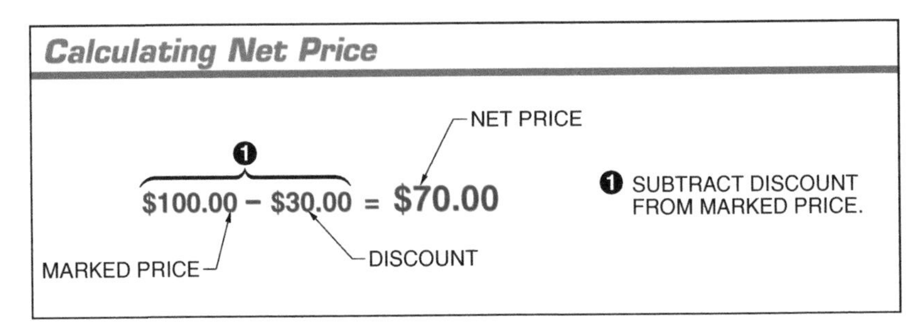

Figure 5-10. The net price of an item is the price paid after the discount has been applied.

Example—Calculating Net Price

1. Calculate the net price of a product with a marked price of $144.00 and a discount of $24.50.
   ANS: **$119.50**
   ❶ Subtract 24.50 from 144.00.
   ❶{ $144.00 – $24.50 = **$119.50**

QUICK REFERENCE

- *Subtract the discount from the marked price.*

Calculating Net Price after Multiple Discounts

A multiple discount is the discount found by applying multiple percent discounts to the price of an item. For example, an item may have a marked price of $75.00, a percent discount of 10%, and an additional percent discount of 20%. Several repeated steps are needed to find the net price of the item.

To find the net price, multiply the marked price by the first percent discount (10%) and subtract the discount from the marked price. The result is the discount price. Multiply this discount price by the second percent discount (20%), and so on, until the net price is found. **See Figure 5-11.**

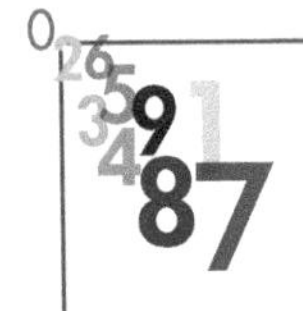

When dealing with discounts, the marked price is the base amount (the amount something originally costs), for example, $200.00. The percent discount is the percent rate (the "percent off"), for example 40%. The discount is the percentage (the actual amount saved), in this case, $80.00.

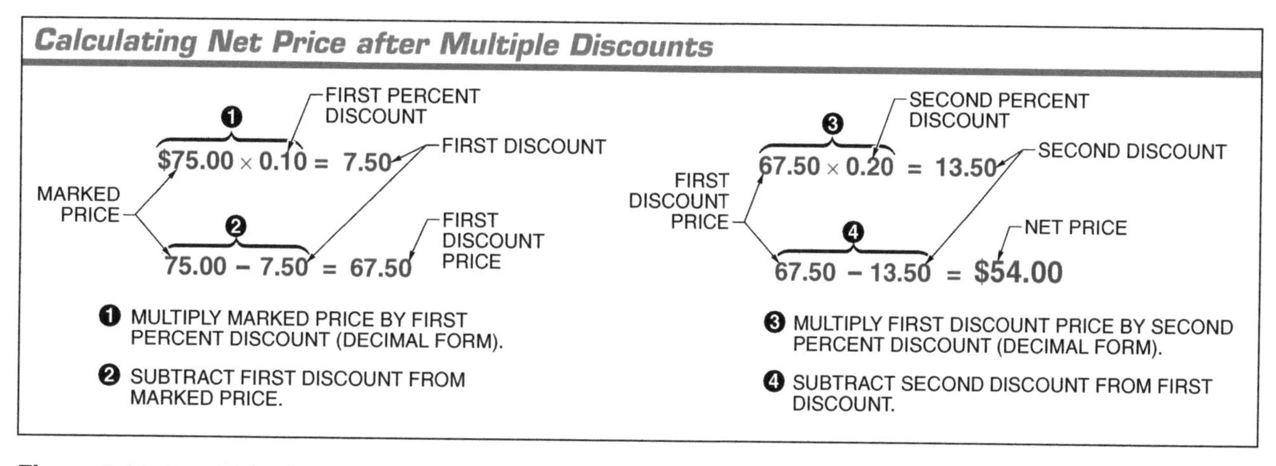

Figure 5-11. A multiple discount is found by applying multiple percent discounts to the price of an item.

Example—Calculating Net Price after Multiple Discounts

1. Find the net price of a product with a marked price of $650.00 and discounts of 25% and 10%.

   ANS: **$438.75**

   ❶ Multiply 650.00 by 0.25 to find the first discount (650.00 × 0.25 = 162.50).

   ❷ Subtract 162.50 from 650.00 to find the first discount price (650.00 – 162.50 = $487.50).

   ❸ Multiply 487.50 by 0.10 to find the second discount (487.50 × 0.10 = 48.75).

   ❹ Subtract 48.75 from 487.50 to find the net price.

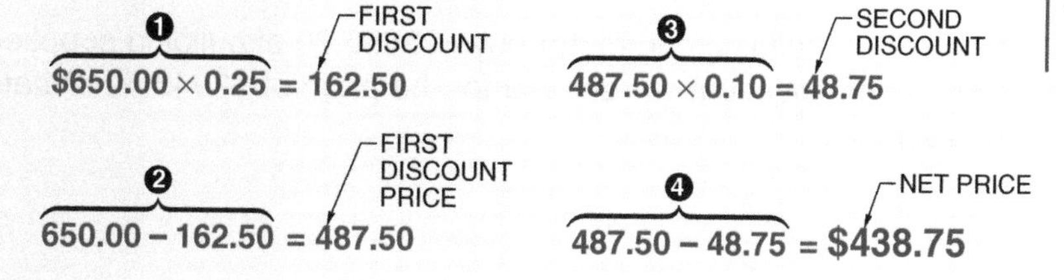

QUICK REFERENCE

- *Multiply the marked price by the first percent discount (decimal form).*
- *Subtract the first discount from the marked price.*
- *Multiply the first discount price by the second percent discount (decimal form).*
- *Subtract the second discount from the first discount price.*

MATH EXERCISES—Calculating Costs

_______________ **1.** Find the discount of a product with a marked price of $22,000.00 at a percent discount of 5%.

_______________ **2.** Find the net price of a product with a marked price of $250.00 and a multiple discount of 10% and 5%.

_______________ **3.** Which is the best price for a buyer on a \$3500.00 purchase, a single discount of 15%, or a multiple discount of 10% and 5%?

_______________ **4.** A purchase of \$8920.00 has a discount of \$312.20. What is the percent discount?

PRACTICAL APPLICATIONS—Calculating Costs

_______________ **5. HVAC:** HVAC supplies are purchased for \$1250.00. The merchant is offered a 3½% discount for cash. What is the cash discount?

_______________ **6. Construction:** What is the percent discount of the bricks?

_______________ **7. Construction:** A building contractor buys \$3550.00 of building supplies from a building supply discount warehouse for the price of \$2946.50. What is the percent discount?

Calculating Tax

A *tax* is a charge paid on income, products, and services. Taxes are a means of raising money to pay the expenses of city, county, state, and federal governments.

When calculating tax, the tax on an item is a percentage of the marked price (the base amount), and the tax rate is the percent rate. To find the tax paid on an item, multiply the marked price by the tax rate (decimal form). **See Figure 5-12.**

Calculating Tax

MARKED PRICE, TAX RATE, TAX

❶ {$10.00 × 0.70 = $.70

❶ MULTIPLY MARKED PRICE BY TAX RATE (DECIMAL FORM).

Figure 5-12. To find the tax, multiply the marked price by the tax rate in decimal form.

Example—Calculating Tax

1. Find the tax paid on a $5200.00 product with an 8% tax rate.
   ANS: **$416.00**
   ❶ Multiply 5200.00 by 0.08.

   ❶ {$5200.00 × 0.08 = $416.00

QUICK REFERENCE

- *Multiply the marked price by the tax rate (decimal form).*

Calculating Total Cost

To find the total cost of a taxed product, find the tax and add it to the marked price. **See Figure 5-13.**

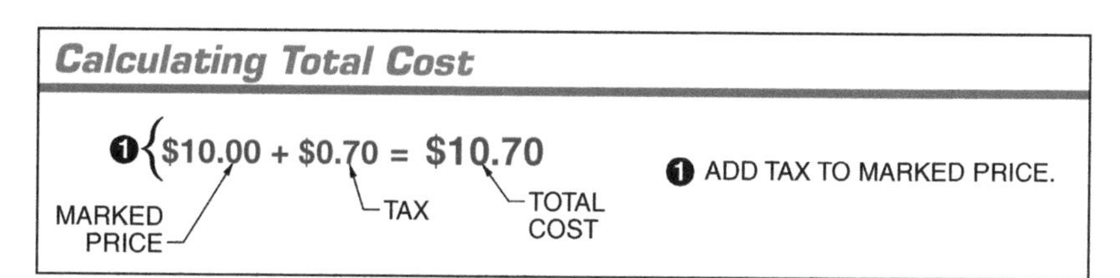

Figure 5-13. Total cost is the sum of the marked price and the tax.

Example—Calculating Total Cost

1. Find the total cost of a $56.00 product with a 5% tax rate.
   ANS: **$58.80**
   ❶ Multiply 56.00 by 0.05 to find the tax (56.00 × 0.05 = 2.80).
   ❷ Add 56.00 to 2.80.

   ❶ {$56.00 × 0.05 = 2.80 (TAX)

   ❷ {56.00 + 2.80 = $58.80

QUICK REFERENCE

- *Multiply the marked price by the tax rate (decimal rate).*
- *Add the tax to the marked price.*

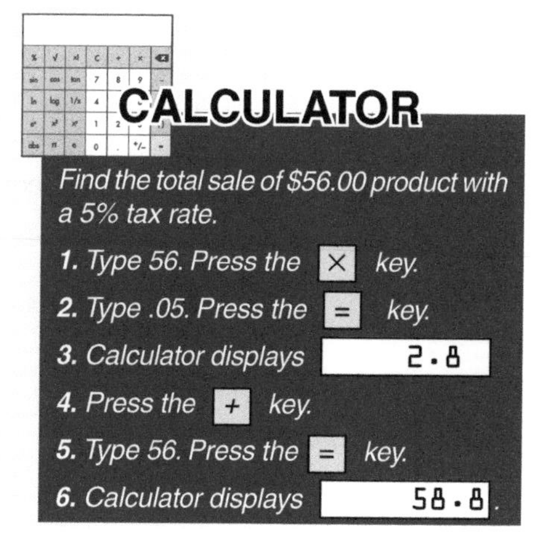

MATH EXERCISES—Calculating Tax and Total Cost

________ 1. A customer purchases three cans of primer at $1.89 per can. The sales tax is 6%. What will the total cost be?

________ 2. What will the tax be on a $250.00 product at an 11% tax rate?

________ 3. Find the total cost of a $134,000.00 product with a 6% tax rate.

________ 4. At 7%, how much tax will be paid on a $140.00 purchase?

________ 5. Find the tax to be paid on a $60.00 purchase. The tax rate is 4½%.

PRACTICAL APPLICATIONS—Calculating Tax and Total Cost

________ 6. **Electrical:** To finish a job, an electrician purchases $223.00 of electrical supplies with a tax rate of 7%. What is the total cost of the supplies?

________ 7. **Construction:** What is the total cost of drywall with a marked price of $439.00 and a tax rate of 5%?

______ **8. Construction:** Find the tax to be paid on the table saw.

$1950.00 WITH AN 8% TAX RATE

TABLE SAW

______ **9. Culinary Arts:** A group of four food orders totals $115.35, and the sales tax is 7.5% of the total. What amount of tax will be charged on this order?

______ **10. Agriculture:** A landscaping project requires 4000 lb of gravel. A discount of 20% can be gained by hauling the gravel to the site instead of requesting delivery. The marked price of the gravel is $13.25 per pound. How much will be saved by using the discount?

For an interactive review of the concepts covered in Chapter 5, refer to the corresponding Quick Quiz® included on the Digital Resources.

Working with Percentages 5
Review

Name ______________________ Date ______________

Math Problems

______ **1.** Change 84% to a decimal.

______ **2.** Change 0.14 to a percent.

______ **3.** Change 71% to a decimal.

______ **4.** Change 40% to a fraction and reduce.

______ **5.** Eight percent of what number is 240?

______ **6.** Change $\frac{5}{7}$ to a percent. *(Round to 2 places.)*

______ **7.** Change $\frac{1}{4}$ to a percent.

______ **8.** Change $\frac{3}{5}$ to a percent.

______ **9.** Thirty-five percent of what number is 350?

Practical Applications

______ **10. Plumbing:** A pump discharges 4500 L of water/hr running at 75% of its capacity. How many liters per hour does the pump discharge at full capacity?

5 Review (continued)

______________ 11. **Agriculture:** One feeding trough has a width of 36″. Another feeding trough has a width of 96″. What is the percentage difference between the widths of the two troughs?

______________ 12. **Welding:** An 8″ fillet weld is increased in length by 50%. What is the total length of the fillet weld?

______________ 13. **Manufacturing:** A 0.750″ drill is what percent larger than a 0.500″ drill?

______________ 14. **Manufacturing:** A 3.00″ block is milled to 2.75″. What percent is removed by milling? *(Round to 2 places.)*

______________ 15. **Electrical:** Voltage readings on a 230 V motor fluctuate between 215 V and 230 V. What is the percent of fluctuation? *(Round to 2 places.)*

Working with Percentages 5
Test

Name ______________________________ Date ______________

Math Problems

__________ **1.** Change $\frac{4}{5}$ to a percent.

__________ **2.** Change 14% to a fraction.

__________ **3.** How much tax is paid on a product with a marked price of $125.00 and a tax rate of 5%?

__________ **4.** Thirty-two percent of what number is 384?

__________ **5.** Forty-five percent of what number is 360?

__________ **6.** What percent of 130 is 19.50?

__________ **7.** Change 2% to a fraction.

__________ **8.** Change 35% to a fraction.

__________ **9.** Change $\frac{3}{4}$ to a percent.

__________ **10.** Forty percent of what number is 3600?

__________ **11.** What percent of 300 is 20? *(Round to the hundredths place.)*

5 Test (continued)

______ 12. What percent of 80 is 16?

Practical Applications

______ 13. **Electrical:** If an electrician buys $380.00 worth of electrical supplies at a tax rate of 8%, what will the total cost of the electrical supplies be?

______ 14. **Construction:** A contractor estimated the cost of a building to be $122,000.00. If 15% is added for profit and overhead, what is the total estimated cost?

______ 15. **Plumbing:** If the marked price of a kitchen faucet is $170.00 and the discount is $20.00, what is the percent discount? *(Round to 2 places.)*

______ 16. **Manufacturing:** A boring machine has a speed of 1200 rpm. In order to increase the output, the speed is raised $8\frac{1}{2}$%. What is the new speed?

5 Test (continued)

_______________ **17. Electrical:** A circuit has an ampacity of 20 A and is protected by a fuse rated at 125% of that ampacity. What fuse rating is required?

_______________ **18. Construction:** If the marked price of a 5 gal. bucket of primer is $50.00 and the sale price is $37.00, what is the percent discount?

_______________ **19. Maintenance:** A steel car is loaded to 70% capacity with 98,000 lb of coal. What is the full capacity of the car?

_______________ **20. Boiler Operation:** If 20 problems are correct on a boiler operator's licensing examination consisting of 25 problems, what is the percentage correct?

_______________ **21. Mechanics:** An engine has an output of 50 HP, which is 80% of the input. What is the input?

5 Test (continued)

_________________ **22. Plumbing:** A plumber estimates that 500′ of pipe is needed for a job. If 12% is added for waste, how much pipe should be ordered?

_________________ **23. HVAC:** An HVAC supplier store advertises "1/5 off on all stock." What is the percent discount?

_________________ **24. Construction:** If there is a 7% tax on all retail sales, and a circular saw costs $150.00 retail, how much tax will be added?

_________________ **25. Welding:** If 2000 lb of iron ore is actually 70% iron, how many pounds of iron are there in the iron ore?

_________________ **26. Electrical:** An electrician bought 800′ of copper conductor, and 55% of it was used for a wiring job. How many feet were used?

5 Test (continued)

_______________ **27. Manufacturing:** Because of one worn cavity in a compression molding die, a factory lost 10 out of every 40 plastic components. What percent of components were lost?

_______________ **28. Maintenance:** The original price of a forklift is increased 15% ($3750.00). What was the original price?

_______________ **29. Alternative Energy:** An ethanol producer purchased 10,000 bushels of corn at $3.90 a bushel. The tax was 6%. What was the total cost of the corn?

_______________ **30. HVAC:** An HVAC component has a marked price of $159.00 and a discount price of $119.25. What is the percent discount?

5 Test (continued)

______ **31. Mechanics:** A motor has a rated speed of 1725 rpm and a synchronous speed of 1800 rpm. Slip is the percent difference between rated speed and synchronous speed. What is the slip of the motor? *(Round to 2 places.)*

______ **32. Agriculture:** A chicken coop measures 4′ (l) × 6′ (w) x 6′10″ (h) and has two window openings that measure 18″ (w) × 27″ (h). If a gallon of wood stain will cover 150 sq ft and has a waste factor of 10%, how many gallons of stain are needed to coat the outside walls of the coop?

______ **33. Culinary Arts:** A recipe calls for 8 lb of diced onions. If the as-purchased unit cost of onions is $0.30/lb and the yield percentage is 95%, what is the total cost of the onions for the recipe?

6 Working with Measurements

Measurements are used in every workplace. It is often necessary to convert units of measure in order to add, subtract, multiply, and divide measurements. Many times a conversion factor or a conversion table may need to be used.

OBJECTIVES

1. Define unit of measure.
2. List the basic units of measure used in the U. S. customary system of measurement.
3. List the basic units of measure used in the metric system of measurement.
4. Explain how length is measured.
5. Explain how area is measured.
6. Explain how volume is measured.
7. Explain how capacity is measured.
8. Explain how weight and mass are measured.
9. List four scales used to measure temperature.
10. Convert measurements to lower units of measure.
11. Convert measurements to higher units of measure.
12. Convert measurements between two systems of measurement.
13. Convert temperature between two systems of measurement.
14. Add and subtract units of measure.
15. Multiply and divide units of measure.

KEY TERMS

- unit of measure
- length
- area
- acre
- hectare
- volume
- capacity
- weight
- mass
- temperature
- absolute zero
- conversion factor

Digital Resources
ATPeResources.com/QuickLinks
Access Code: 764460

SECTION 6-1 UNDERSTANDING UNITS OF MEASURE

Measurements may be composed of single or multiple units of measure. A *unit of measure* is a standard by which a quantity is measured. **See Figure 6-1.**

Units of Measure

SINGLE UNIT OF MEASURE	MULTIPLE UNITS OF MEASURE
4 mi	2 hr, 40 min
144 sq in.	3 gal., 2 qt, 1 pt

Figure 6-1. Measurements may be composed of single or multiple units of measure.

A single unit of measure can be converted to different units of measure without changing value. For example, 15 in. = 1 ft, 3 in. In addition, conversions can be made from one measurement system to another.

Two common measurement systems are the U.S. customary system and the metric system. The U.S. customary system uses the following as basic units of measure: inch (in. or ″), foot (ft or ′), pint (pt), quart (qt), gallon (gal.), ounce (oz), and pound (lb). **See Figure 6-2.**

U.S. Customary System of Measurement

		Unit	Abbr	Equivalents
LENGTH		mile	mi	5280′, 1760 yd
		yard	yd	3′, 36″
		foot	ft *or* ′	12″, 0.333 yd
		inch	in. *or* ″	0.083′, 0.028 yd
AREA		square mile	sq mi *or* mi²	640 A
		acre	A	4840 sq yd, 43,560 sq ft
		square yard	sq yd *or* yd²	1296 sq in., 9 sq ft
		square foot	sq ft *or* ft²	144 sq in., 0.111 sq yd
		square inch	sq in. *or* in²	0.0069 sq ft, 0.00077 sq yd
VOLUME		cubic yard	cu yd *or* yd³	27 cu ft, 46,656 cu in.
		cubic foot	cu ft *or* ft³	1728 cu in., 0.0370 cu yd
		cubic inch	cu in. *or* in³	0.00058 cu ft, 0.000021 cu yd
CAPACITY	*U.S. liquid measure*	gallon	gal.	4 qt, 128 fl oz (231 cu in.)
		quart	qt	2 pt, 32 fl oz (57.75 cu in.)
		pint	pt	16 fl oz (28.875 cu in.)
		fluid ounce	fl oz	0.0625 pt (1.805 cu in.)
	U.S. dry measure	bushel	bu	4 pk (2150.42 cu in.)
		peck	pk	8 qt (537.605 cu in.)
		quart	qt	2 pt (67.201 cu in.)
		pint	pt	0.5 qt (33.600 cu in.)
WEIGHT	*avoirdupois*	ton	t	2000 lb
		pound	lb *or* #	16 oz
		ounce	oz	0.0625 lb
MASS	*troy*	troy pound	lb t	12 oz t
		troy ounce	oz t	20 dwt or 480 gr
		pennyweight	dwt or pwt	24 gr
		grain	gr	0.04 dwt

Figure 6-2. The U.S. customary system of measurement uses inches, feet, pints, quarts, gallons, ounces, and pounds.

The metric system is the most commonly used measurement system in the world. This system is based on the meter (m), liter (L), and kilogram (kg). **See Figure 6-3.** Prefixes are used in the metric system to represent multipliers. For example, when the prefix kilo (k), meaning 1000, is added to the root word "meter," the result is 1000 meters, or 1 kilometer.

Metric System of Measurement

	Unit	Abbr	Equivalents
LENGTH	kilometer	km	1000 m
	meter	m	100 cm
	centimeter	cm	10 mm
	millimeter	mm	0.1 cm
AREA	square kilometer	sq km *or* km^2	1,000,000 m^2
	hectare	ha	10,000 m^2
	square meter	sq m *or* m^2	10,000 cm^2
	square centimeter	sq cm *or* cm^2	100 mm^2
VOLUME	cubic meter	m^3	1,000,000 cm^3
	cubic centimeter	cu cm, cm^3, *or* cc	0.000001 m^3
CAPACITY	kiloliter	kL	1000 L
	liter	L	1000 mL
	milliliter	mL	0.001 L
MASS	kilogram	kg	1000 g
	gram	g	100 cg, 1000 mg
	centigram	cg	10 mg
	milligram	mg	0.1 cg, 0.001 g

Figure 6-3. The metric system of measurement uses meters, liters, and grams.

The U.S. customary system of measurement and the metric system of measurement use different standard units for determining length, area, volume, weight or mass, and temperature.

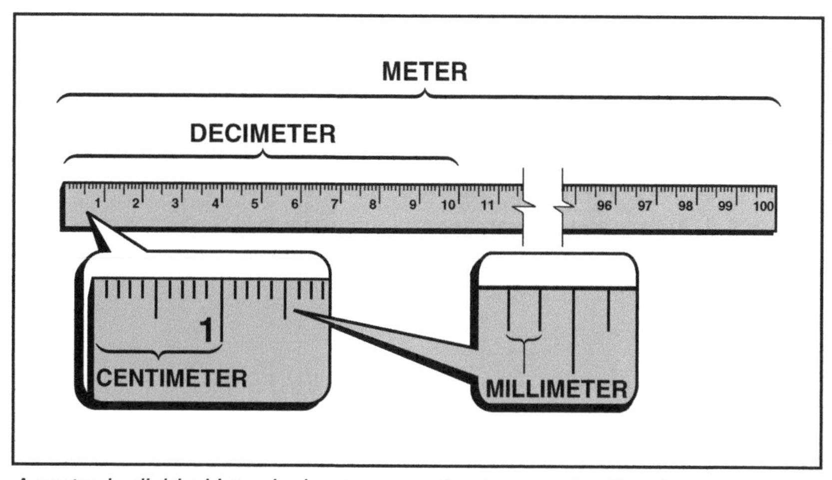

A meter is divided into decimeters, centimeters, and millimeters.

Length

Length is distance expressed in linear measure. Length is used to measure an object or a distance between objects. In the U.S. customary system, common units for expressing length are the inch (in. or ″), foot (ft or ′), and mile (mi). In the metric system, the basic unit for expressing length is the meter (m). Smaller units of measure such as the millimeter (mm), centimeter (cm), and kilometer (km) are also used. **See Figure 6-4.**

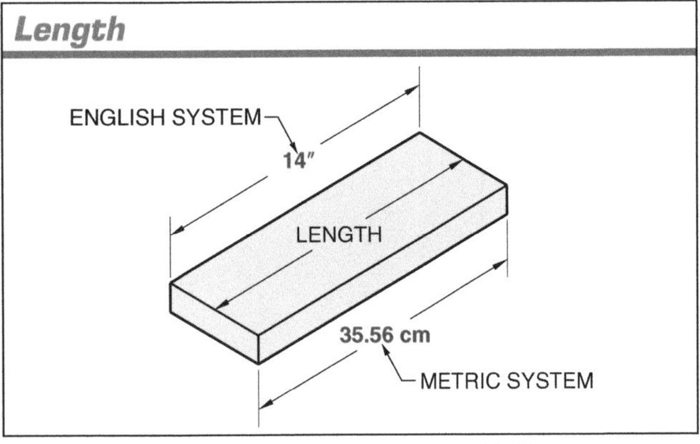

Figure 6-4. Length is distance expressed in linear measure.

Area

Area is space as expressed in square units. **See Figure 6-5.**

In the U.S. customary system, the common units for expressing area are the square inch (sq in.), square foot (sq ft), and square mile (sq mi). To express area in the metric system, a superscript "2" is added to the unit used for expressing length, such as in square meters (m^2).

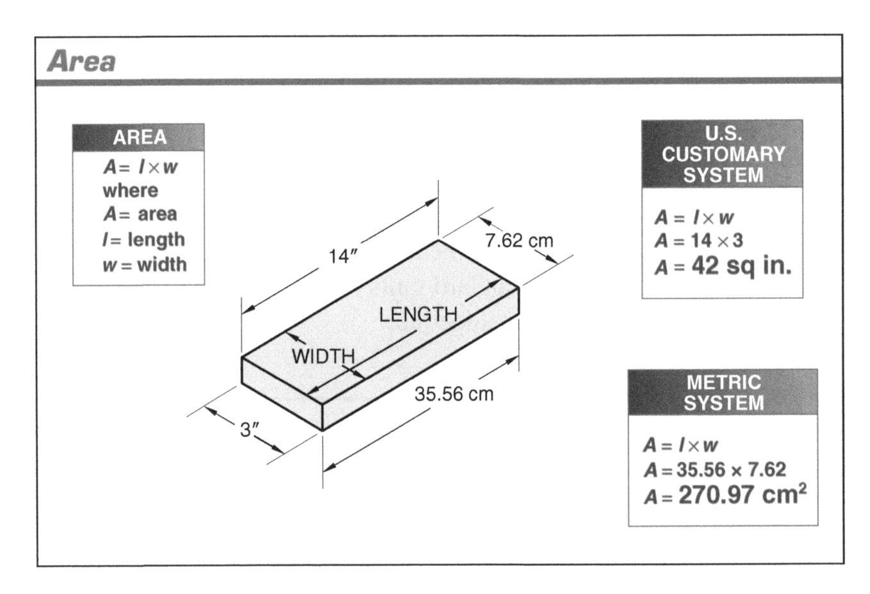

Figure 6-5. Area is space expressed in square units.

In the U.S. customary system, the acre is used to measure large areas of land. An *acre* is an area of land containing 43,560 sq ft. In the metric system, the hectare is used. A *hectare* is an area of land containing 10,000 m^2.

Volume

Volume is space as expressed in cubic measure. It represents three dimensions of an object: length, width, and height. **See Figure 6-6.**

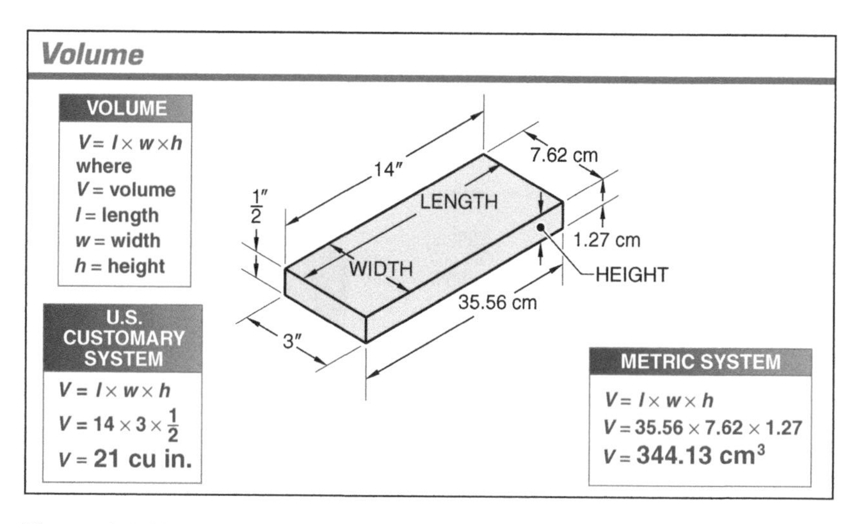

Figure 6-6. Volume is space expressed in cubic measure.

In the U.S. customary system, the common units for expressing volume are cubic inch (cu in.), cubic foot (cu ft), and cubic yard (cu yd). To express volume in the metric system, a superscript "3" is added to the unit used for expressing length, such as in cubic meters (m^3).

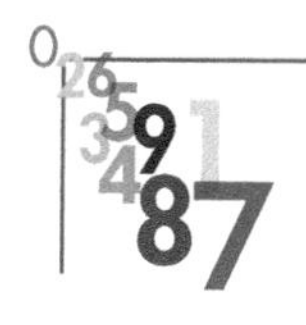

In the construction trades, the term "yard" is often used to mean "cubic yard", such as when measuring the volume of concrete or soil.

Copper tubing can carry a specific volume of water depending on its diameter and length.

Capacity

Capacity is the maximum volume that a container can hold. **See Figure 6-7.** The U.S. customary system uses two types of measurement for capacity: liquid measure and dry measure. Liquid measure consists of measurements such as gallons (gal.), quarts (qt), and pints (pt) and pecks (pk). Dry measure consists of measurements such as bushels (bu). In the metric system, there is only one measurement for capacity, liters (L).

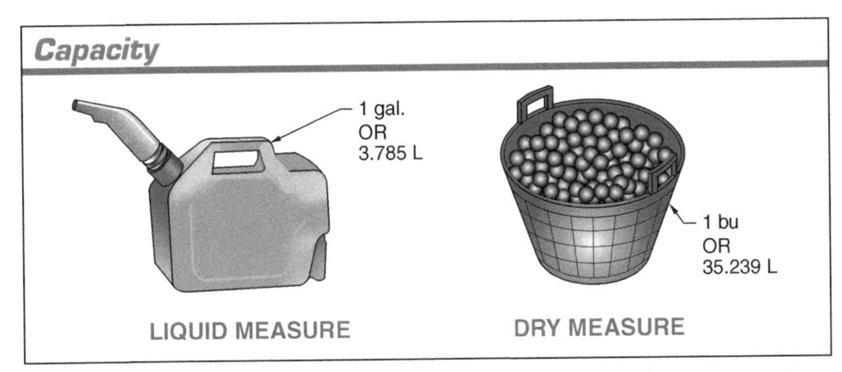

Figure 6-7. Capacity is the maximum volume that a container can hold.

Weight and Mass

Weight is a measurement that indicates the heaviness of an object. **See Figure 6-8.** The U.S. customary system uses avoirdupois weight for objects such as grain, livestock, meats, and groceries. Common units of avoirdupois weight are the ton (t), pound (lb), and ounce (oz). Precious materials such as diamonds, gold, and silver are measured by mass. *Mass* is the amount of matter contained in an object. The troy system uses grains, penny weights, troy ounces, and troy pounds to measure precious metals. The metric system uses mass, not weight, to measure objects. Common units are the gram (g) and kilogram (kg).

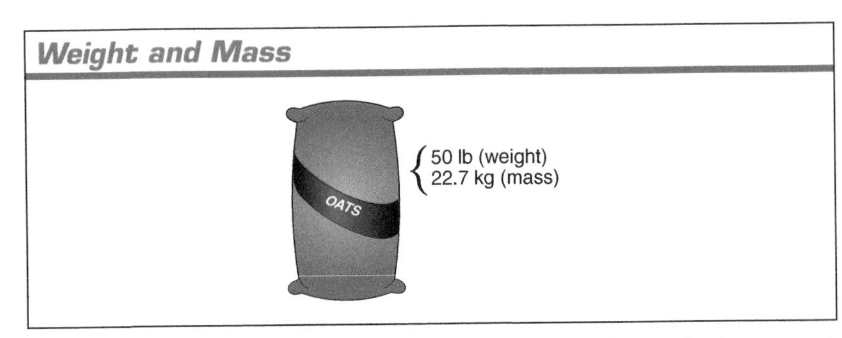

Figure 6-8. Weight is the heaviness of an object, and mass is the amount of matter contained in an object.

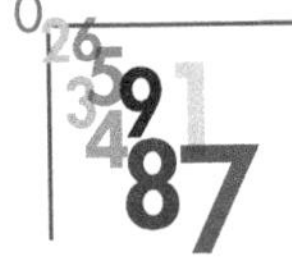

A troy ounce is 10% heavier than an avoirdupois ounce. Yet, a troy pound weighs (12 oz) less than an avoirdupois pound (16 oz).

Temperature

A gauge may display both English and metric units.

Temperature is a measurement of an amount of heat expressed in degrees. Fahrenheit (°F) and Celsius (°C) are the two scales most often used for measuring temperature. The Fahrenheit scale is used in conjunction with the U.S. customary system of measurement, and the Celsius scale is used in conjunction with the metric system.

Other types of temperature scales use absolute zero as a base. *Absolute zero* is a theoretical condition where no heat is present. Rankine (°R) is the absolute temperature scale that relates to Fahrenheit. Kelvin (K) is the absolute temperature scale that relates to Celsius. Absolute scales are typically only used in calculations involving gas laws.

SECTION 6-2 CONVERTING UNITS OF MEASURE

Converting units within a measurement system or to units in another system is often necessary. For example, it may be necessary to convert feet to inches, or it may be necessary to convert kilometers to miles. Addition, subtraction, multiplication, and division can only be performed if common units of measure are determined.

Common units of measure are determined through the use of a conversion factor. A *conversion factor* is a number that translates one unit of measure into another unit of measure of the same value. The conversion factor is typically the number of smaller units within 1 larger unit of measure, such as 8 oz in 1 cup.

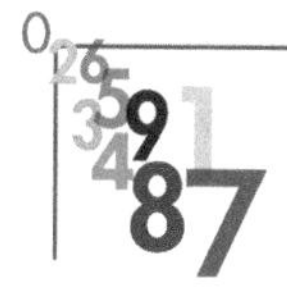

Conversion factors can either be memorized or taken from conversion tables.

Converting to Lower Units of Measure

To convert a higher unit of measure to a lower one requires multiplying the higher unit of measure by the conversion factor. For example, to convert 3 ft, 5 in., 3 is multiplied by the conversion factor 12, the number of inches in 1 ft ($3 \times 12 = 36$). In this case, the remaining inches, because they are common units, can be added (36 in. + 5 in. = 41 in.).

When converting multiple units of measure, each unit must be converted separately. For example, to convert 2 m, 73 cm, 6 mm to millimeters, the number of centimeters in a meter (100) must be used as the first conversion factor, and the number of millimeters in a centimeter (10) as the second conversion factor. **See Figure 6-9.**

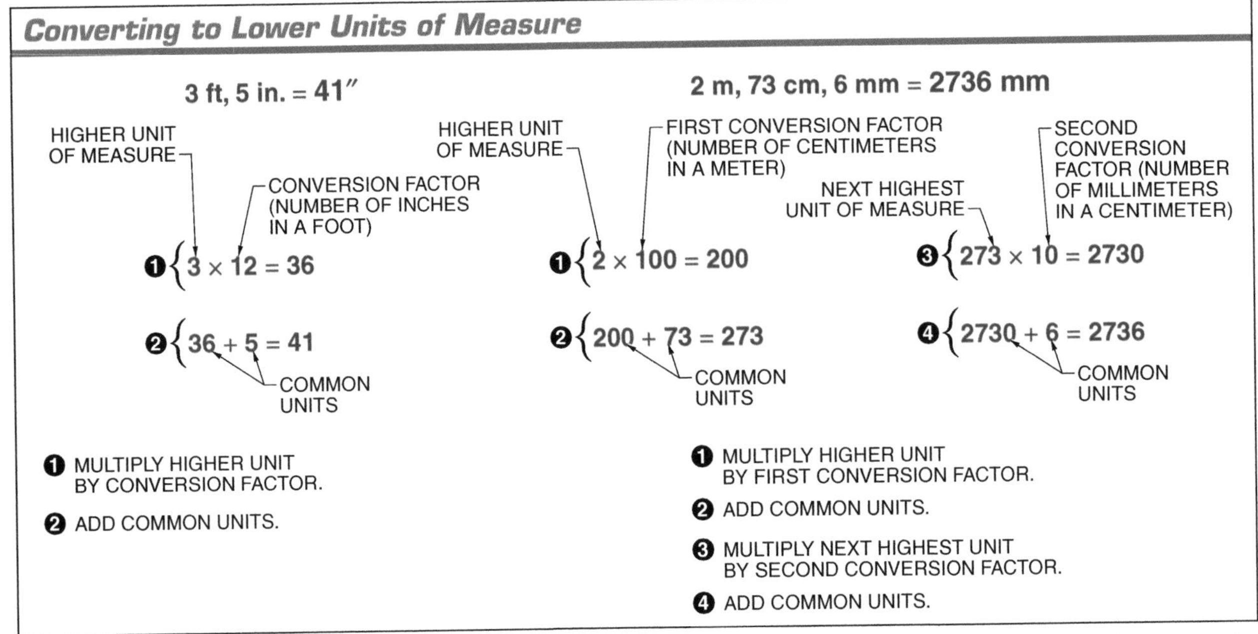

Figure 6-9. Converting a higher unit of measure to a lower one requires multiplying the higher unit of measure by the conversion factor.

Examples—Converting to Lower Units of Measure

1. Convert 6 mi, 116 yd, 2 ft to feet.

ANS: **32,030′**

❶ Multiply 6 by 1760, the number of yards in a mile (6 × 1760 = 10,560).
❷ Add 10,560 and 116 (10,560 + 116 = 10,676).
❸ Multiply 10,676 by 3, the number of feet in a yard (10,676 × 3 = 32,028).
❹ Add 32,028 and 2.

❶{ 6 × 1760 = 10,560 ❷{ 10,560 + 116 = 10,676

❸{ 10,676 × 3 = 32,028 ❹{ 32,028 + 2 = 32,030

QUICK REFERENCE

- *Multiply higher unit of measure by conversion factor.*

2. Convert 7 gal., 3 qt, 1 pt to pints.

ANS: **63 pt**

❶ Multiply 7 by 4, the number of quarts in a gallon (7 × 4 = 28).
❷ Add 28 and 3 (28 + 3 = 31).
❸ Multiply 31 by 2, the number of pints in a quart (31 × 2 = 62 pt).
❹ Add 62 and 1.

❶{ 7 × 4 = 28 ❷{ 28 + 3 = 31 ❸{ 31 × 2 = 62 ❹{ 62 + 1 = 63

3. Convert 134 kg to grams.

ANS: **134,000 g**

❶ Multiply 134 by 1000, the number of grams in a kilogram.

❶{ 134 × 1000 = 34,000

MATH EXERCISES—Converting to Lower Units of Measure

_______________ 1. Convert 244 sq yd to square inches.

_______________ 2. Convert 25 t, 70 lb to pounds.

_______________ 3. Convert 555 cm to millimeters.

PRACTICAL APPLICATIONS—Converting to Lower Units of Measure

_______________ 4. **Agriculture:** How many square feet does ½ acre contain?

_______________ 5. **Construction:** The length of a wall is 6 ft, 6 in. What is the length in inches?

Converting to Higher Units of Measure

Converting a lower unit of measure to a higher one requires dividing the lower unit of measure by the conversion factor. For example, to convert 48 ft to yards, 48 is divided by the conversion factor 3, the number of feet in 1 yd (48 ÷ 3 = 16 yd). **See Figure 6-10.**

Since division is the operation used when converting to higher units of measure, a remainder may occur. In this case, the units are seen more clearly when long division is used. For example, to convert 765 pints to gallons, the number of pints in a quart (2) must be used as the first conversion factor, and the number of quarts in a gallon (4) as the second conversion factor.

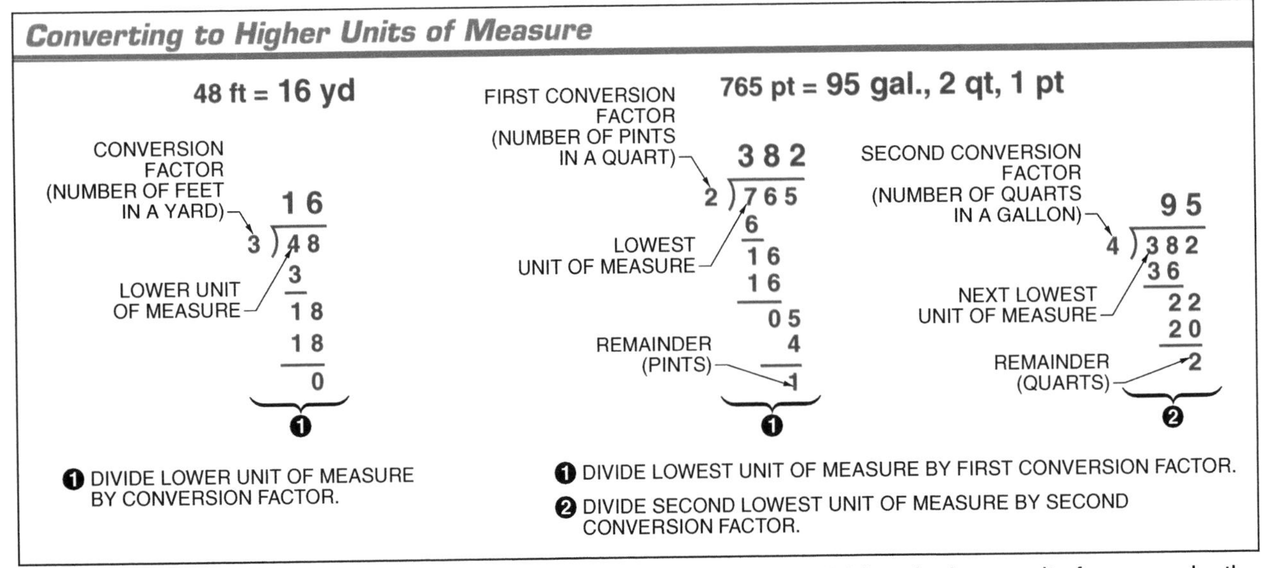

Figure 6-10. Converting a lower unit of measure to a higher unit requires dividing the lower unit of measure by the conversion factor.

Examples—Converting to Higher Units of Measure

1. Convert 2000 cu in. to cubic feet.

ANS: **1 cu ft, 272 cu in.**

❶ Divide 2000 by 1728, the number of cubic inches in a cubic foot (2000 ÷ 1728 = 1 with a remainder of 272 cu in.).

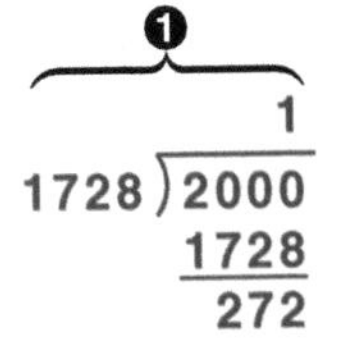

QUICK REFERENCE

- *Divide lower unit of measure by conversion factor.*

2. Convert 49,062 sq ft to acres.

ANS: **1 A, 611 sq yd, 3 sq ft**

❶ Divide 49,062 by 9, the number of square feet in a square yard (49,062 ÷ 9 = 5451 with a remainder of 3 sq ft).

❷ Divide 5451 by 4840, the number of square yards per acre (5451 ÷ 4840 = 1 with a remainder of 611 sq yd).

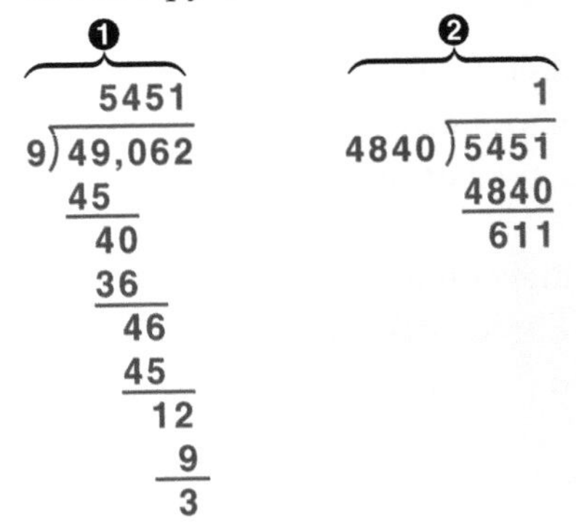

3. Convert 50,000 L to kiloliters.

ANS: **50 kL**

❶ Divide 50,000 by 1000, the number of liters in a kiloliter.

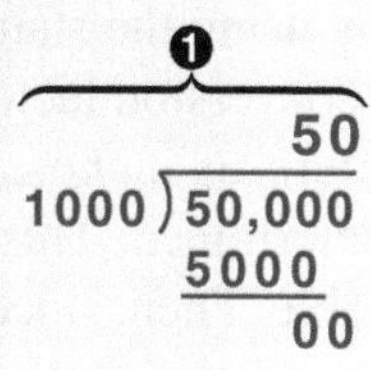

MATH EXERCISES—Converting to Higher Units of Measure

________________ **1.** Convert 1507 pt to gallons.

________________ **2.** Convert 375 mm to centimeters.

________________ **3.** Convert 17,000 ft to miles. (*Round to the tenths place.*)

PRACTICAL APPLICATIONS—Converting to Higher Units of Measure

________________ **4.** **Electrical:** A standard length of rigid metal conduit (RMC) is 120″. What is the standard length of RMC in feet?

________________ **5.** **Electrical:** A print specifies that 250 patch cords, 36″ each, be installed in the telecommunications closet of a building. Because the manufacturer sells the patch cords in 1′ lengths, the installer must convert inches to feet. How many 1′ patch cords does the installer need?

Converting Between Measurement Systems

Converting units between measurement systems is basically the same as converting units within a measurement system. U.S. customary and metric measurements are typically converted by using the standard number of smaller units within 1 larger unit as the conversion factor.

For example, to convert 3 centimeters to inches, the number of centimeters in an inch (2.54) must be determined. This number can be memorized or taken from a conversion table. Then, since a larger unit of measure is sought (inches are larger than centimeters), the 3 is divided by the conversion factor 2.54 ($3 \div 2.54 = 1.18$ in.). **See Figure 6-11.**

Converting Metric Measurements to U.S. Customary Measurements

	Metric Unit	U.S. Customary Equivalent		
LENGTH	kilometer	0.62 mi		
	meter	39.37″, 3.281′		
	centimeter	0.39″		
	millimeter	0.039″		
AREA	square kilometer	0.3861 sq mi		
	hectare	2.47 A		
	square centimeter	0.155 sq in.		
VOLUME	cubic centimeter	0.061 cu in.		
	cubic meter	1.307 cu yd		
CAPACITY		*cubic*	*dry*	*liquid*
	kiloliter	1.31 cu yd	—	—
	liter	61.02 cu in.	0.908 qt	1.057 qt
	milliliter	0.061 cu in.	—	0.27 fl dr
MASS AND WEIGHT	kilogram	2.2046 lb		
	gram	0.03527 oz		

Figure 6-11. A conversion table can be used to find the U.S. customary equivalents for metric measurements.

To convert 5 miles to kilometers, the number of kilometers within each mile (1.609) must be determined. Then, since a smaller unit of measure is sought (kilometers are shorter than miles), the 5 is multiplied by the conversion factor 1.609 ($5 \times 1.609 = 8.045$ km). **See Figure 6-12.**

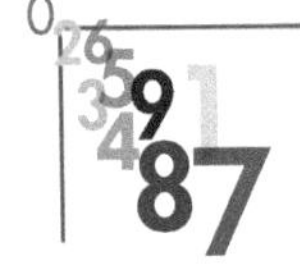

To convert a number to a higher unit of measure, divide. To convert a number to a lower unit of measure, multiply.

Converting U.S. Customary Measurements to Metric Measurements

		U.S. Unit	Metric Equivalent
LENGTH		mile	1.609 km
		yard	0.9144 m
		foot	30.48 cm
		inch	2.54 cm
AREA		square mile	2.590 km²
		acre	0.405 hectare, 4047 m²
		square yard	0.836 m²
		square foot	0.093 m²
		square inch	6.452 cm²
VOLUME		cubic yard	0.765 m³
		cubic foot	0.028 m³
		cubic inch	16.387 cm³
CAPACITY	*U.S. liquid measure*	gallon	3.785 L
		quart	0.946 L
		pint	0.473 L
		fluid ounce	29.573 mL
	U.S. dry measure	bushel	35.239 L
		peck	8.810 L
		quart	1.101 L
		pint	0.551 L
WEIGHT	*avoir-dupois*	pound	0.454 kg
		ounce	28.350 g

Figure 6-12. A conversion table can be used to find the metric equivalents for U.S. customary measurements.

Examples — Converting Between Measurement Systems

1. Convert 12 kg to pounds.

ANS: **26.43 lb**

❶ Divide 12 by 0.454, the number kilograms in a pound.

❶{ $12 \div 0.454 = \mathbf{26.43}$

QUICK REFERENCE

- *Multiply or divide by the conversion factor.*

2. Convert 6 ft, 3 in. to centimeters.

ANS: **190.5 cm**

❶ Multiply 6 ft by 12, the number of inches in a foot (6 × 12 = 72).
❷ Add 72 and 3 (72″ + 3″ = 75″).
❸ Multiply 75 by 2.54, the number of centimeters in an inch.

❶{ $6 \times 12 = 72$ ❷{ $72 + 3 = 75$

❸{ $75 \times 2.54 = \mathbf{190.5}$

3. Convert 6 sq mi to square kilometers.

ANS: **15.54 km²**

❶ Multiply 6 sq mi by 2.590, the number of square kilometers in a square mile.

❶{ $6 \times 2.590 = \mathbf{15.54}$

MATH EXERCISES—Converting Between Measurement Systems

(Round to the tenths place.)

__________ **1.** Convert 6 L to quarts.

__________ **2.** Convert 1500 g to ounces.

__________ **3.** Convert 3 pt to liters.

__________ **4.** Convert 11 oz to grams.

__________ **5.** Convert 5 m to inches.

__________ **6.** Convert 58 km to miles.

PRACTICAL APPLICATIONS—Converting Between Measurement Systems

__________ **7. Electrical:** An electrical engineer specifies that 16′ rigid metallic conduit (RMC) be installed from the control box of a plastic injection molding machine to a plant's power source. Because the molding machine is being installed where the metric system is used, the installer must convert feet to centimeters. How long is the conduit in centimeters? (*Round to the tenths place.*)

8. **HVAC:** The ductwork to a furnace must be reworked. The existing filter (18″ × 24″) is 432 sq in. in size. Replacement racks and filters are rated in square meters. How many square meters must the filter be to offer the same resistance to flow as the original filter? (*Round to the hundredths place.*)

Converting Temperature

When temperature is converted between scales, two things are considered: the difference in a common point on both the scales and the ratio between the scales. The common point chosen between the Fahrenheit and Celsius scales is the freezing point of water, 32°F and 0°C. Thus the difference is 32.

The ratio between the scales is determined from the range of degrees from freezing to boiling on both scales, 32°F to 212°F and 0°C to 100°C. The difference in each range provides the ratio: 180:100, or 1.8. This means there is 1.8 degrees on the Fahrenheit scale for every 1 degree on the Celsius scale.

To convert Fahrenheit to Celsius, subtract 32 from the Fahrenheit reading and divide by 1.8. To convert Celsius to Fahrenheit, multiply the Celsius reading by 1.8 and add 32. **See Figure 6-13.**

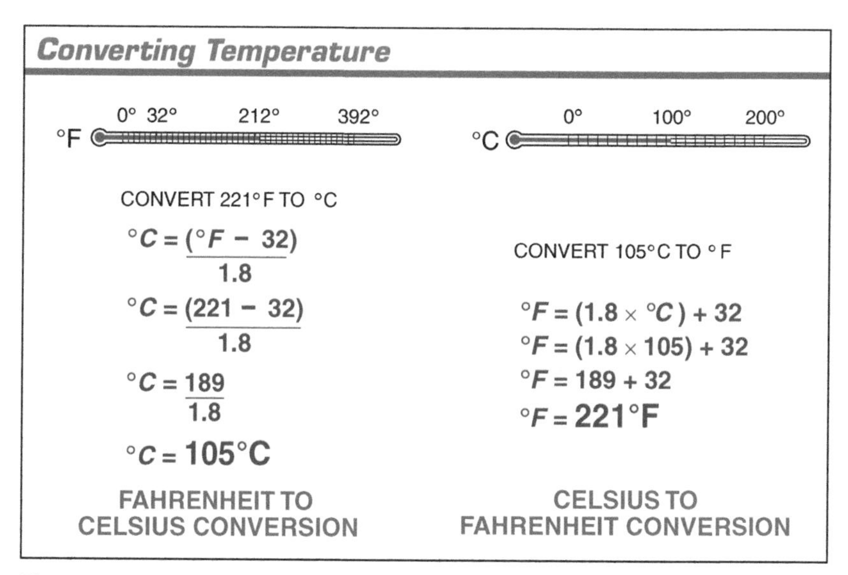

Figure 6-13. Temperature conversion between Fahrenheit and Celsius is frequently performed.

MATH EXERCISES—Converting Temperature

_______________ **1.** Convert 190°C to °F.

_______________ **2.** Convert 167°C to °F.

_______________ **3.** Convert 29°C to °F.

_______________ **4.** Convert 180°F to °C.

_______________ **5.** Convert 37°F to °C.

PRACTICAL APPLICATIONS—Converting Temperature

_______________ **6.** **Electrical:** NEC® Table 310.15(B)(16) permits THW Cu conductor to be installed in an ambient temperature of 75°C. What is the temperature in °F?

_______________ **7.** **HVAC:** The recommended comfort zone for one-family dwellings is 68°F to 72°F during the daytime. What is the comfort range in °C?

SECTION 6-3 ADDING AND SUBTRACTING UNITS OF MEASURE

For units of measure to be added and subtracted, the units must be common. If the units of measure are not common, they must be converted. For example, 6 centimeters and 2 inches cannot be added, but 6 centimeters and 5.08 centimeters can, or 2.36 inches and 2 inches can.

Multiple units of measure are added or subtracted one unit at a time. For example, to add 3 gal., 2 pt, 12 gal., and 1 pt, the gallons are added first, then the pints, for a sum of 15 gal., 3 pt. **See Figure 6-14.**

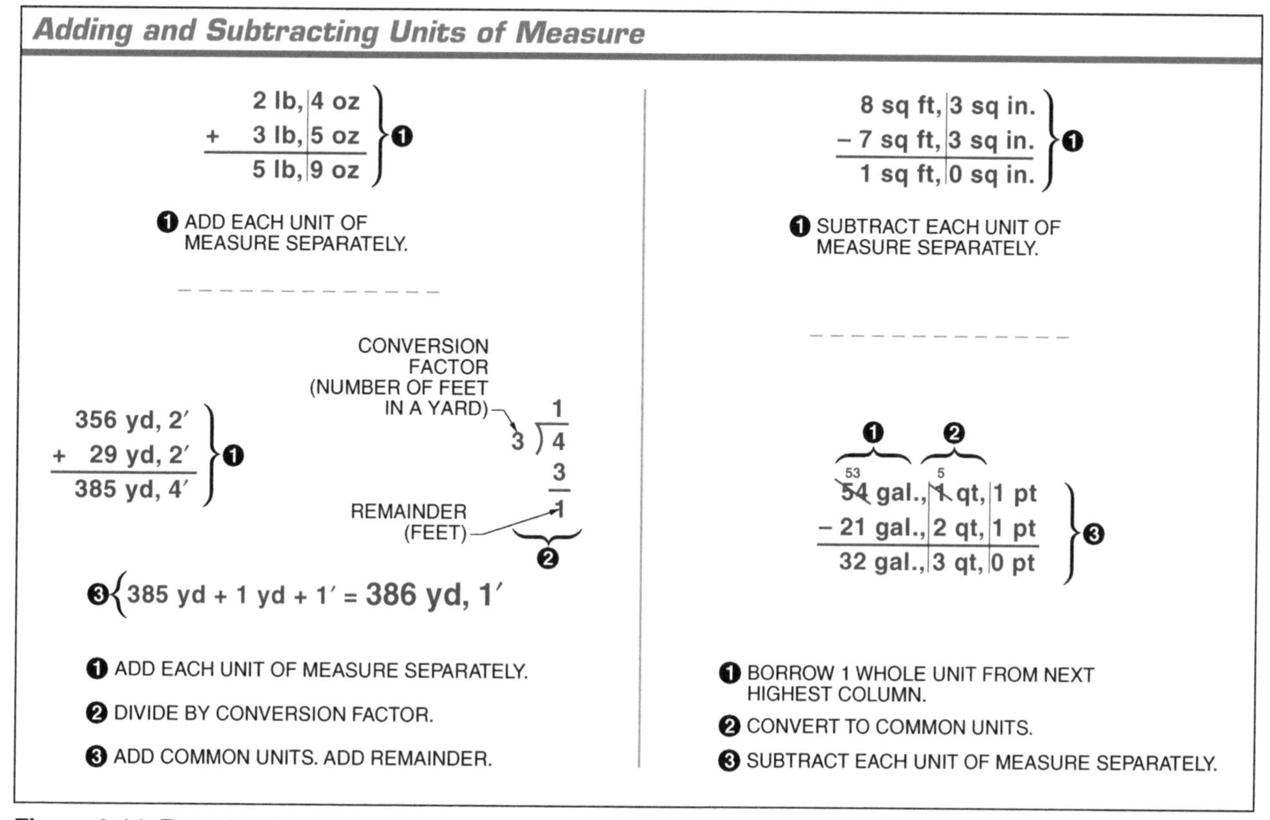

Figure 6-14. For units of measure to be added and subtracted, the units must be common.

Examples — Adding and Subtracting Units of Measure

1. Add 15 mi, 45 yd, 2′ and 3 mi, 89 yd, 2′.

ANS: **18 mi, 135 yd, 1′**

❶ Add the miles.
❷ Add the yards.
❸ Add the feet.
❹ Convert the 4′ to 1 yd, 1′.
❺ Add common units.

15 mi, 45 yd, 2′
+ 3 mi, 89 yd, 2′
18 mi, 134 yd, 4′

❹ $4 \div 3 = 1$, remainder 1

❺ 18 mi, 134 yd + 1 yd, 1′ = **18 mi, 135 yd, 1′**

QUICK REFERENCE

- *Add or subtract each unit separately.*
- *Convert as required.*

2. Subtract 1 gal., 2 qt from 6 gal., 1 qt.

ANS: **4 gal., 3 qt**

❶ Borrow 1 gal. from the 6 gal. (6 gal., 1 qt = 5 gal., 5 qt).

❷ Subtract the gallons.

❸ Subtract the quarts.

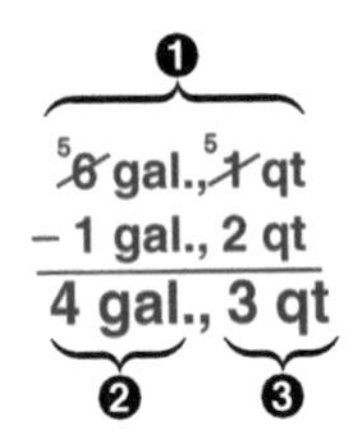

MATH EXERCISES—Adding and Subtracting Units of Measure

____________________ **1.** Subtract 4 yd, 2′, 10″ from 15 yd, 2′, 7″.

____________________ **2.** Add 2′-10″, 9″, 2′-7″, and 11′-6″.

PRACTICAL APPLICATIONS—Adding and Subtracting Units of Measure

____________________ **3. Mechanics:** How much oil was removed from the second drum?

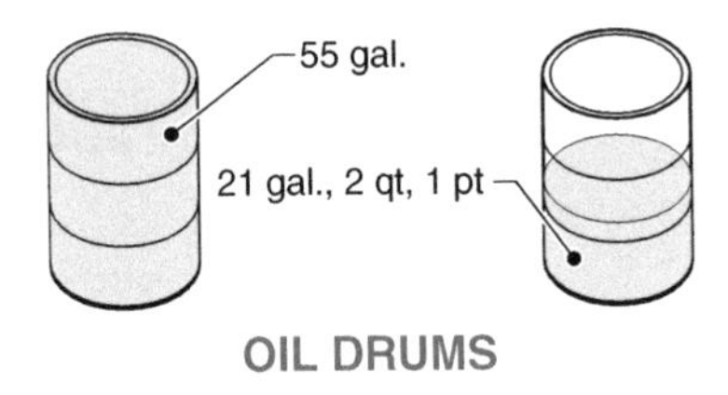

OIL DRUMS

____________________ **4. Agriculture:** What is the perimeter of the stable?

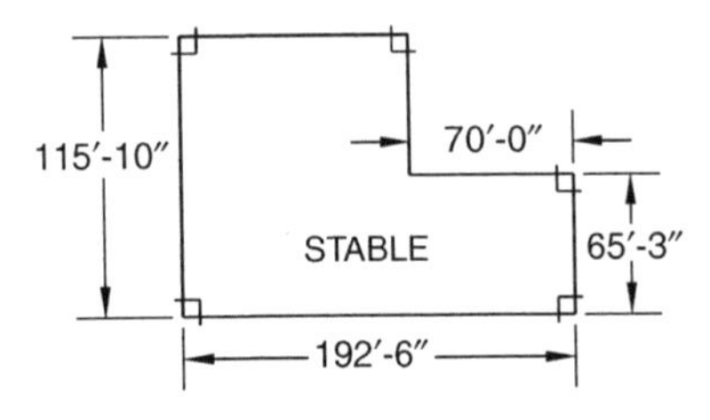

SECTION 6-4 MULTIPLYING AND DIVIDING UNITS OF MEASURE

Units of measure are easy to multiply and divide. In calculations involving multiple units of measure, each unit is multiplied or divided separately. Conversion is performed as necessary. **See Figure 6-15.**

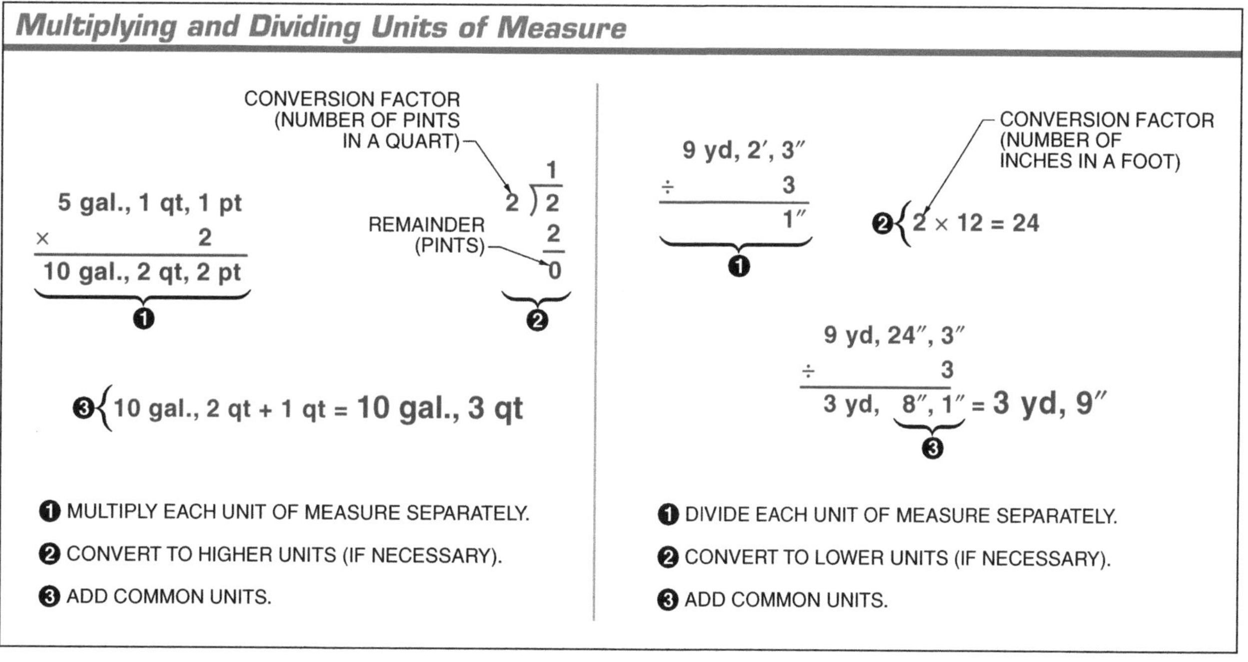

Figure 6-15. In multiple units of measure, each unit is multiplied or divided separately, and conversion is performed as necessary.

Examples — Multiplying and Dividing Units of Measure

1. Multiply 12 cu yd, 24 cu ft by 3.

ANS: **38 cu yd, 18 cu ft**

❶ Multiply 12 cu yd by 3.
❷ Multiply the 24 cu yd by 3.
❸ Convert 72 cu ft to cubic yards (72 cu ft ÷ 27 = 2 cu yd, 18 cu ft).
❹ Add common units.

12 cu yd, 24 cu ft
× 3
36 cu yd (❶), 72 cu ft (❷)

❸ $27\overline{)72}$ = 2, 54, 18

❹ 36 cu yd + 2 cu yd, 18 cu ft = **38 cu yd, 18 cu ft**

QUICK REFERENCE

- *Multiply or divide each unit of measure separately.*
- *Convert as required.*

2. Divide 12 yd, 2′, 6″ by 3.

ANS: **4 yd, 10″**

❶ Divide 12 yd by 3.
❷ Convert 2′ to inches (2 × 12 = 24).
❸ Divide 24″ by 3.
❹ Divide 6″ by 3.
❺ Add common units.

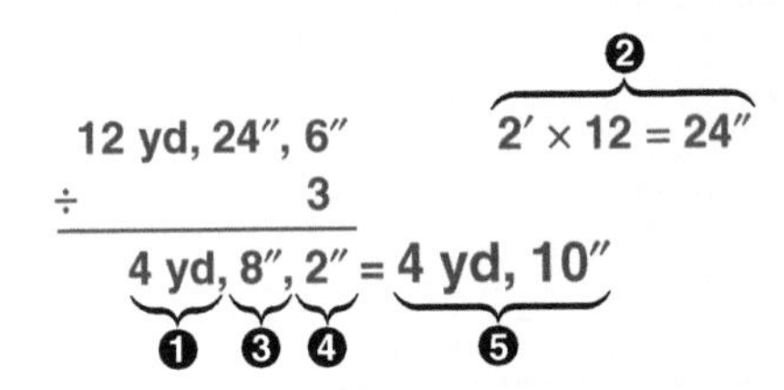

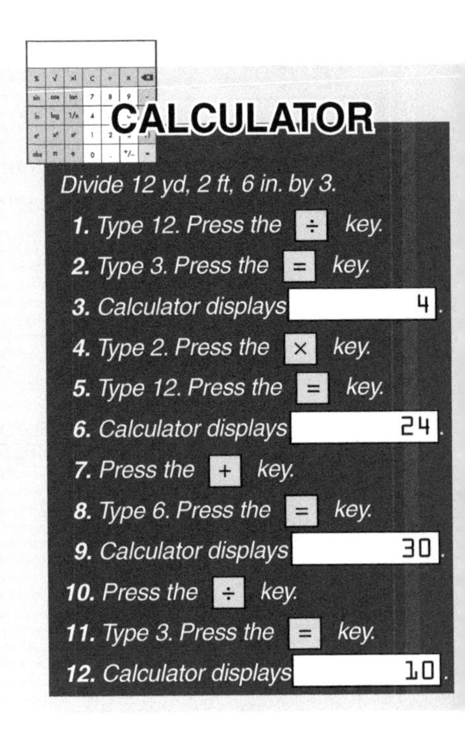

MATH EXERCISES—Multiplying and Dividing Units of Measure

________________ 1. Multiply 4 yd, 2′, 8″ by 5.

________________ 2. Divide 3 gal., 3 qt by 6.

________________ 3. Multiply 18 cu ft, 9 cu in. by 2.

________________ 4. Divide 8 cm, 7 mm by 2.

PRACTICAL APPLICATIONS — Multiplying and Dividing Units of Measure

5. Construction: How many cubic feet of sand can be moved in five trips with the wheelbarrow filled to a 4½ cu ft capacity each trip?

MAXIMUM CAPACITY = $4\frac{1}{2}$ CU FT

6. Construction: A carpenter has used five 8′-long studs for the first wall of a room. Three more walls that are identical to the first wall are built. What is the total linear feet of studs used?

7. Maintenance: Three oil reservoirs for the plows of sugar centrifugals require 4 gal., 2qt of hydraulic fluid. How much hydraulic fluid is needed to fill one reservoir?

8. Maintenance: A school custodian has 12 gal., 3 qt of paint, enough to paint one classroom. How much paint is needed for four classrooms, all the same size?

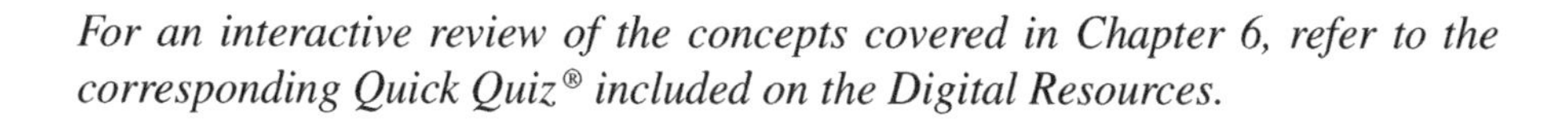

For an interactive review of the concepts covered in Chapter 6, refer to the corresponding Quick Quiz® included on the Digital Resources.

Working with Measurements 6

Review

Name ______________________ Date ____________

Math Problems

______________ 1. Convert 17′-6″ to inches.

______________ 2. Convert 7 m to millimeters.

______________ 3. Convert 20 m to kilometers.

______________ 4. Convert 40 cu yd to cubic feet.

______________ 5. Convert 3 sq mi to square kilometers.

______________ 6. Convert 46 cm to inches.

______________ 7. Convert 12 qt to liters.

______________ 8. Convert 22°C to Fahrenheit.

______________ 9. Convert 81°F to Celsius.

6 Review (continued)

_______________ **10.** Add 5 gal., 3 pt and 1 gal., 4 pt.

_______________ **11.** Subtract 2 L, 900 mL from 17 L, 320 mL.

_______________ **12.** Multiply 6 g, 15 mg by 20.

_______________ **13.** Divide 6 lb, 3 oz by 8.

Practical Applications

_______________ **14.** **Construction:** A roofer needs to lay shingles on a roof that measures 30 m × 10 m. How many bundles of shingles must be ordered to complete the job? *Note:* 1 square or 100 sq. ft is equal to 3 bundles of shingles.

_______________ **15.** **Welding:** Three welds of 1′-8″ each need to be made. What is the total length of the welds?

Working with Measurements 6

Test

Name ______________________ Date ______________

Math Problems

______________ **1.** Convert 512 cm to millimeters.

______________ **2.** Convert 17 t to pounds.

______________ **3.** Convert 1½ acres to square feet.

______________ **4.** Convert 12 qt, 6 pt to pints.

______________ **5.** Convert 473 km to meters.

______________ **6.** Convert 8 mi, 27 yards, 3 ft to feet.

______________ **7.** Convert 1188 sq ft to square yards.

______________ **8.** Convert 121.5 cu ft to cubic yards.

______________ **9.** Convert 13,550 mm to centimeters.

6 Test (continued)

______________ **10.** Convert 128 fl oz to quarts.

______________ **11.** Convert 110 km to miles.

______________ **12.** Convert 19″ to centimeters.

______________ **13.** Convert 16 m to inches.

______________ **14.** Convert 21 m^3 to cubic yards.

______________ **15.** Convert 160 A to hectares.

______________ **16.** Convert 10 g to ounces.

______________ **17.** Convert 88°C to Fahrenheit.

______________ **18.** Convert 185°F to Celsius.

______________ **19.** Convert 102°F to Celsius.

______________ **20.** Convert 510°C to Fahrenheit.

6 Test (continued)

________________ **21.** Add 3 mi, 302′, 7″ and 4 mi, 19′, 6″.

________________ **22.** Add 14 t, 900 lb and 27 t, 1300 lb.

________________ **23.** Subtract 3 yd, 2′, 8″ from 12 yd, 2′, 6″.

________________ **24.** Subtract 3 t, 460 lb from 16 t, 305 lb.

________________ **25.** Multiply 3 yd, 2′, 6″ by 4.

________________ **26.** Multiply 8 pt, 3 fl oz by 20.

________________ **27.** Divide 46′-6″ by 9.

________________ **28.** Divide 25 cm, 2 mm by 12.

Practical Applications

________________ **29.** **Welding:** A 7 ft, 6 in. length of steel angle stock will be cut into 9 in. long pieces. How many pieces can be cut from this stock?

6 Test (continued)

_______________ **30. HVAC:** A popular size of a supply plenum is 51.5 cm × 51.5 cm × 77 cm. Convert the size of the plenum to inches.

_______________ **31. Alternative Energy:** The output of solar photovoltaic modules is typically rated at 25°C. What is this temperature in degrees Fahrenheit?

_______________ **32. Agriculture:** How many square feet does a ¾ acre lot contain?

_______________ **33. Culinary Arts:** A quart of cookie dough makes 2 dozen cookies. How many cookies can be made from 1½ gal. of cookie dough?

7 Working with Exponents

Exponents are important when dealing with large numbers. Both scientific and engineering notation make use of exponents. Exponents provide a shorter way to do calculations. Roots are factors of numbers, and the square root of a number is the factor of that number before it is raised to the power of 2.

OBJECTIVES

1. Define exponent and base number.
2. Raise a positive base number to a power.
3. Raise a negative base number to a power.
4. Raise a fraction base number to a power.
5. Raise a decimal base number to a power.
6. Find the power of a power using the base exponent method.
7. Find the power of a power using the new exponent method.
8. Find the value of a base number raised to a negative power.
9. Multiply two or more base numbers with exponents.
10. Divide two or more base numbers with exponents.
11. Differentiate between square root and perfect square.
12. Use a calculator to find square roots.

KEY TERMS

- exponent
- base number
- square root
- radical sign
- perfect square

Digital Resources
ATPeResources.com/QuickLinks
Access Code: 764460

SECTION 7-1 UNDERSTANDING EXPONENTS

An *exponent* is a number that indicates how many times a base number is to be multiplied. A *base number* is a factor that is multiplied a given number of times according to an exponent. For example, in the problem $3 \times 3 \times 3 \times 3 = 81$, the repeated factor 3 is the base number and the exponent is 4, which indicates that the base number is to be multiplied 4 times. The problem is expressed as $3^4 = 81$, or 3 to the fourth power equals 81. **See Figure 7-1.**

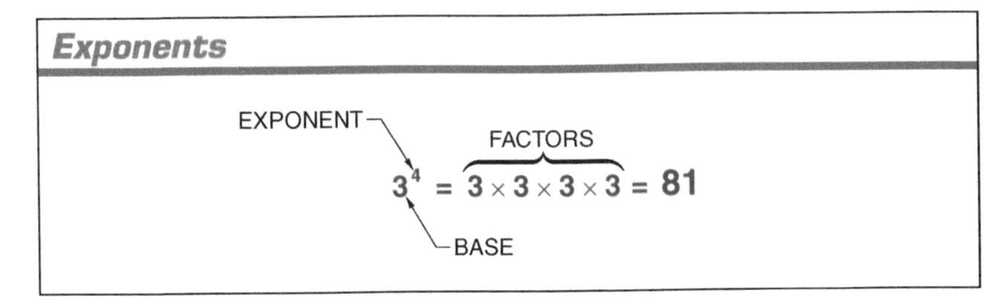

Figure 7-1. Exponents raise a base number to a power.

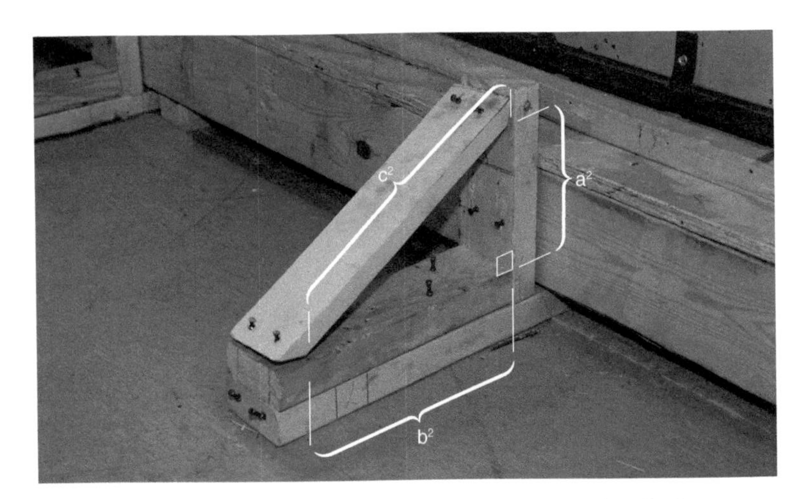

The length of the brace can be calculated using the Pythagorean theorem, ($c^2 = a^2 + b^2$).

When a base is used twice as a factor, the product is to the second power or squared. For example, the square of 4 is 4^2, or $4 \times 4 = 16$. The square is used to find the area of flat surfaces. When the base is used three times as a factor, the product is to the third power or cubed. For example, the cube of 5 is 5^3, or $5 \times 5 \times 5 = 125$. The cube is used to find the volume of solids. When the base is used as a factor more than three times, the product is to the fourth power, fifth power, sixth power, etc. and is indicated by an exponent.

With exponents, a base can be a positive number, a negative number, a fraction, or a decimal. An exponent can be a positive number, a negative number, or a zero. A power can be raised to a power, and equivalent bases raised with the same or different exponents can be multiplied and divided.

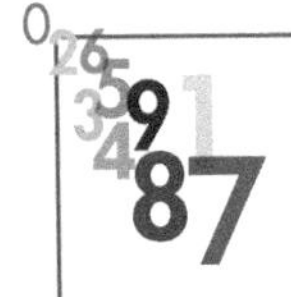

Scientific notation and engineering notation are used to simplify large or small numbers using the powers of 10. For example, 24,100,000 in scientific notation is written as 2.41×10^7 and 0.0009 is written as 9×10^{-4}.

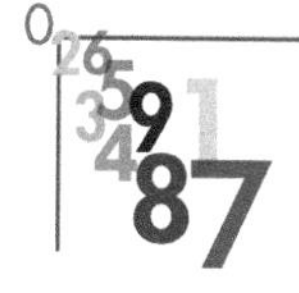

In scientific notation and engineering notation, a positive exponent means a decimal shift to the right and a negative exponent means a decimal shift to the left.

Positive Base Numbers

To raise a positive base number to a power, multiply the base number by itself as many times as shown by the exponent. **See Figure 7-2.**

Positive Base Numbers

Figure 7-2. A positive base number raised to a power is multiplied as many times as shown by the exponent.

Example—Positive Base Numbers

1. Find the value of 23^4, or the fourth power of 23.
ANS: **279,841**

❶ Multiply the positive base number 23 by itself four times.

$$23^4 = \underbrace{23 \times 23 \times 23 \times 23}_{❶} = 279,841$$

QUICK REFERENCE

- *Multiply the base as many times as shown by the exponent.*

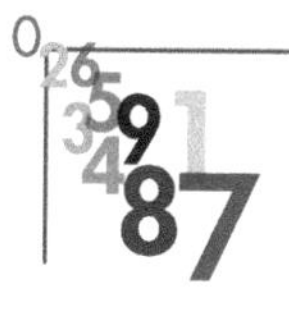

Any base raised to the power of zero is equal to 1. For example, $1^0 = 1$, $14^0 = 1$, and $(-3)^0 = 1$. The only exception is 0^0, which is undefined.

Negative Base Numbers

To raise a negative base number to a power, multiply the base number by itself as many times as shown by the exponent. **See Figure 7-3.** If the negative base number is raised to an even power, the answer is a positive number. For example, to find the value of $(-2)^4$, multiply –2 by itself four times ($-2 \times -2 \times -2 \times -2 = 16$). If the negative base number is raised to an odd power, the answer is a negative number. For example, to find the value of $(-2)^5$, multiply –2 by itself five times ($-2 \times -2 \times -2 \times -2 \times -2 = -32$).

Negative Base Numbers

EVEN EXPONENT

$(-2)^4 = -2 \times -2 \times -2 \times -2$

$= 16$

POSITIVE NUMBER

NEGATIVE BASE NUMBER

ODD EXPONENT

$(-2)^5 = -2 \times -2 \times -2 \times -2 \times -2$

$= -32$

NEGATIVE NUMBER

NEGATIVE BASE NUMBER

Figure 7-3. A negative base number raised to a power is multiplied as many times as shown by the exponent.

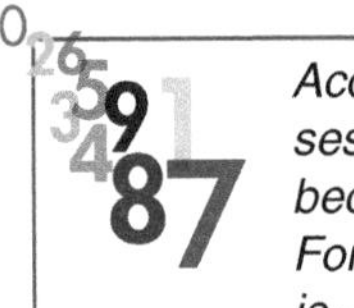

According to the order of operations, calculations within parentheses are made first. Parentheses are important in exponentiation because they demonstrate to what base the exponent applies. For example, the difference between $(-4)^2$ and $-(4)^2$ is that $(-4)^2$ is $-4 \times -4 = 16$ and $-(4)^2$ is $-(4 \times 4) = -16$.

Examples—Negative Base Numbers

1. Find the value of $(-4)^4$, or the fourth power of -4.

ANS: **256**

❶ Multiply the base number -4 by itself four times.

EVEN EXPONENT

POSITIVE NUMBER

$$(-4)^4 = \underbrace{-4 \times -4 \times -4 \times -4}_{❶} = 256$$

2. Find the value of $(-3)^5$.

ANS: **–243**

❶ Multiply the base number -3 by itself five times.

ODD EXPONENT

$$(-3)^5 = \underbrace{-3 \times -3 \times -3 \times -3 \times -3}_{❶} = -243$$

NEGATIVE NUMBER

QUICK REFERENCE

- *Multiply the base as many times as shown by the exponent.*

Fraction Bases

To raise a fraction to a power, multiply the numerator and the denominator separately according to the exponent. **See Figure 7-4.** A proper fraction raised to a positive exponent has an answer of lesser value than the base fraction. For example, $(3/5)^2 = 9/25$, which is less than $3/5$. An improper fraction raised to a positive exponent has an answer of greater value than the base fraction. For example, $(4/3)^2 = 16/9$, which is greater than $4/3$.

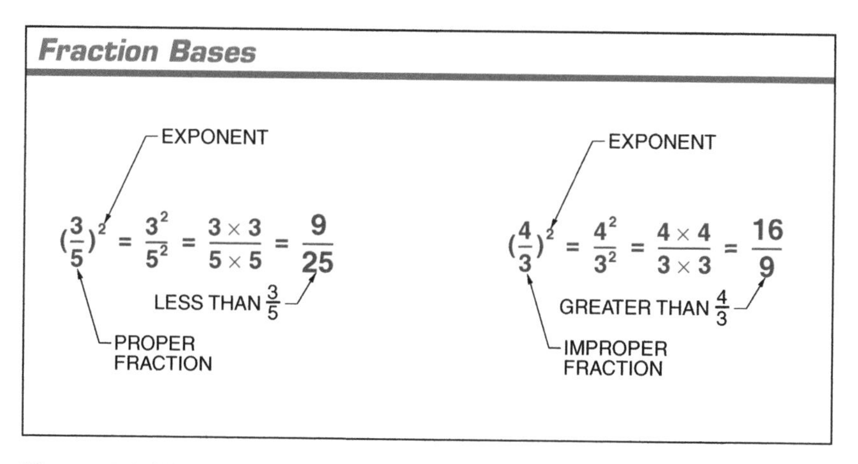

Figure 7-4. With a fraction base, both the numerator and denominator are multiplied as many times as shown by the exponent.

Examples—Fraction Bases

1. Find the value of $(3/4)^5$.

ANS: $243/1024$

❶ Raise the 3 to the fifth power and raise the 4 to the fifth power.

$$❶\left\{(\frac{3}{4})^5 = \frac{3^5}{4^5} = \frac{3 \times 3 \times 3 \times 3 \times 3}{4 \times 4 \times 4 \times 4 \times 4} = \frac{243}{1024}\right.$$

LESS THAN $\frac{3}{4}$

PROPER FRACTION

QUICK REFERENCE

- *Multiply both the numerator and the denominator separately as many times as shown by the exponent.*

2. Find the value of $(5/3)^3$.

ANS: $125/27$

❶ Raise the 5 to the third power and raise the 3 to the third power.

$$❶\left\{(\frac{5}{3})^3 = \frac{5^3}{3^3} = \frac{5 \times 5 \times 5}{3 \times 3 \times 3} = \frac{125}{27}\right.$$

GREATER THAN $\frac{5}{3}$

IMPROPER FRACTION

Decimal Bases

To raise decimal bases or mixed decimals to a power, multiply the decimal by itself as many times as indicated by the exponent. **See Figure 7-5.** Decimals are similar to proper fractions in that decimals raised to a positive exponent have lesser value than the base decimal. For example, $0.2^2 = 0.04$, which is less than 0.2.

Decimal Bases

EXPONENT

$(0.25)^3 = 0.25 \times 0.25 \times 0.25 = 0.015625$

DECIMAL BASE

NUMBER OF DECIMAL PLACES EQUALS DECIMAL PLACES IN BASE × EXPONENT

Figure 7-5. With a decimal base, the decimal is multiplied by itself as many times as shown by the exponent.

Example—Decimal Bases

1. Find the value of 0.03^3, or the third power of 0.03.

ANS: **0.000027**

❶ Multiply 0.03 by 0.03 by 0.03.

❶ $(0.03)^3 = 0.03 \times 0.03 \times 0.03 = 0.000027$

NUMBER OF DECIMAL PLACES EQUALS DECIMAL PLACES IN BASE × EXPONENT

QUICK REFERENCE

- *Multiply the decimal base as many times as shown by the exponent.*

Power of a Power

The power of a power may be found by the base exponent method or the new exponent method. **See Figure 7-6.** To use the base exponent method, solve the base with the first exponent and raise it to the second exponent. For example, to raise 2^2 to the third power, solve 2^2 ($2 \times 2 = 4$). Raise 4 to the third power ($4^3 = 4 \times 4 \times 4 = 64$).

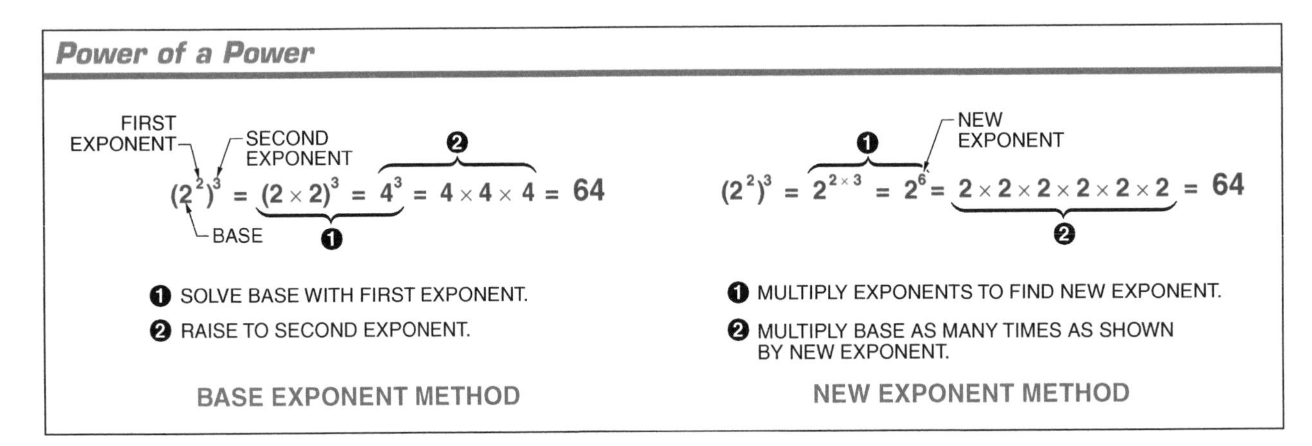

Figure 7-6. A power of a power is found by solving the base with the first exponent and raising it to the second exponent, or by multiplying the exponents to find a new exponent.

To use the new exponent method, multiply the two exponents to find the new exponent. For example, to raise 2^2 to the third power, multiply 2 (first exponent) by 3 (second exponent) to get 6 (new exponent). Raise 2 to the sixth power ($2^6 = 2 \times 2 \times 2 \times 2 \times 2 \times 2 = 64$).

Examples — Power of a Power

1. Find the value of $(10^3)^4$ using the base exponent method.

ANS: **1,000,000,000,000**

❶ Solve the base with the first exponent by multiplying 10 by 10 by 10 ($10 \times 10 \times 10 = 1000$).

❷ Raise the 1000 to the fourth power.

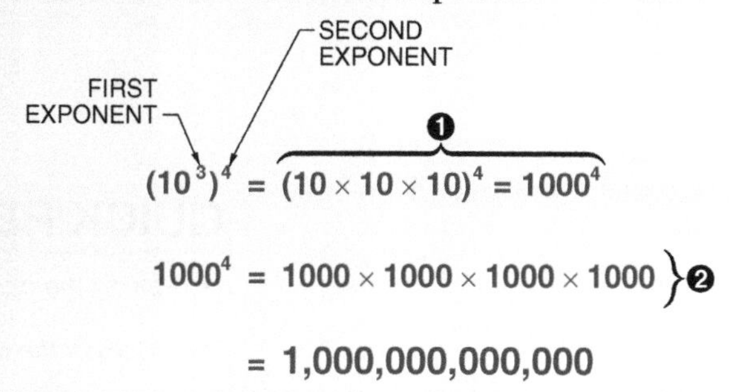

QUICK REFERENCE

- *Solve the base with the first exponent.*
- *Raise the result to the second exponent.*

2. Find the value of $(10^3)^4$ using the new exponent method.

ANS: **1,000,000,000,000**

❶ Multiply the exponents to find the new exponent ($10^{3 \times 4} = 10^{12}$).

❷ Raise 10 to the twelfth power.

NEW EXPONENT

❶ $(10^3)^4 = 10^{3 \times 4} = 10^{12}$

❷ $10^{12} = 10 \times 10 \times 10 \times 10 \times 10 \times 10 \times 10 \times 10 \times 10 \times 10 \times 10 \times 10$

$= 1,000,000,000,000$

QUICK REFERENCE

- *Multiply the exponents to find the new exponent.*
- *Multiply the base as many times as shown by the new exponent.*

Negative Exponents

A base number raised to a negative power is the reciprocal of the base with a positive exponent. **See Figure 7-7.** To find the value of a base raised to a negative power, invert the base to its reciprocal and multiply the denominator according to the exponent.

Negative Exponents

NEGATIVE EXPONENT

BASE NUMBER

$$10^{-4} = \frac{1}{10^4} = \frac{1}{10} \times \frac{1}{10} \times \frac{1}{10} \times \frac{1}{10} = \frac{1}{10{,}000}$$

POSITIVE EXPONENT

❶ INVERT BASE TO RECIPROCAL.

❷ MULTIPLY DENOMINATOR.

Figure 7-7. A base number with a negative exponent is the reciprocal of the base with a positive exponent.

Example—Negative Exponents

1. Find the value of 8^{-3}.

ANS: $\frac{1}{512}$

❶ Invert the base and make the exponent positive ($8^{-3} = \frac{1}{8^3}$).

❷ Multiply the denominator.

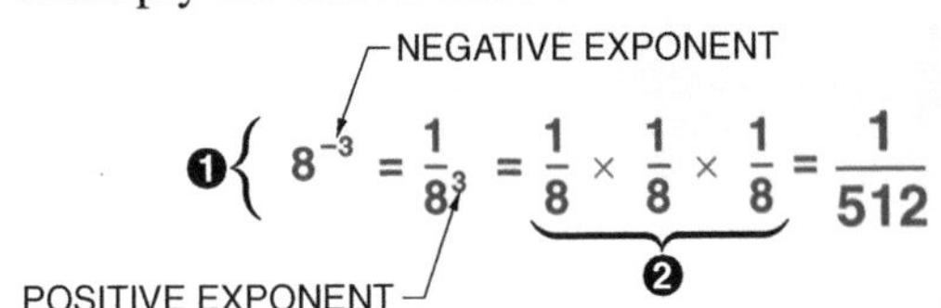

QUICK REFERENCE

- *Invert the base to its reciprocal.*
- *Multiply the denominator.*

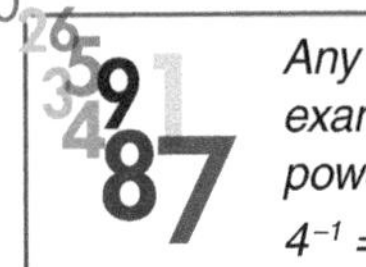

Any number raised to the power of 1 equals the base number. For example, $33^1 = 33$ and $(-87)^1 = -87$. Any number raised to the power of -1 equals the reciprocal of the base number. For example, $4^{-1} = \frac{1}{4}$ and $(-\frac{2}{3})^{-1} = -\frac{3}{2}$.

Multiplying Base Numbers

To multiply two or more like bases with exponents, add the exponents to find the new exponent. **See Figure 7-8.**

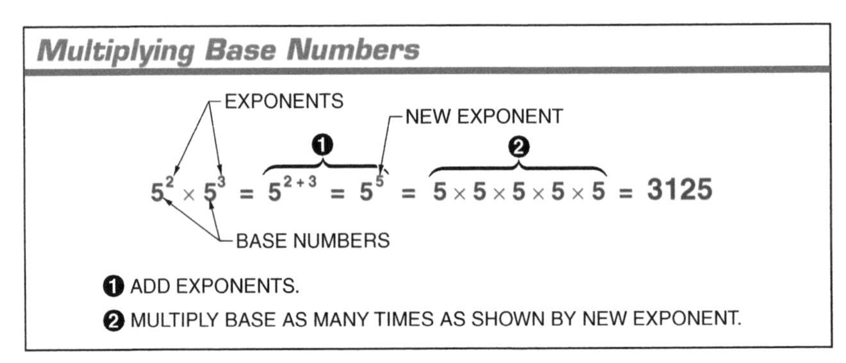

Figure 7-8. When multiplying two or more like bases that have the same or different exponents, the exponents are added to find the new exponent.

Example — Multiplying Base Numbers

1. Find the value of $4^2 \times 4^2 \times 4^3$.
ANS: **16,384**
❶ Add exponents (2 + 2 + 3 = 7).
❷ Raise 4 to the seventh power.

$$4^2 \times 4^2 \times 4^3 = \overbrace{4^{2+2+3}}^{❶} = 4^7 = \underbrace{4 \times 4 \times 4 \times 4 \times 4 \times 4 \times 4}_{❷} = 16{,}384$$

NEW EXPONENT

QUICK REFERENCE

- *Add exponents.*
- *Multiply the base as many times as shown by the new exponent.*

Dividing Base Numbers

To divide two like bases with exponents, subtract the exponent of the divisor from the exponent of the dividend to find the new exponent. **See Figure 7-9.**

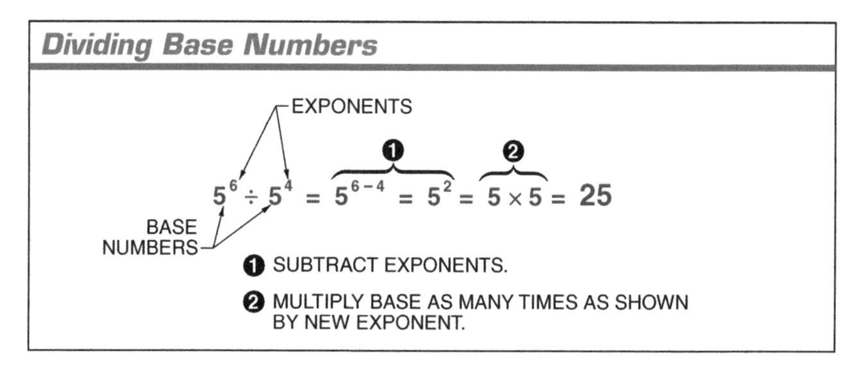

Figure 7-9. When dividing two or more like bases that have the same or different exponents, the second exponent is subtracted from the first to find the new exponent.

Example — Dividing Base Numbers

1. Find the value of $3^9 \div 3^5$.
ANS: **81**
❶ Subtract exponents (9 – 5 = 4).
❷ Raise 3 to the fourth power.

$$3^9 \div 3^5 = \overbrace{3^{9-5}}^{❶} = 3^4 = \underbrace{3 \times 3 \times 3 \times 3}_{❷} = 81$$

NEW EXPONENT

QUICK REFERENCE

- *Subtract exponents.*
- *Multiply the base as many times as shown by the new exponent.*

MATH EXERCISES—Exponents

__________ **1.** Find the value of 2^6.

__________ **2.** Find the fourth power of 21.

__________ **3.** Find the third power of $\frac{2}{9}$.

__________ **4.** Find the value of $(2^4)^3$.

__________ **5.** Find the value of $(3^3)^{-2}$.

__________ **6.** Find the value of 144^0.

__________ **7.** Find the value of $(-2)^{-2}$.

__________ **8.** Find the value of $3^3 \times 3^4$.

__________ **9.** Find the value of $0.2^3 \times 0.2^3$. (*Round to 6 places.*)

__________ **10.** Find the value of $2.2^5 \div 2.2^2$. (*Round to 3 places.*)

__________ **11.** Find the value of $750^8 \div 750^6$.

__________ **12.** Find the value of $(\frac{2}{5})^2 \times (\frac{2}{5})^3$.

PRACTICAL APPLICATIONS—Exponents

_______________ **13. Construction:** Find the area of the floor in the supply room. *Hint:* The *area* (A) of a square equals one side (s) raised to the second power ($A = s^2$).

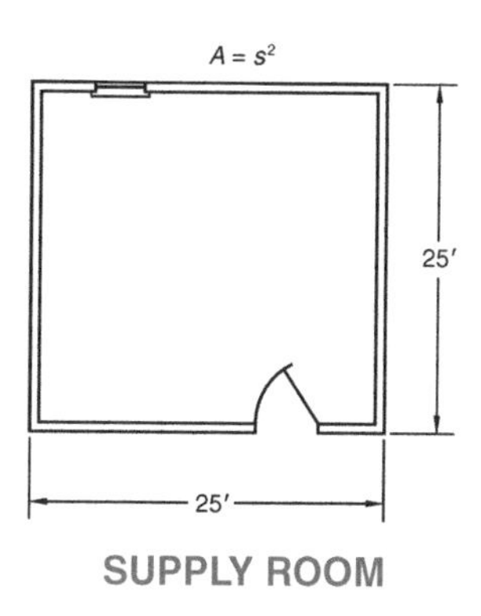

SUPPLY ROOM

_______________ **14. Boiler Operation:** The heating surface area of a boiler tube is calculated as *area* = 2 × 3.1416 × *radius* × *length*. What is the area of a tube with a radius of 1.5″ and a length of 120″?

_______________ **15. Electrical:** The electrical power in watts used in a circuit is calculated as $P = (current)^2 \times resistance$. What is the power if the current is 2.2 amps and the resistance is 100 ohms?

SECTION 7-2 UNDERSTANDING SQUARE ROOTS

A *square root* is one of the two equal factors, or roots, used to obtain a number. For example, 4 and 4 are the two equal factors of 16 ($4^2 = 4 \times 4 = 16$). Therefore, 4 is a square root of 16. The *radical sign* ($\sqrt{\ }$) is the symbol used to indicate the square root of a number. Thus, $\sqrt{16}$ is 4. **See Figure 7-10.** A *perfect square* is a number whose square root is a whole number.

Square Roots

$4^2 = 4 \times 4 = 16$

RADICAL SIGN

$\sqrt{16} = 4$

SQUARE ROOT OF 16

FIRST 12 PERFECT SQUARES

$1^2 = 1$	$7^2 = 49$
$2^2 = 4$	$8^2 = 64$
$3^2 = 9$	$9^2 = 81$
$4^2 = 16$	$10^2 = 100$
$5^2 = 25$	$11^2 = 121$
$6^2 = 36$	$12^2 = 144$

Figure 7-10. A square root is one of the two equal factors of a number.

Most numbers, however, do not have perfect squares so a calculator is used to find the square root. **See Figure 7-11.** The exact key sequence depends on the type of calculator. In most cases, a number is entered and the $\sqrt{\ }$ key is pressed. For example, to find the square root of 9.61, type 9.61 and press the $\sqrt{\ }$ key. The answer, 3.1, is displayed. To find the square root of a fraction, convert the fraction to a decimal and calculate the square root.

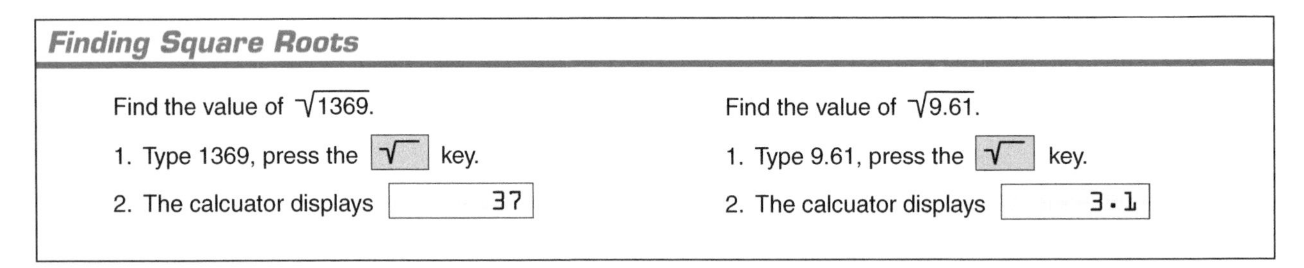

Figure 7-11. A calculator is normally used to find square roots.

Example—Finding Square Roots

1. Find the square root of 64, or $\sqrt{64}$.
   ANS: **8**
   ❶ Type 64 into a calculator.
   ❷ Press the $\sqrt{\ }$ key.

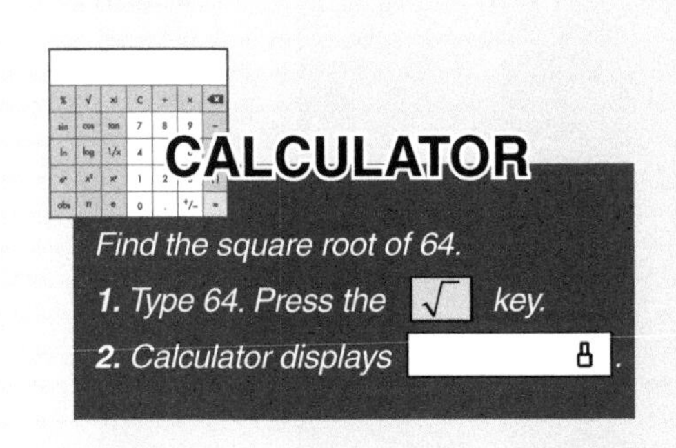

The cube root is one of the three equal factors, or roots, used to obtain a number. For example, $27 = 3 \times 3 \times 3$, so 3 is the cube root of 27. When the cube root is to be found, a superscript figure 3 is placed by the radical sign indicating that $\sqrt[3]{27}$ is the cube root of 27.

MATH EXERCISES—Square Roots

________________ **1.** Find the value of $\sqrt{36}$.

________________ **2.** Find the value of $\sqrt{121}$.

________________ **3.** Find the square root of 49.

________________ **4.** Find the square root of $\frac{1}{9}$.

________________ **5.** Find the value of $\sqrt{0.1849}$.

________________ **6.** Find the square root of 1444.

________________ **7.** Find the square root of 22.5625.

________________ **8.** Find the value of $\sqrt{85}$. (*Round to 5 decimal places.*)

______ **9.** Find the value of $\sqrt{289/225}$.

______ **10.** Find the square root of 24. (*Round decimal to nearest tenth.*)

PRACTICAL APPLICATIONS—Square Roots

______ **11. Construction:** A contractor knows that the total area of a square lot equals 4225 sq ft. What is the length of one of the sides of the lot?

______ **12. Boiler Operation:** A safety valve is used to protect a boiler by relieving excess pressure. The required diameter of a safety valve is calculated as $diameter = \sqrt{\frac{area}{0.7854}}$. What is the required diameter of a safety valve with an area of 9.56 sq in.?

For an interactive review of the concepts covered in Chapter 7, refer to the corresponding Quick Quiz® included on the Digital Resources.

QUICK QUIZ®
Working with Exponents

7 Working with Exponents
Review

Name ____________________ Date ____________

Math Problems

__________ **1.** Find the value of 7^4.

__________ **2.** Find the value of $(-2)^8$.

__________ **3.** Find the value of $(-6)^3$.

__________ **4.** Find the third power of $\frac{2}{3}$.

__________ **5.** Find the value of 0.3^2.

__________ **6.** Find the value of $(3^2)^2$.

__________ **7.** Find the value of 9^{-2}.

__________ **8.** Multiply 4^3 by 4^2 by 4^2.

__________ **9.** Divide 3^7 by 3^5.

7 Review (continued)

_________________ **10.** Find the fourth power of $\frac{5}{8}$.

_________________ **11.** Find the value of $\sqrt{4}$.

_________________ **12.** Find the square root of 10.89.

_________________ **13.** Find the square root of $\frac{9}{25}$.

Practical Applications

_________________ **14. Boiler Operation:** The amount of force in pounds developed by a reciprocating feedwater pump is calculated as *force* = 3.1416 × (*radius*)^2 × *pressure.* What is the force developed when the radius is 3″ and the pressure is 250 lb/sq in.?

_________________ **15. HVAC:** A custom right triangle diverter must be fabricated from sheet metal. The Pythagorean theorem ($c^2 = a^2 + b^2$) is used to find the length of each side of the triangle. If side a is 5″ long and side b is 7″ long, how long is side c? (*Hint: The answer is the square root of* c^2.)

Working with Exponents 7

Test

Name ______________________ Date ____________

Math Problems

______ 1. Find the value of 14^4.

______ 2. Find the value of $(-5)^3$.

______ 3. Find the fourth power of –5.

______ 4. Find the third power of $\frac{1}{2}$.

______ 5. Find $(\frac{3}{4})^3$.

______ 6. Find 0.25^3.

______ 7. Find the value of $(6^3)^2$ using the base exponent method.

______ 8. Find the value of $(7^3)^2$ using the new exponent method.

______ 9. Find the value of $(0.5^3)^2$.

7 Test (continued)

____________ **10.** Find the value of 3^3.

____________ **11.** Multiply 3^3 by 3^3 by 3^4.

____________ **12.** Multiply the fourth power of 4 by the square of 4.

____________ **13.** Divide the fifth power of 4 by the square of 4.

____________ **14.** Divide the third power of 12 by the cube of 12.

____________ **15.** Find the value of $(2^3)^4$.

____________ **16.** Find the square root of 1024.

____________ **17.** Find the square root of 22. (*Round to 5 places.*)

____________ **18.** Find the square root of 1369.

____________ **19.** Find the value of 14^4.

____________ **20.** Find the value of $-(5)^4$.

7 Test (continued)

_______________ **21.** Find the third power of $\frac{3}{16}$.

_______________ **22.** Find the third power of 0.15.

_______________ **23.** Find the value of $(2^4)^3$ using the base exponent method.

_______________ **24.** Find the value of $(10^3)^2$ using the new exponent method.

_______________ **25.** Multiply 5^3 by 5^3.

_______________ **26.** Multiply 4^2 by 4^3 by 4^3.

_______________ **27.** Divide 47 by 4^3.

_______________ **28.** Divide 35 by 3^2.

_______________ **29.** Find the square root of 1156.

_______________ **30.** Find the square root of 1764.

7 Test (continued)

Practical Applications

____________ **31. Construction:** Panels with tongue-and-groove edges are to be used as a subfloor in a square bedroom that measures 484 sq ft. What is the length of one side of the subfloor?

____________ **32. Mechanics:** The amount of force created by a 6″ diameter pneumatic cylinder is found using the formula *force* = *pressure* × *area*. If the pressure to the cylinder is 85 psi, what is the force created by the cylinder? (*Hint: area* = 3.1416 × r^2.)

____________ **33. Boiler Operation:** The heating surface area of a boiler tube is calculated as *area* = 2 × 3.1416 × *radius* × *length.* What is the area of a tube with a radius of 1.25″ and a length of 240″?

Working with Ratios and Proportions

8

A ratio demonstrates the mathematical relationship between two quantities. A proportion is an expression of equality between two ratios. Knowing how to find the ratio of two quantities and their proportions is essential to all trades. Tasks such as mixing cement and determining shingle quantity all require being able to solve ratio and proportion problems.

OBJECTIVES

1. Demonstrate how to express ratios.
2. Explain inverse ratios.
3. Find the compound ratio of two or more ratios.
4. Solve ratios that include fractions.
5. Demonstrate how to express proportions.
6. Use cross-multiplication to verify a proportion.
7. Find a missing term of a proportion.
8. Solve direct proportions.
9. Solve inverse proportions.

KEY TERMS

- ratio
- inverse ratio
- compound ratio
- proportion
- direct proportion
- inverse proportion

Digital Resources
ATPeResources.com/QuickLinks
Access Code: 764460

SECTION 8-1 USING RATIOS

A ratio is a mathematical way to represent the relationship between two or more numbers, or terms. A ratio is the mathematical method of making a comparison. The symbol used to indicate the relationship between terms in a ratio is the colon (:). A ratio can also be expressed as a fraction where the first term is the numerator and the second term is the denominator. **See Figure 8-1.**

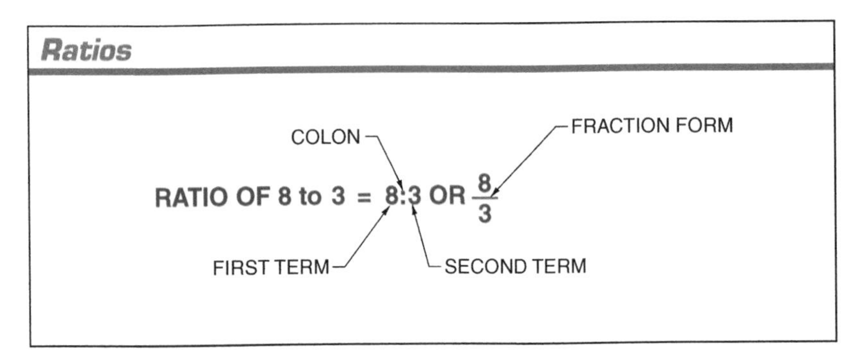

Figure 8-1. A colon can be used to separate the terms in a ratio or the terms can be expressed as a fraction.

Inverse Ratios

An *inverse ratio* is the ratio of the reciprocals of two quantities. To find an inverse ratio, simply invert the terms of the fraction. An inverse ratio can be used when two elements work opposite one another. **See Figure 8-2.**

In the following problem, fluid flow and time are inversely proportional because time increases when fluid flow decreases and time decreases when flow increases.

If it takes 14 hours to fill a barrel with a flow of 18 liters per minute, how long would it take to fill the same barrel if the flow were reduced to 7 liters per minute?

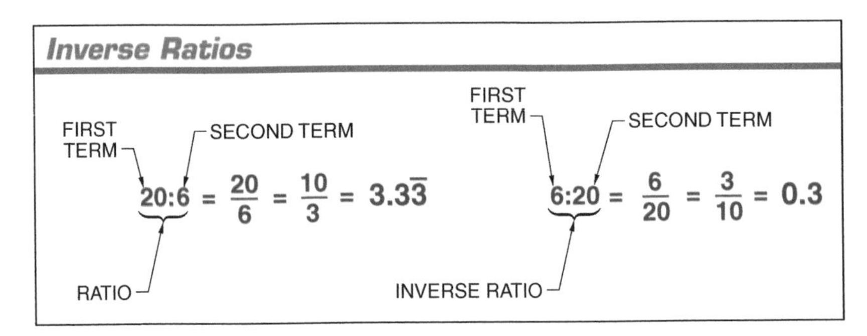

Figure 8-2. In an inverse ratio, the terms of the ratio are inverted and the ratio becomes the reciprocal of the two quantities.

Examples — Inverse Ratios

1. Reduce the ratio 21:3.
   ANS: 7⁄1 or **7:1**
   ❶ Put ratio in fraction form.
   ❷ Reduce.

   $$21{:}3 = \underbrace{\frac{21}{3}}_{❶} = \underbrace{\frac{7}{1}}_{❷} = 7{:}1$$

2. Invert the ratio 36:9 and reduce.
   ANS: 1⁄4 or **1:4**
   ❶ Put ratio in fraction form.
   ❷ Invert.
   ❸ Reduce.

   $$36{:}9 = \underbrace{\frac{36}{9}}_{❶} \qquad \underbrace{\frac{9}{36}}_{❷} = \underbrace{\frac{1}{4}}_{❸} = 1{:}4$$

QUICK REFERENCE

- *Put the ratio in fraction form.*
- *Reduce as required.*

Compound Ratios

A *compound ratio* is the product of two or more ratios. To find a compound ratio of two ratios, change the ratios to fractions, and multiply the numerators (first terms) and denominators (second terms). The product is a compound ratio in fraction form. **See Figure 8-3.**

Compound Ratios

$$8{:}3 \text{ and } 7{:}2 = \overbrace{\frac{8}{3} \times \frac{7}{2}}^{❶} = \frac{56}{6} = \frac{28}{3} = \underbrace{28{:}3}$$

Figure 8-3. A compound ratio is the product of two simple ratios.

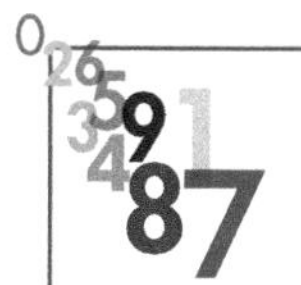

When the terms include units of measure, a ratio can be formed only if the units of measure are common. It may be possible to convert one term to match the unit of measure of the other term. For example, 6 ft can be converted to yards to compare it to 3 yd (2 yd:3 yd).

Example — Compound Ratios

1. Find the compound ratio of 6:1 and 3:2.

   ANS: ⁹⁄₁ or **9:1**

   ❶ Put ratios in fraction form. Multiply the numerators (6 and 3), then multiply the denominators (1 and 2).

   ❷ Reduce ¹⁸⁄₂ to ⁹⁄₁.

$$6{:}1 \text{ and } 3{:}2 = \underbrace{\frac{6}{1} \times \frac{3}{2} = \frac{18}{2}}_{❶} = \underbrace{\frac{9}{1}}_{❷} = 9{:}1$$

QUICK REFERENCE

- *Change the ratios to fraction form and multiply both numerators by numerators and denominators by denominators.*
- *Reduce as required.*

Ratios with Fractions

Sometimes a ratio expression can have a fraction as the first term, second term, or both terms. To solve ratios with fractions, any whole number needs to be changed to a fraction. When both ratios are fractions, they can be divided. To do this, invert the second ratio, multiply the numerators, and multiply the denominators. **See Figure 8-4.**

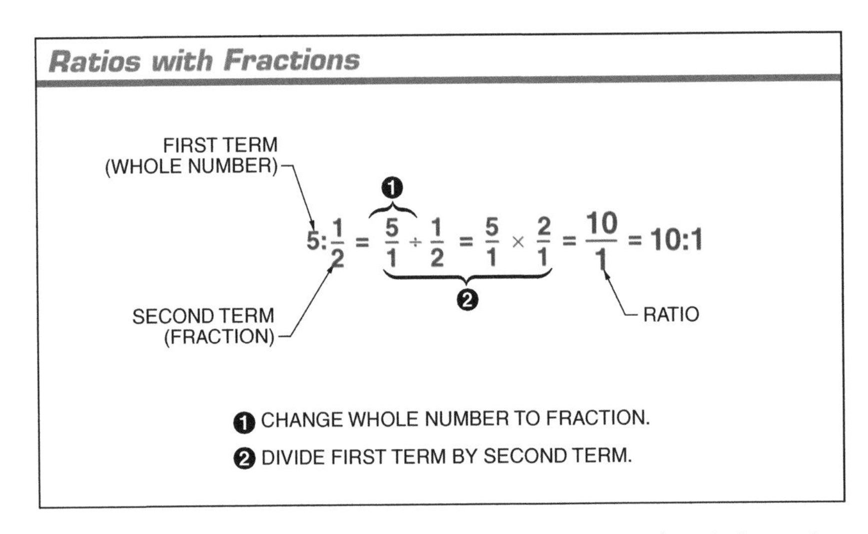

Figure 8-4. To find ratios that include a fraction, change the whole number to a fraction and divide the first term by the second term.

When using a calculator to solve ratios with fractions, convert each fraction to a decimal number and then divide the first term by the second term.

Examples—Ratios with Fractions

1. Find the ratio of ¾ : 8.

ANS: **3 : 32**

❶ Change whole number 8 to a fraction.

❷ Divide fractions by inverting second fraction (8⁄1 becomes ⅛) and multiplying.

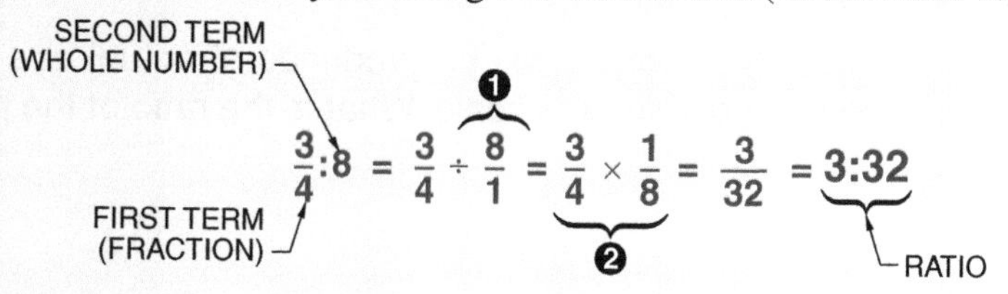

QUICK REFERENCE

- *Change a whole number to a fraction.*
- *Divide the fractions by inverting the second fraction and multiplying.*
- *Reduce as required.*

2. Find the ratio of ⅖ : ⅔.

ANS: **3 : 5**

❶ Divide fractions by inverting second fraction (⅔ becomes 3⁄2) and multiplying.

❷ Reduce to ⅗.

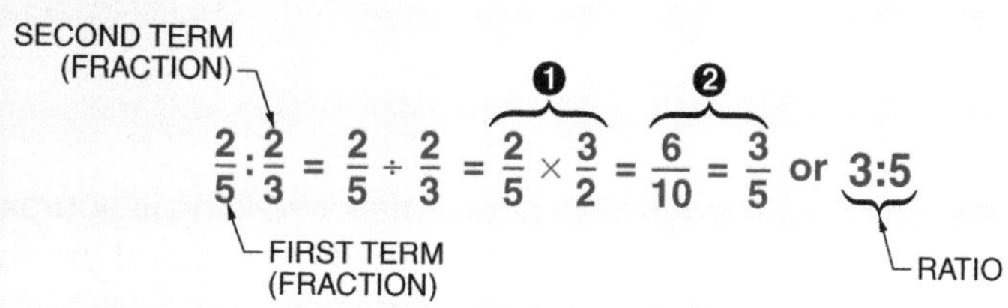

MATH EXERCISES—Ratios

_______________ **1.** Reduce the ratio 36:4.

_______________ **2.** Find the compound ratio of 9:2 and 10:3 and reduce.

_______________ **3.** Find the inverse ratio of 3 to 9 and reduce.

_______________ **4.** Find the ratio of ⅔ to 5.

_______________ **5.** Find the compound ratio of 22:1 and 3:4 and reduce.

_______________ **6.** Find the inverse ratio of 24 to 48 and reduce.

_______________ **7.** Find the ratio of 3⅓ to 16⅔.

_______________ **8.** Find the compound ratio of 10:3 and 6:4 and reduce.

PRACTICAL APPLICATIONS—Ratios

_______________ **9. Manufacturing:** A prototype will be produced at actual size while the print for the prototype is drawn at ¼ scale. What is the ratio of the prototype to the print?

_______________ **10. Welding:** The ratio of oxygen to acetylene for welding is approximately 1:1. For cutting operations, the oxygen is increased by one and one-half times. What is the ratio of oxygen to acetylene for cutting?

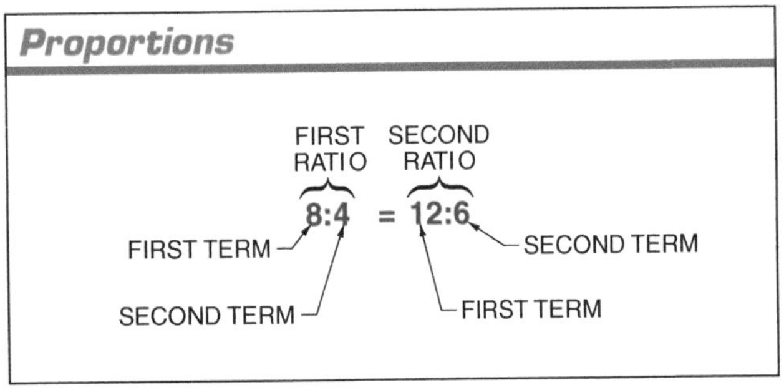

Proportions

FIRST RATIO SECOND RATIO

8:4 = 12:6

FIRST TERM SECOND TERM SECOND TERM FIRST TERM

Figure 8-5. All proportions are composed of two equal ratio expressions.

SECTION 8-2 USING PROPORTIONS

A *proportion* is an expression of equality between two ratios. **See Figure 8-5.** For example, the ratio expressions 8:4 and 12:6 can both be simplified as 2:1, and are therefore equal. The proportion is read, "8 is to 4 as 12 is to 6."

The cross-product rule states that the product of the two inner numbers of the proportions (the means) equals the product of the two outer numbers of the proportions (the extremes). This means that in any proportion, multiplying the means will yield the same product as when multiplying the extremes. For example, in the ratios 8:4 and 12:6, the product of the means, 4 × 12 (the inner numbers) and the product of the extremes 8 × 6 (the outer numbers) are both equal to 48. **See Figure 8-6.**

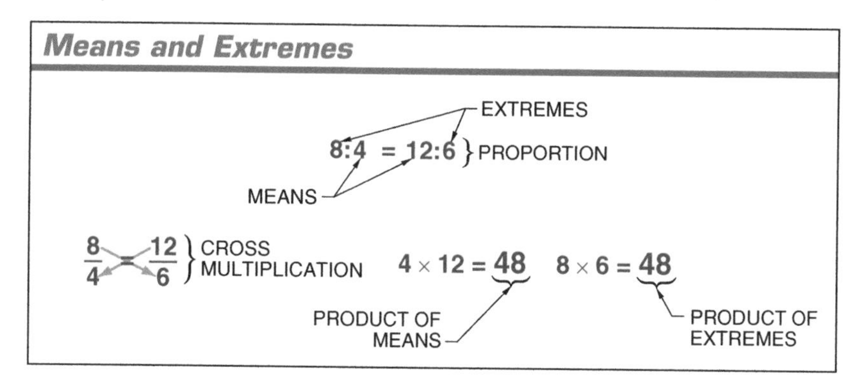

Figure 8-6. Cross-multiplication can be applied to a proportion to prove its validity.

Sometimes it may be necessary to find an unknown term of a proportion. To find the term, the ratios should be written as fractions and the cross-product rule applied. **See Figure 8-7.**

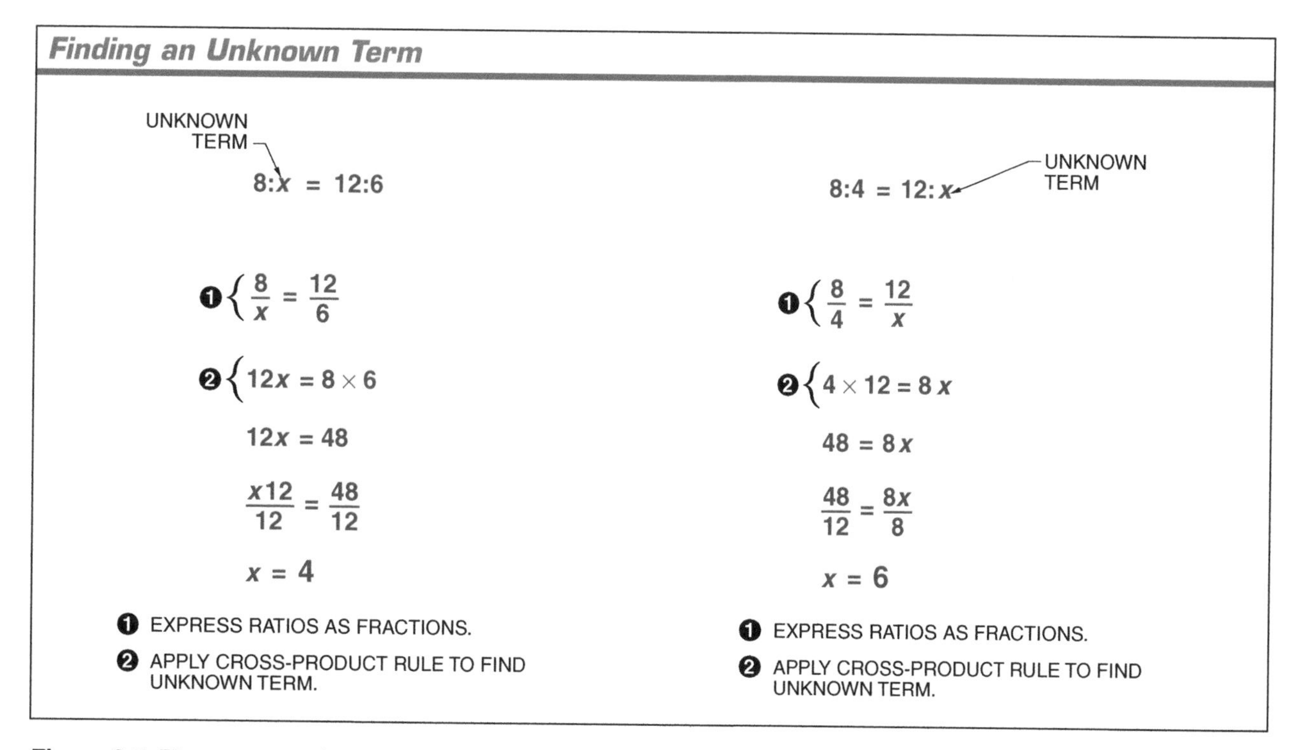

Figure 8-7. The cross-product rule is applied when finding an unknown term.

Examples—Finding an Unknown Term

1. Find the unknown term in $36{:}x = 12{:}17$.

   ANS: **51**

   ❶ Set proportion up as fractions.
   ❷ Cross multiply (36 by 17, x by 12).
   ❸ Solve for x.

   UNKNOWN MEAN — KNOWN MEAN — EXTREME — EXTREME

   $$36{:}x = 12{:}17$$

   $$❶\left\{\frac{36}{x} = \frac{12}{17} \quad ❷\left\{36 \times 17 = x12 \quad ❸\left\{\frac{612}{12} = \frac{x12}{12} \quad x = 51\right.\right.\right.$$

2. Find the unknown term in $x{:}120 = 8{:}192$.

   ANS: **5**

   ❶ Set proportion up as fractions.
   ❷ Cross multiply (8 by 120, x by 192).
   ❸ Solve for x.

   UNKNOWN EXTREME — KNOWN EXTREME — MEANS

   $$x{:}120 = 8{:}192$$

   $$❶\left\{\frac{x}{120} = \frac{8}{192} \quad ❷\left\{8 \times 120 = x192 \quad ❸\left\{\frac{960}{192} = \frac{x192}{192} \quad x = 5\right.\right.\right.$$

QUICK REFERENCE

- *Set the proportions up as fractions.*
- *Cross multiply the fractions.*
- *Solve for x by dividing.*

MATH EXERCISES—Proportions

Find each unknown term.

_______________ **1.** $17{:}x = 51{:}54$

_______________ **2.** $19/x = 209/143$

_______________ **3.** $x{:}3.9 = 40{:}78$

_______________ **4.** $6/0.6 = x/20$

_______________ **5.** $35/7 = x/91$

_______________ **6.** $48:20 = x:50$

_______________ **7.** $x:300 = 20:100$

_______________ **8.** $1:x = 7:84$

_______________ **9.** $48:x = 67.25:201.75$

_______________ **10.** $12:5 = x:40$

_______________ **11.** $16:24 = x:15$

_______________ **12.** $4:4\frac{2}{3} = 9\frac{3}{7}:x$

_______________ **13.** $48:x = 240:6$

_______________ **14.** $x:9 = 13:10$

_______________ **15.** $2.76:3.45 = 2.28:x$

Solving Direct Proportions

A *direct proportion* is a proportion where an increase in one quantity leads to a proportional increase in the related quantity, and a decrease leads to a proportional decrease. For example, in the proportion 8:4 = 12:6, if the 8 is increased twice, the 12 must be increased twice to keep the proportion equal, 16:4 = 24:6. **See Figure 8-8.**

Concrete is composed of cement, aggregate, and water, all in specific proportions.

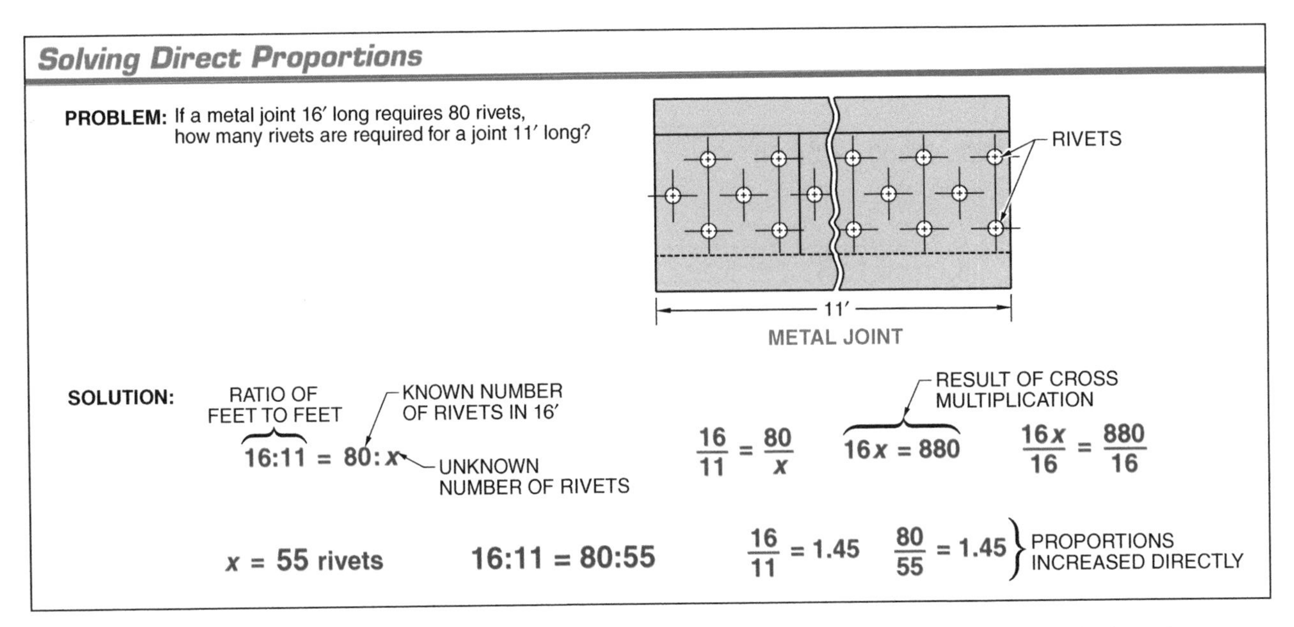

Figure 8-8. In a direct proportion, a change in one ratio results in a directly proportional change in the other.

Example—Solving Direct Proportions

1. If 15 ironworkers can build a form 12′ high in one day, how many ironworkers are required to build a form 20′ high in one day?

   ANS: **25**

   ❶ Set proportion up as fractions.
   ❷ Cross multiply (12 by x, 15 by 20).
   ❸ Solve for x.

KNOWN EXTREME — UNKNOWN EXTREME — MEANS

$$12:20 = 15:x$$

❶ $\left\{ \frac{12}{20} = \frac{15}{x} \right.$ ❷ $\left\{ 12x = 15 \times 20 \right.$ ❸ $\left\{ \frac{12x}{12} = \frac{300}{12} \right.$

x = 25 iron workers

12:20 = 15:25

QUICK REFERENCE

- *Set the proportion up as fractions.*
- *Cross multiply the fractions.*
- *Solve for x by dividing.*

Solving Inverse Proportions

An *inverse proportion* is a proportion where an increase in one quantity leads to a proportional decrease in the related quantity, and a decrease leads to a proportional increase. In an inverse proportion, the second ratio is inverted to make the proportion equal. **See Figure 8-9.**

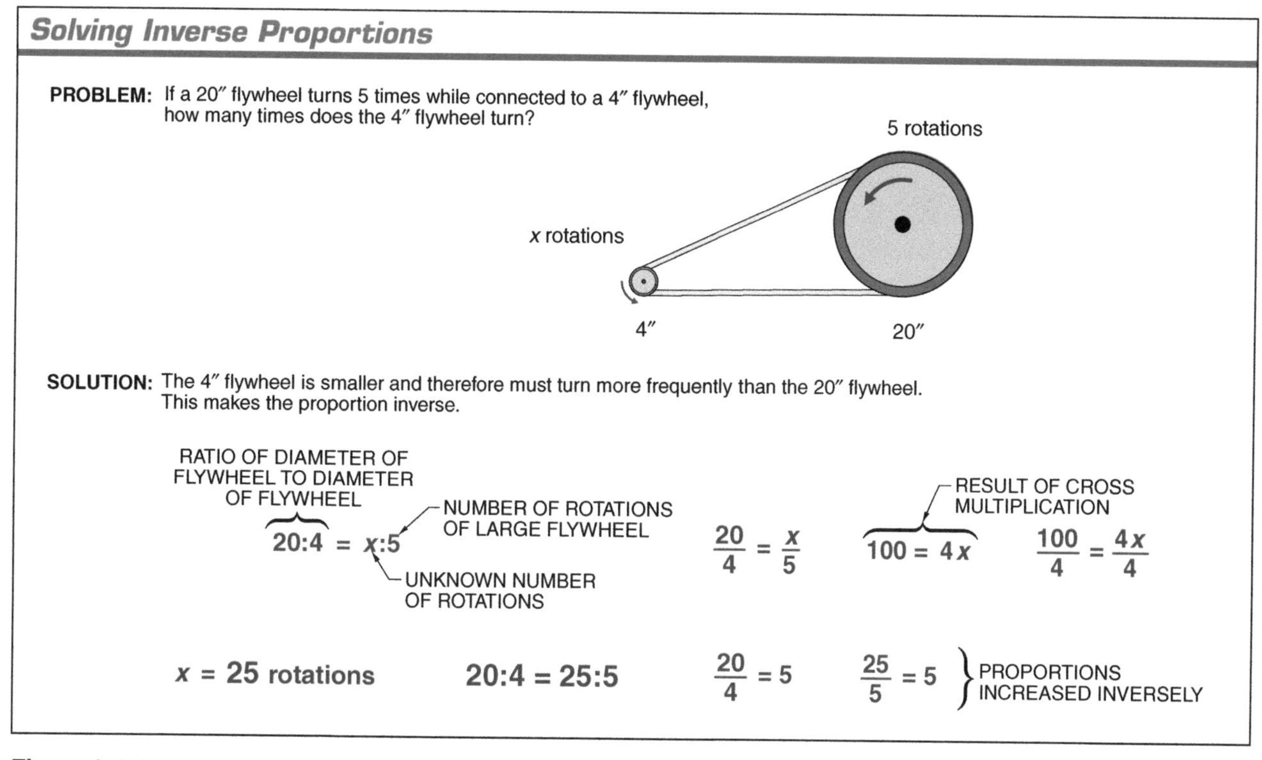

Figure 8-9. In an inverse proportion, a change in one ratio results in an inversely proportional change in the other.

Example—Solving Inverse Proportions

1. When levers are used for mechanical work, their forces are inversely proportional to the distances the forces are applied over. Find the weight (force) that can be lifted by the lever.

   ANS: **500 lb**

   ❶ Set proportion up as inverse proportion.
   ❷ Set proportion up as fractions.
   ❸ Cross multiply (200 by 10, 4 by x).
   ❹ Solve for x.

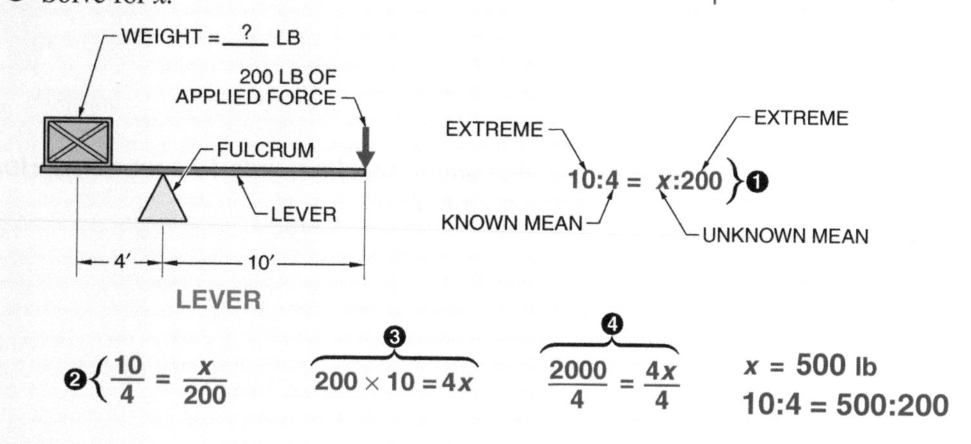

QUICK REFERENCE

- *Set the proportion up as an inverse proportion.*
- *Set the proportion up as fractions.*
- *Cross multiply the fractions.*
- *Solve for x by dividing.*

PRACTICAL APPLICATIONS—Proportions

_________________ 1. **Boiler Operation:** If burning 50 gal. of #2 fuel oil releases 300 lb of carbon into the atmosphere, how much carbon is released by burning 125 gal. of #2 fuel oil?

_________________ 2. **Welding:** The pressure on 10 cu ft of gas in a tank is 20 psi. What would the pressure be if the gas were compressed to 5 cu ft? (*Hint: This proportion is inverse.*)

_________________ 3. **Construction:** If 10 drywallers can hang 40 pieces of drywall in 1 hour, how long should it take 2 drywallers to hang 40 pieces of drywall?

_________________ 4. **Alternative Energy:** One fuel cell uses 3 liters per minute (Lpm) of hydrogen to produce 175 W of electricity. At this rate, how much hydrogen is required to produce 525 W of electricity?

_________________ 5. **Welding:** One panel of sheet metal requires 13 spot welds. How many spot welds are required on 7 panels?

________________ 6. **Boiler Operation:** Boiler water treatment systems may use a flow ratio controller to control the amount of treatment chemicals added to a boiler with the makeup feedwater. If a flow ratio controller adds 2.3 lb of water treatment chemicals for every 920 gal. of makeup feedwater, how much water treatment chemical is used with 200 gal. of makeup feedwater?

________________ 7. **Pipefitting:** A 415 ft run of 4 in. pipe requires 43 hangers. How many hangers are needed for 280 ft of 4 in. pipe?

________________ 8. **Mechanics:** A lever is 20 ft long on the applied force side and 4 ft long on the load side. How many pounds of applied force are required to lift a 500 lb load? *(Hint: This is an inverse proportion.)*

For an interactive review of the concepts covered in Chapter 8, refer to the corresponding Quick Quiz® included on the Digital Resources.

Working with Ratios and Proportions 8

Review

Name ______________________ Date ____________

Math Problems

________ **1.** Reduce the ratio 24:4.

________ **2.** Find the inverse ratio of 32:8 and reduce.

________ **3.** Find the compound ratio of 3:1 and 5:3 and reduce.

________ **4.** Find the ratio of $\frac{1}{4}$:16.

________ **5.** Find the unknown term in $x:4 = 3:1$.

________ **6.** Find the unknown term in $42:x = 14:1$.

________ **7.** Find the inverse ratio of 64:16 and reduce.

________ **8.** Find the unknown term in $x:6 = 0.5:1$.

________ **9.** Find the unknown term in $128:x = 4:1$.

________ **10.** Find the unknown mean in $6:0.5 = x:5$.

________ **11.** Find the unknown extreme in $3:4 = 16:x$.

8 Review (continued)

Practical Applications

________________ 12. **Alternative Energy:** An array of 8 solar thermal panels produces about 24,000 Btu of heat energy per hour. How much heat energy would a similar array of 6 panels produce per hour?

________________ 13. **Construction:** If it takes 10 hours for three painters to paint 1200 sq ft of drywall, how much drywall can be painted in 16 hours?

________________ 14. **Manufacturing:** If a pulley has a 12 in. diameter wheel that turns 5 times per minute, what diameter is the second wheel if it turns 10 times per minute?

________________ 15. **Construction:** If 2 carpenters can build the formwork for a foundation in 12 hours, how long should it take 3 carpenters to build the same foundation?

Working with Ratios and Proportions 8

Test

Name ______________________ Date ____________

Math Problems

_______ 1. Reduce the ratio 44:11.

_______ 2. Find the whole number ratio of 0.50:0.10.

_______ 3. Find the inverse ratio of 36:3 and reduce.

_______ 4. Find the inverse ratio of 0.75:0.25 and reduce.

_______ 5. Find the compound ratio of 5:1 and 5:4.

_______ 6. Find the compound ratio of 6:1 and 7:1.

_______ 7. Find the unknown term in x:16 = 0.4:1.

_______ 8. Find the unknown term in x:8 = 0.25:1.

_______ 9. Find the unknown term in 12:x = 0.3:1.

8 Test (continued)

Practical Applications

_______________ **10. Construction:** What is the ratio of jobs accepted for a construction firm that wins 4 jobs for every 10 jobs it bids?

_______________ **11. Maintenance:** Twenty-four shims (no stacking) are used to level the 720 sq in. bed of a wood turning lathe. Using the same leveling requirements, how many shims should be required to level the 1260 sq in. bed of a horizontal boring mill?

_______________ **12. Mechanics:** Using a pulley, a load is raised 1 ft when the lead line on a four-part reeve is pulled 4 ft. If the force required by the lead line is 20 lb, what is the actual weight of the load? (*Hint: This proportion is inverse.*)

_______________ **13. Boiler Operation:** The horsepower (HP) of a boiler determines the amount of steam produced. A 100 HP boiler produces 3450 lb/hr of steam. How many HP are required to produce 5175 lb/hr of steam?

8 Test (continued)

_______________ **14. Welding:** A team of three production welders using gas metal arc welding can weld a total of 720 in. in 1 hour. How long should it take two welders to weld the same amount?

_______________ **15. Electrical:** If an electrician can pull 100 ft of No. 14 Cu conductor in 0.7 hr, how many feet can be pulled in 4 hr? (*Round to the nearest whole foot.*)

_______________ **16. Plumbing:** A plumber installs 100 ft of 4 in. Schedule 40 black pipe in 6 hr. How many hours should be required to install 240 ft?

_______________ **17. Boiler Operation:** Steam piping expands when heated. If a 200 ft section expands 5 in., how much does a 240 ft section expand?

_______________ **18. Mechanics:** A gear train consists of a gear with 18 teeth that rotates at a speed of 100 rpm and a gear containing 50 teeth. At what speed does the gear with 50 teeth rotate? (*Hint: This proportion is inverse.*)

8 Test (continued)

_______________ **19. Construction:** Two tractor operators backfill a foundation with 100 cu yd of earth in 0.9 hr. In how many hours can the two tractor operators backfill 800 cu yd of earth?

_______________ **20. Electrical:** An electrical circuit with a constant voltage has a resistance of 10 Ω and a current of 11 A. What is the current if the resistance is 55 Ω? (*Hint: This proportion is inverse.*)

_______________ **21. Alternative Energy:** A system's efficiency is the ratio of its output to the input it receives. A solar photovoltaic panel receives 1200 W (input) of solar power on its surface and converts it to 150 W (output) of electrical power. What is the module's efficiency?

Working with Plane and Solid Figures

Plane figures are two-dimensional figures, such as squares, rectangles, triangles, and circles. The ability to determine the area of a plane figure is important in many trades. Solid figures are three-dimensional figures, such as cubes, rectangular prisms, and cylinders. The ability to determine both the volume and the surface area of a solid figure is important in many trades.

OBJECTIVES

1. Explain how the area of a plane figure is measured.
2. Identify seven types of lines that can form a perimeter of a plane figure.
3. Differentiate among straight, right, acute, and obtuse angles.
4. Describe what all polygons have in common.
5. Find the areas of squares and rectangles.
6. Describe what all triangles have in common.
7. Find the areas of triangles using the Pythagorean theorem.
8. Describe what all circles have in common.
9. Find the areas of circles.
10. Explain how the volume and the surface area of a solid figure are measured.
11. Find the volumes and surface areas of rectangular solids.
12. Find the volumes and surface areas of cylinders.
13. Find the volumes and surface areas of square pyramids.
14. Find the volumes and surface areas of cones.
15. Find the volumes and surface areas of spheres.

KEY TERMS

- plane figure
- perimeter
- perpendicular line
- parallel lines
- angle
- right angle
- polygon
- quadrilateral
- right triangle
- isosceles triangle
- equilateral triangle
- scalene triangle
- hypotenuse
- circumference
- diameter
- radius
- solid figure
- prism
- cylinder
- pyramid
- cone
- sphere

Digital Resources
ATPeResources.com/QuickLinks
Access Code: 764460

SECTION 9-1 UNDERSTANDING PLANE FIGURES

A *plane figure* is a two-dimensional figure. The two dimensions are length and width, which create the area of the figure. The area of a plane figure is measured in square units such as square inches, square feet, square millimeters, or square meters.

All plane figures are composed of points connected by straight or curved lines. A *point* is a specific location in space. A point is shown as a dot and designated by a single letter. **See Figure 9-1.**

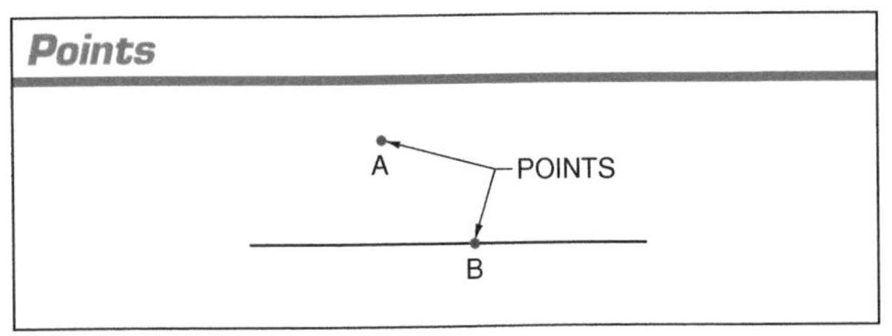

Figure 9-1. A point is a specific location in space.

Lines

A *line* is a one-dimensional figure that appears as a long, narrow band. **See Figure 9-2.** In plane geometry, lines form the perimeter of a figure. The *perimeter* is the sum of the lengths of the sides of a closed plane figure. Lines are measured in units such as inches, feet, yards, millimeters, or meters.

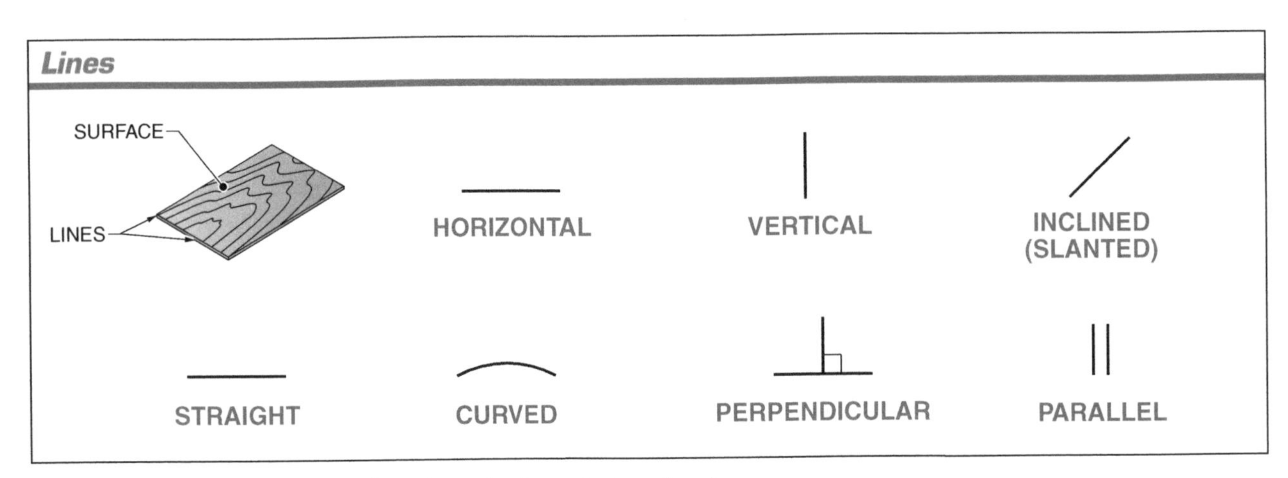

Figure 9-2. In plane geometry, lines form the boundaries of surfaces.

A *horizontal line* is a line that is parallel to the horizon. It may be referred to as a level line. A *vertical line* is a line that is perpendicular to the horizon. It is often referred to as a plumb line. An *inclined line*, or slanted line, is a line that is neither horizontal nor vertical and can incline in any direction. Lines can be straight or curved.

Perpendicular lines are two lines that form a 90° angle. The symbol for perpendicular is ⊥. *Parallel lines* are two or more lines that never intersect. The symbol for parallel lines is ||.

Angles

An *angle* is a figure created by two intersecting lines. A *vertex* is the point of intersection of two or more lines. To identify angles, letters are placed at the end of each line and at the vertex.

In plane geometry, any angle can be thought of as having its vertex at the center of a circle. Since a circle measures 360°, an angle can measure any number of degrees less than 360°. **See Figure 9-3.**

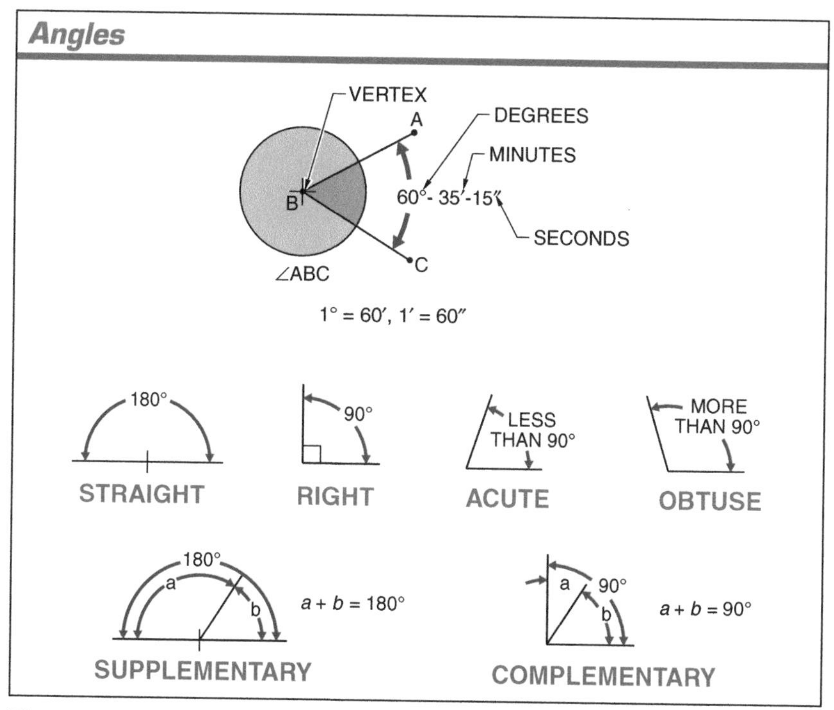

Figure 9-3. The vertex of any angle can be thought of as the centerpoint of a circle.

A *straight angle* is an angle that measures 180° and forms a straight line. If a line intersects a straight angle, it creates two angles called supplementary angles that, added together, equal 180°.

A *right angle* is an angle that measures 90°. It is one-fourth of a circle, or 90°. If a line intersects a right angle, it creates two angles called complementary angles that, added together, equal 90°.

An *acute angle* is an angle that measures less than 90°. Angles of 30°, 45°, and 60° are acute angles. An *obtuse angle* is an angle that measures more than 90° but less than 180°. Angles of 105°, 120°, and 175° are obtuse angles.

Polygons

A *polygon* is a multiple-sided plane figure that has a perimeter of straight lines. All polygons have a base or bases and altitude (height). The *base* is the bottom of a polygon. If the opposite side is parallel to the base, it can also be referred to as a base. *Height* is the length measurement from the base of a polygon to its other base or to the vertex on the opposite side.

Polygons include triangles, quadrilaterals, pentagons, hexagons, heptagons, and octagons. **See Figure 9-4.** A *quadrilateral* is a four-sided polygon with four interior angles. The sum of the four angles of a quadrilateral is always 360°. Quadrilaterals include squares, rectangles, rhomboids, rhombuses, trapezoids, and trapeziums. **See Figure 9-5.**

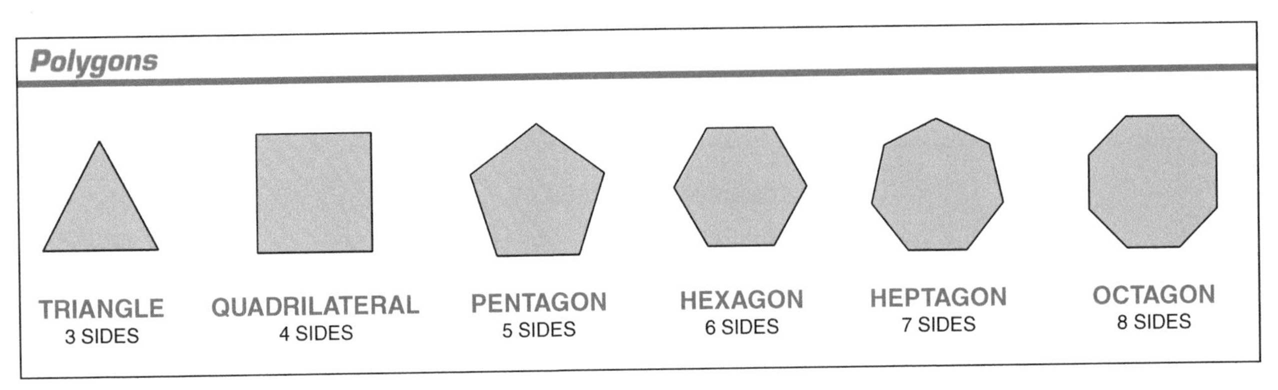

Figure 9-4. All polygons are bounded by straight lines.

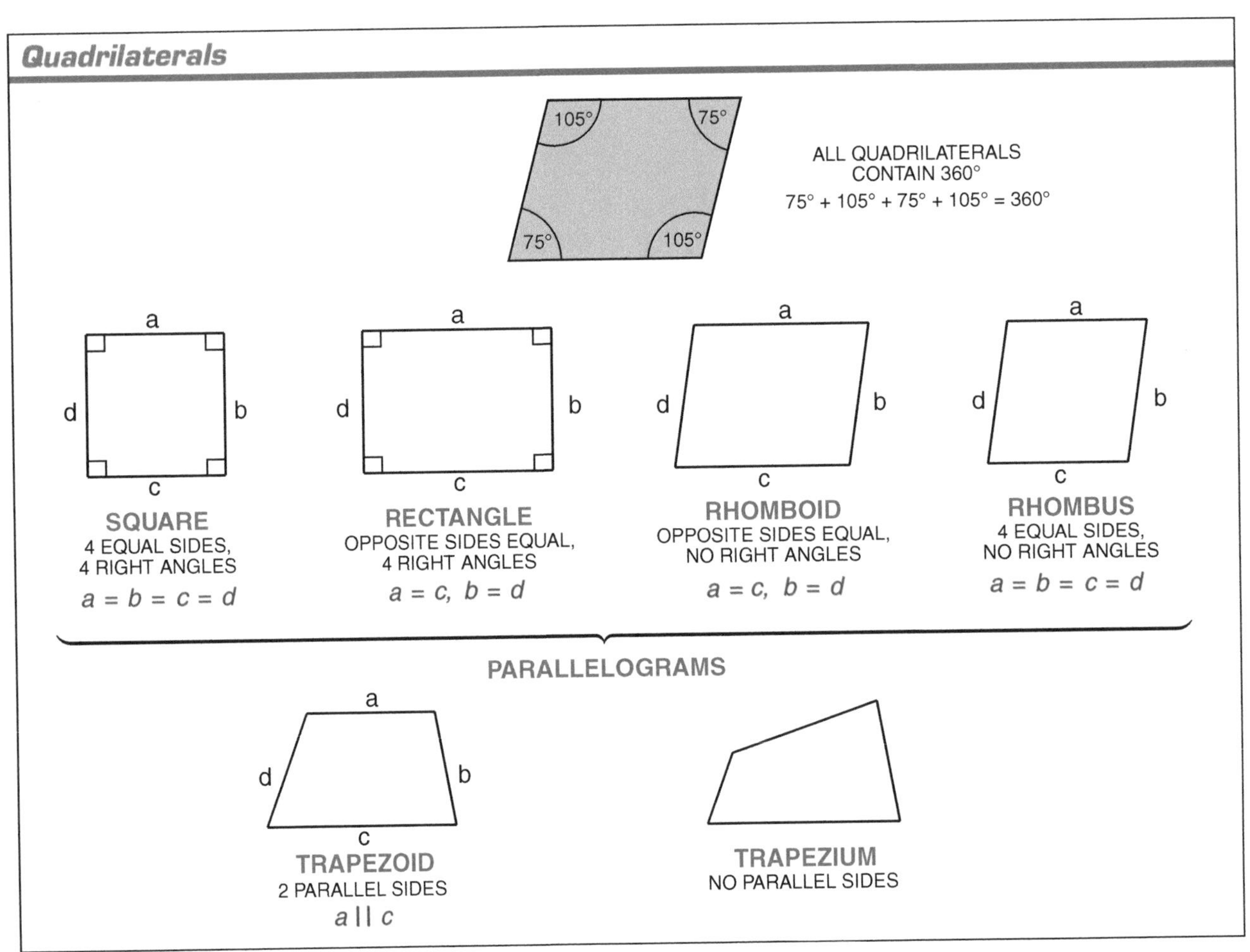

Figure 9-5. Squares, rectangles, rhomboids, rhombuses, trapezoids, and trapeziums are quadrilaterals and each contains 360°.

A *square* is a quadrilateral with four equal sides and four right angles. A *rectangle* is a quadrilateral with opposite sides of equal length and four right angles. The square, rectangle, rhombus, and rhomboid are parallelograms. Trapezoids and trapeziums are not parallelograms because not all opposite sides are parallel.

The area of a square or rectangle is found by applying the following formula:

$A = l \times w$

where

A = area

l = length

w = width

Example—Squares and Rectangles

1. Find the area of a 16″ × 7″ rectangle.

ANS: **112 sq in.**

❶ Determine the length (16) and width (7) of the rectangle.

❷ Multiply 16 by 7 to find the area of the rectangle.

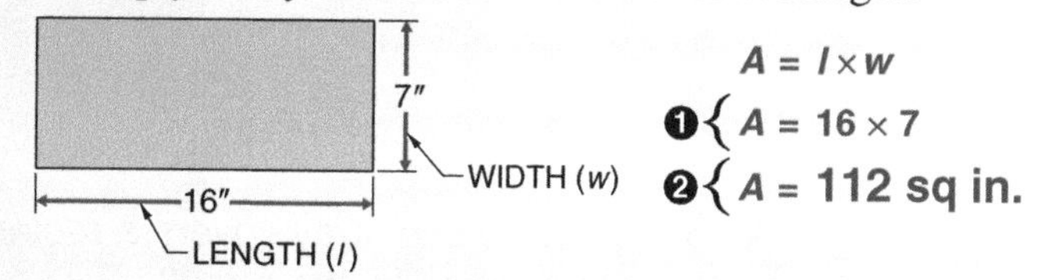

QUICK REFERENCE

- *The formula for the area of a square or rectangle is A = l × w.*

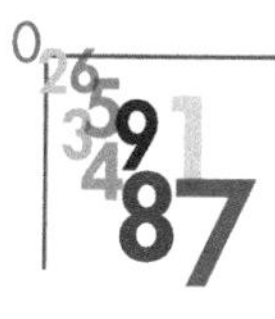

Since all sides are equal in a square, the formula for finding its area can be simplified to $A = l^2$.

MATH EXERCISES—Squares and Rectangles

________________ **1.** What is the area of a 14′-6″ × 16′-0″ rectangle?

________________ **2.** What is the area of 12′ × 18′ carpet in square yards?

________________ **3.** How many square inches are in a square having 2.75″ sides? *(Round decimals to nearest hundredth.)*

4. What is the area of Lot 312?

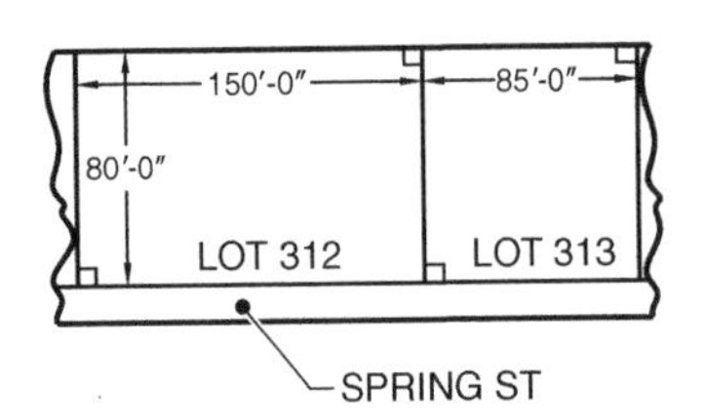

5. Lot 312 contains how many more square feet than Lot 313?

PRACTICAL APPLICATIONS—Squares and Rectangles

6. **Construction:** A print for a ten-story building with a telecommunications closet (TC) on each floor requires 3½ sheets of Type A plywood. Each sheet measures 4′-0″ × 8′-0″. What is the total area of the plywood required?

7. **Alternative Energy:** A solar photovoltaic array covers a rectangular area of 8′-6″ by 21′-0″. What is the area covered by the array?

8. **Construction:** A carpenter has 2 pieces of 3′ × 5′ plywood. How many square feet of plywood does the carpenter have?

Triangles. A *triangle* is a three-sided polygon with three interior angles. The sum of the three angles of a triangle is always 180°. The sign (Δ) indicates a triangle. Any side of a triangle can be its base. The height of a triangle is the perpendicular dimension from its base to the vertex opposite the base.

The angles of a triangle are named by uppercase letters. The sides of a triangle are named by lowercase letters. For example, a triangle may be named ΔABC and contain sides *d, e,* and *f*. Triangles include right triangles, isosceles triangles, equilateral triangles, and scalene triangles. **See Figure 9-6.**

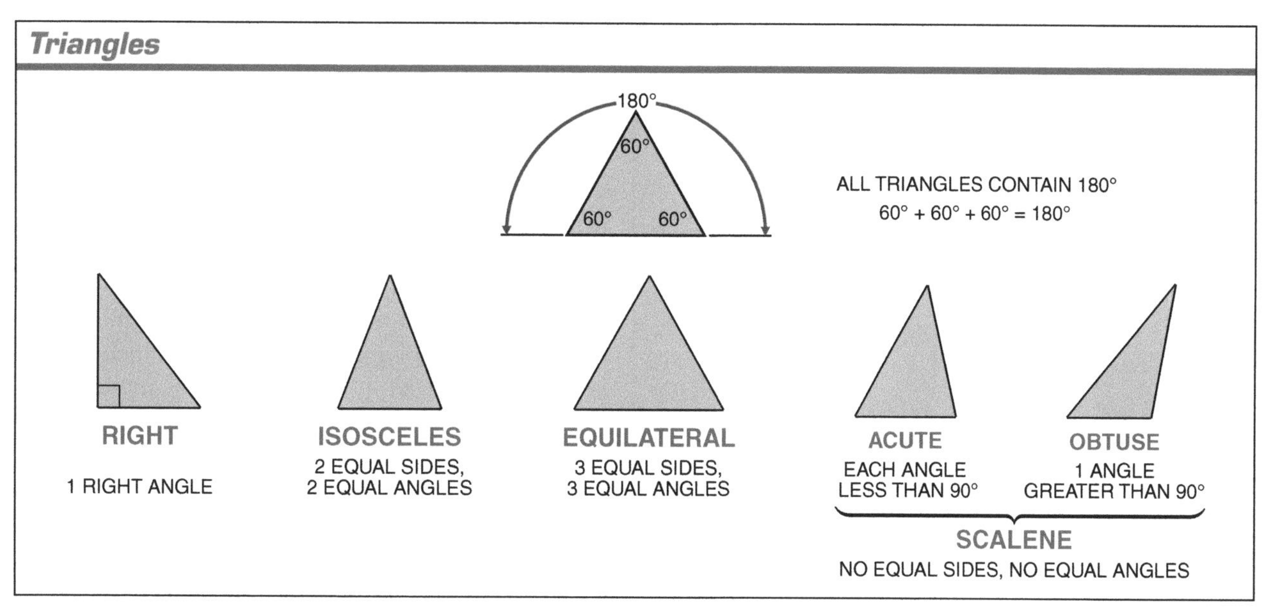

Figure 9-6. A triangle is a three-sided polygon.

A *right triangle* is a triangle with one 90° angle. An *isosceles triangle* is a triangle with two equal angles and two equal sides. An *equilateral triangle* is a triangle with three equal angles and three equal sides. Each angle of an equilateral triangle is 60°. A *scalene triangle* is a triangle with no equal angles or equal sides. A scalene triangle may be acute or obtuse.

The area of a triangle is found by applying the following formula:

$A = \frac{1}{2}bh$

where

A = area

b = base

h = height

Example—Triangles

1. Find the area of a triangle with a 4 in. base and a 5 in. height.

ANS: **10 sq in.**

❶ Determine the base (4) and height (5) of the triangle.

❷ Multiply 4 by 5 ($4 \times 5 = 20$).

❸ Multiply ½ by 20.

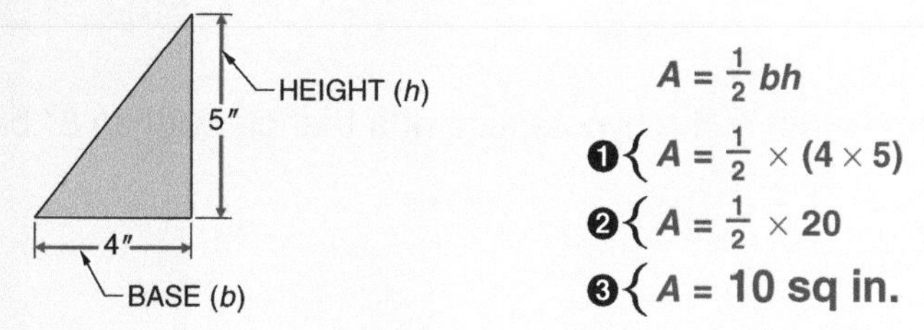

$A = \frac{1}{2}bh$

❶ $A = \frac{1}{2} \times (4 \times 5)$

❷ $A = \frac{1}{2} \times 20$

❸ $A = 10$ **sq in.**

QUICK REFERENCE

- *The formula for the area of a triangle is A = ½bh.*

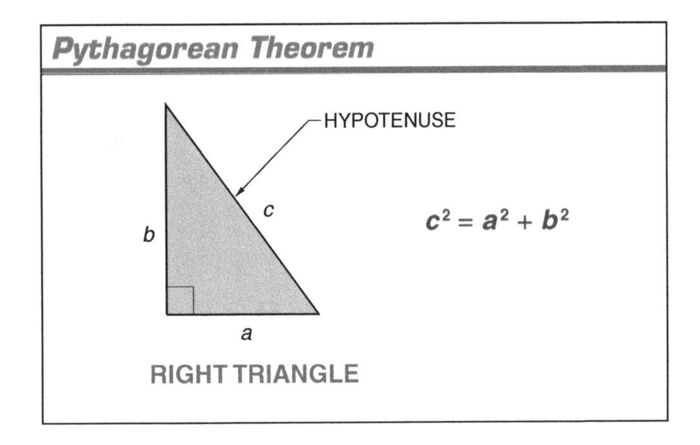

Figure 9-7. The Pythagorean theorem can be used to find the length of any side of a right triangle if the lengths of the other two sides are known.

Pythagorean Theorem. The Pythagorean theorem can be used to find the length of any side of a right triangle if the lengths of the other two sides are known. The Pythagorean theorem states that the square of the hypotenuse of a right triangle is equal to the sum of the squares of the other two sides, or $c^2 = a^2 + b^2$. The *hypotenuse* is the side of a right triangle opposite the right angle. **See Figure 9-7.**

The length of the hypotenuse of a right triangle is found by applying the Pythagorean theorem:

$$c^2 = a^2 + b^2$$

where

c = length of hypotenuse

a = length of one side

b = length of other side

Example — Pythagorean Theorem

1. Find the length of the hypotenuse of a right triangle with sides of 3 ft and 4 ft.

   ANS: **5 ft**

   ❶ Find the square root of c^2 (c) and determine the length of the sides of the triangle (3 and 4).

   ❷ Find the values of 3^2 and 4^2 and add them together (9 + 6 = 25).

   ❸ Find the square root of 25.

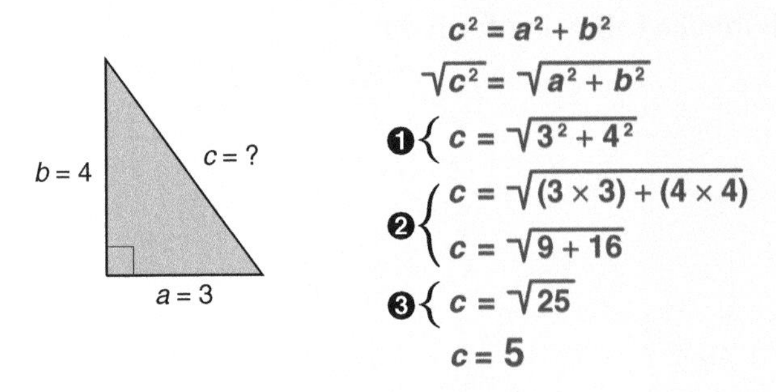

QUICK REFERENCE

- *The Pythagorean theorem states that $c^2 = a^2 + b^2$.*

MATH EXERCISES—Triangles

__________ 1. What is the area of a triangle with a 16″ base and a 22″ height?

__________ 2. What is the length of the hypotenuse of a triangle with an 8″ base and a 12″ height?

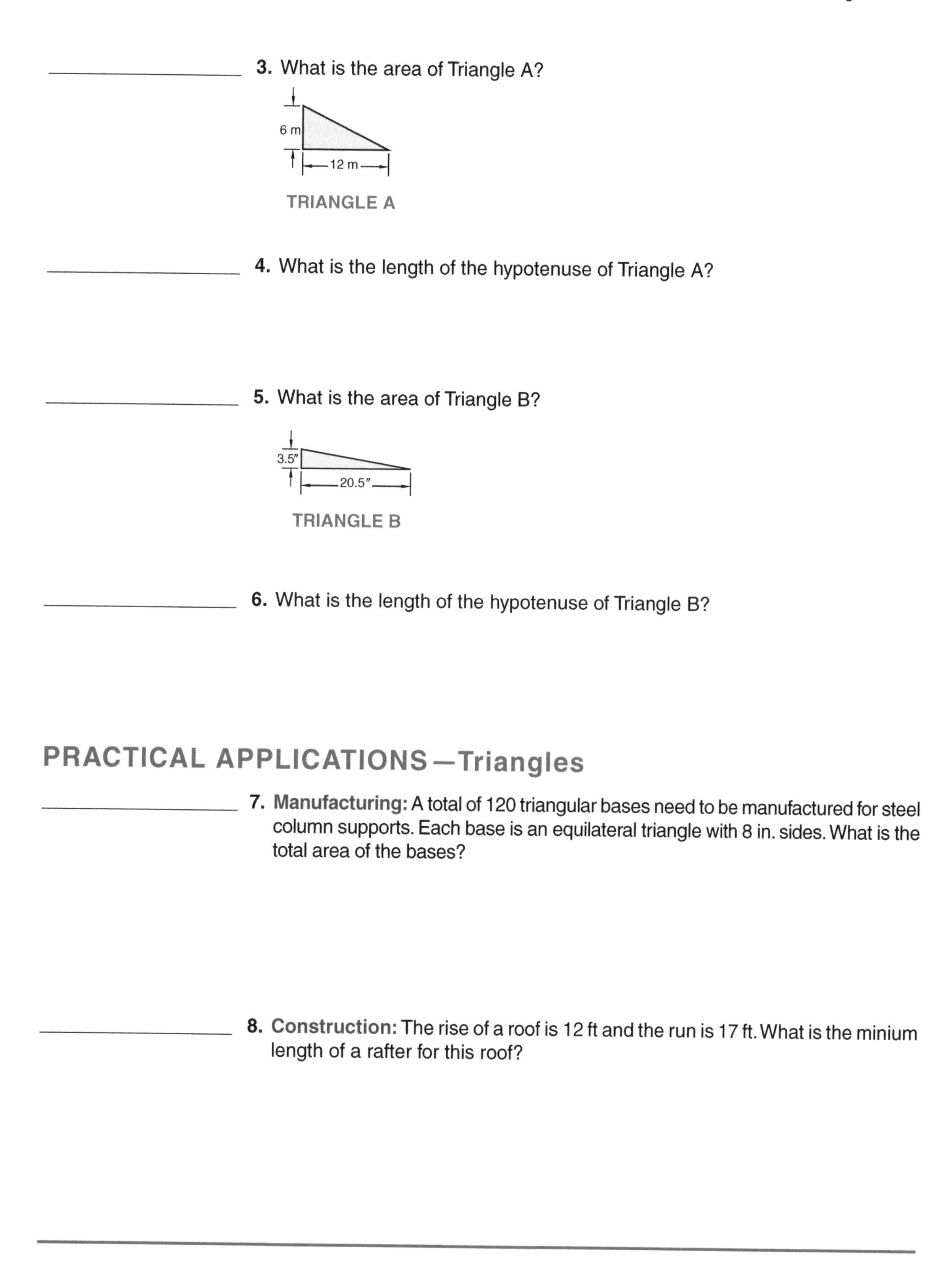

______________________ **3.** What is the area of Triangle A?

TRIANGLE A

______________________ **4.** What is the length of the hypotenuse of Triangle A?

______________________ **5.** What is the area of Triangle B?

TRIANGLE B

______________________ **6.** What is the length of the hypotenuse of Triangle B?

PRACTICAL APPLICATIONS — Triangles

______________________ **7. Manufacturing:** A total of 120 triangular bases need to be manufactured for steel column supports. Each base is an equilateral triangle with 8 in. sides. What is the total area of the bases?

______________________ **8. Construction:** The rise of a roof is 12 ft and the run is 17 ft. What is the minium length of a rafter for this roof?

Circles

A *circle* is a round plane figure made of a curved line in which all points on the curve are the same distance from a centerpoint. The *circumference* is the boundary of a circle. All circles contain 360°. The *diameter* is the length of a line from one edge of a circle, through the centerpoint, and to the opposite edge. The *radius* is the distance from the centerpoint of a circle to the edge. The radius is always half the length of the diameter. **See Figure 9-8.**

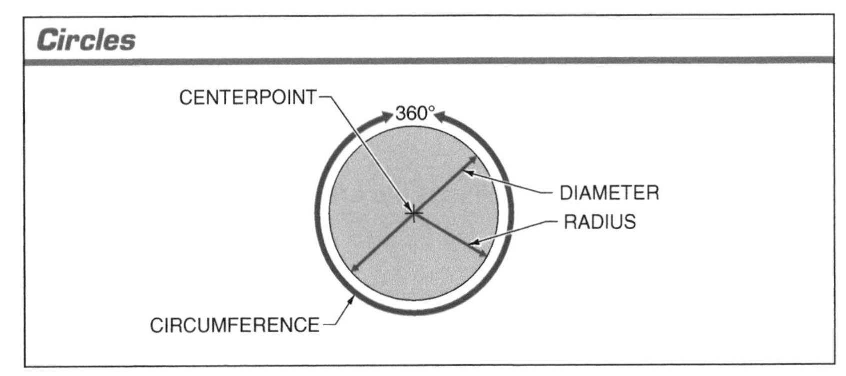

Figure 9-8. A circle is a 360° plane figure generated around a centerpoint.

When the radius is known, the circumference of a circle is found by applying the following formula:

$C = 2\pi r$

where

C = circumference

$\pi = 3.1416$

r = radius

When the radius is known, the area of a circle is found by applying the following formula:

$A = \pi r^2$

where

A = area

$\pi = 3.1416$

r = radius

Examples — Circles

1. Find the circumference of a circle with a 10″ radius.

   ANS: **62.83″**

   ❶ Determine the length of the radius (10).

   ❷ Multiply 2 by π by 10.

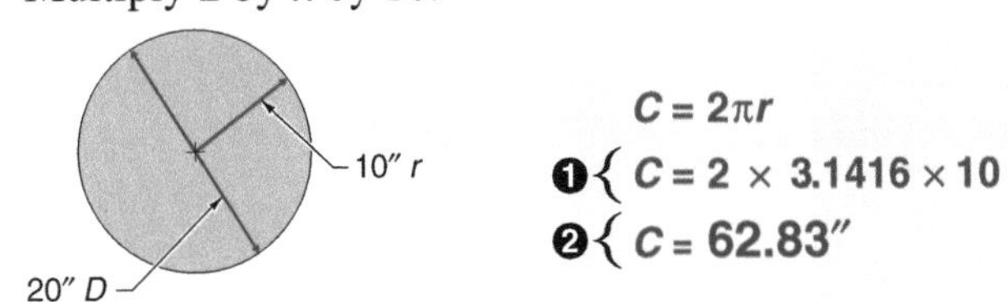

QUICK REFERENCE

- *The formula for the circumference of a circle is $C = 2\pi r$.*

2. Find the area of a circle with a 10″ radius.

***ANS:* 314.16 sq in.**

❶ Determine the length of the radius (10).

❷ Find the value of 10^2 ($10 \times 10 = 100$).

❸ Multiply π by 100 ($3.1416 \times 100 = 314.16$).

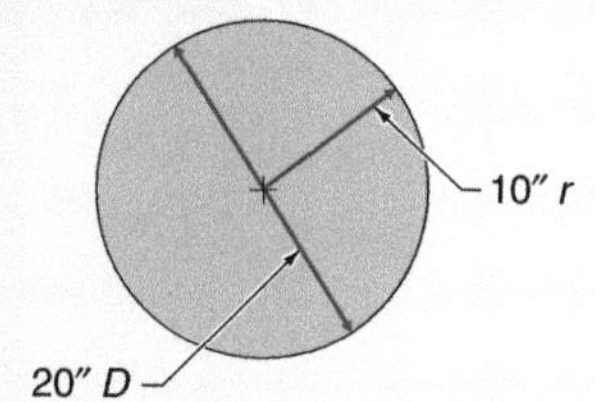

$$A = \pi r^2$$

❶ { $A = 3.1416 \times 10^2$

❷ { $A = 3.1416 \times 100$

❸ { $A = 314.16$ **sq in.**

QUICK REFERENCE

- *The formula for the area of a circle is $A = \pi r^2$.*

MATH EXERCISES — Circles

Round decimals to nearest hundredth.

________________ **1.** What is the circumference of a circle with a 4″ radius?

________________ **2.** What is the circumference of a circle with a 15″ diameter?

________________ **3.** What is the circumference of Circle A?

17″

CIRCLE A

________________ **4.** What is the area of Circle A?

________________ **5.** What is the circumference of Circle B?

18 cm

CIRCLE B

________________ **6.** What is the area of Circle B?

PRACTICAL APPLICATIONS—Circles

_______________ 7. **Boiler Operation:** The amount of steam flow through an orifice meter depends on the area of a circular hole in the meter. What is the area of a hole with a radius of 1.5 in.?

_______________ 8. **Construction:** A carpenter must cut a hole in a wooden door for a round window that has a radius of 9″. What is the circumference of the hole?

SECTION 9-2 UNDERSTANDING SOLID FIGURES

A *solid figure* is a three-dimensional figure that has length, width, and height. As with plane figures, solid figures have bases and altitudes. The most common solid figures are prisms (which include cubes and other rectangular solids), cylinders, pyramids, cones, and spheres.

Prisms

A *prism* is a solid figure with two identical bases and lateral faces (sides) which are parallelograms. The *base* is one of the two parallel polygons of a prism. A prism can be triangular, rectangular, pentagonal, hexagonal, octagonal, etc. according to the shape of its bases. **See Figure 9-9.** Rectangular solids, including cubes, are the most commonly seen prisms.

The *altitude* is the perpendicular distance between the two bases. When the bases are perpendicular to the faces, as with rectangular solids, the altitude equals the length of the edge of a lateral face.

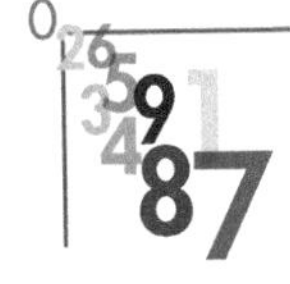

A right prism has bases that are parallel. With an oblique prism, the bases are not parallel.

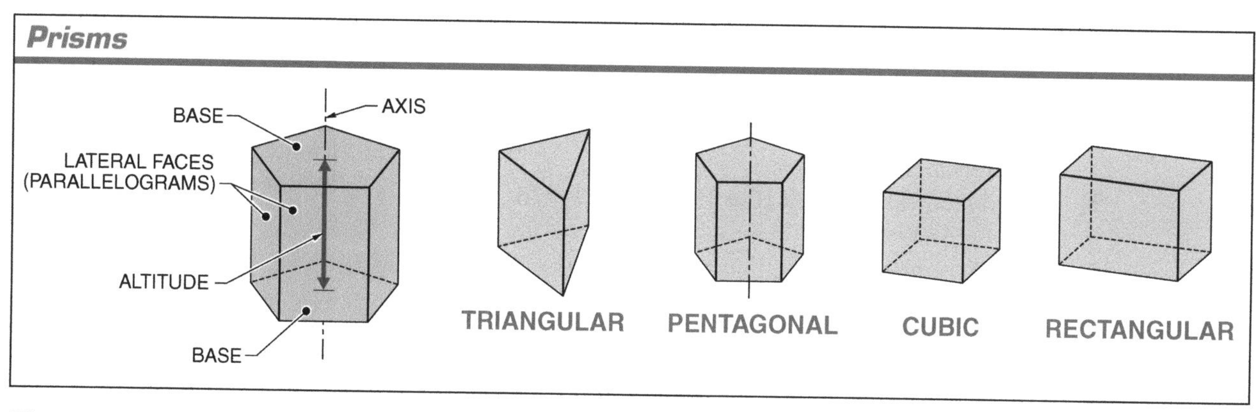

Figure 9-9. Square and rectangular solids are prisms.

The volume of a rectangular solid is found by applying the following formula:

$V = l \times w \times h$

where

V = volume

l = length

w = width

h = height

The surface area of a rectangular solid is found by applying the following formula:

$A = 2(lw + wh + lh)$

where

A = total surface area

l = length

w = width

h = height

Examples—Rectangular Solids

1. Find the volume of a 19″ × 10″ × 7″ rectangular solid.

   ANS: **1330 cu in.**

   ❶ Determine the length (19), width (10), and height (7) of the rectangular solid.

   ❷ Multiply 19 by 10 by 7 to find the volume.

QUICK REFERENCE

- *The formula for the volume of a rectangular solid or cube is $V = l \times w \times h$.*

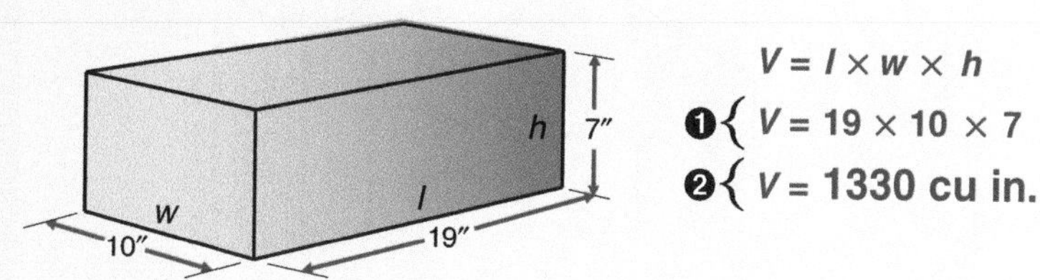

$V = l \times w \times h$

❶{ $V = 19 \times 10 \times 7$

❷{ $V = 1330$ cu in.

2. Find the surface area of a 19″ × 10″ × 7″ rectangular solid.

ANS: **786 sq in.**

❶ Determine the length (19), width (10), and height (7) of the rectangular solid.

❷ Multiply the values (19 × 10 = 190), (10 × 7 = 70), and (19 × 7 = 133).

❸ Add the products of the values together (190 + 70 + 133 = 393).

❹ Multiply 2 by 393.

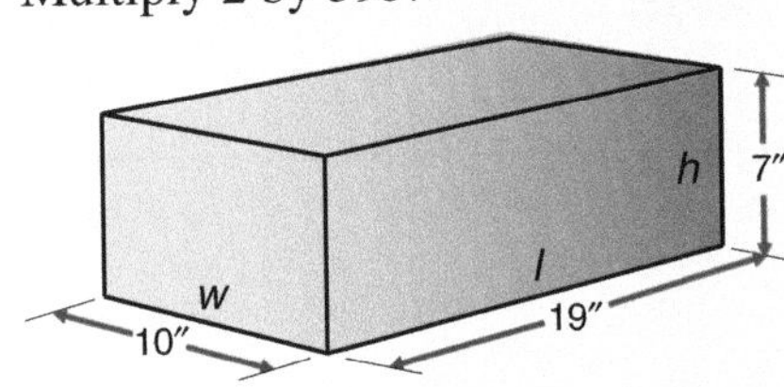

$$A = 2(lw + wh + lh)$$

❶ $A = 2([19 \times 10] + [10 \times 7] + [19 \times 7])$

❷ $A = 2(190 + 70 + 133)$

❸ $A = 2(393)$

❹ $A = 786$ sq in.

QUICK REFERENCE

- *The formula for the surface area of a rectangular solid or cube is $A = 2(lw + wh + lh)$.*

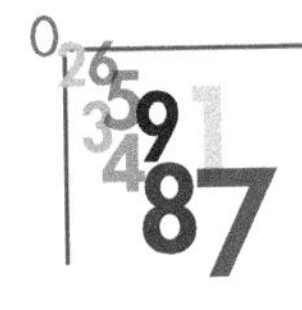

The formula for finding the volume and surface area of a rectangular solid is also used to find the volume and surface area of a cube.

MATH EXERCISES—Rectangular Solids

_______________ **1.** What is the volume of a 4″ × 8″ × 36″ rectangular solid?

_______________ **2.** What is the surface area of a 4″ × 8″ × 36″ rectangular solid?

_______________ **3.** What is the volume of a 3.5″ × 6.25″ × 18″ rectangular solid?

_______________ **4.** Which of the three cartons will hold the most material?

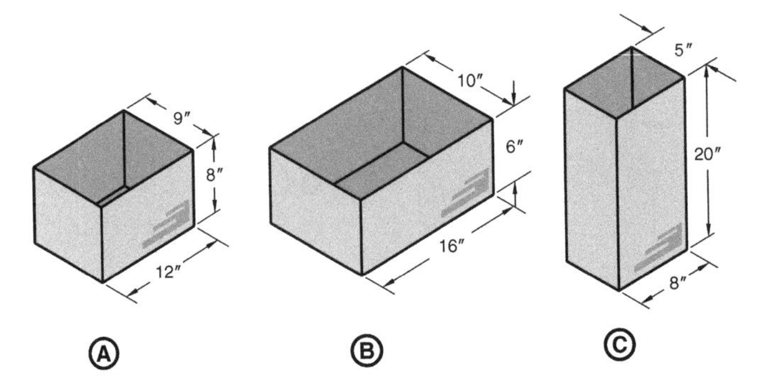

PRACTICAL APPLICATIONS—Rectangular Solids

__________ 5. **Construction:** A formwork contractor must order concrete for four walls. All of the walls are to be 6″ (0.5′) thick. Two of the walls are 8′ × 20′. The other two walls are 8′ × 30′. How much concrete does the contractor need to order?

__________ 6. **Maintenance:** A process tunnel in a manufacturing plant needs to be painted with waterproof paint on all six sides. The tunnel is 6′ × 8′ × 15′. What is the area that needs to be painted?

Cylinders

A *cylinder* is a solid figure with two circular bases. **See Figure 9-10.** The curved surface of the cylinder follows the circumference of the bases. The diameter or radius of the cylinder is the same as the diameter or radius of its bases. The *altitude* is the perpendicular distance between the bases.

When the radius and altitude are known, the volume of a cylinder is found by applying the following formula:

$V = \pi r^2 \times h$

where

V = volume

π = 3.1416

r = radius

h = altitude

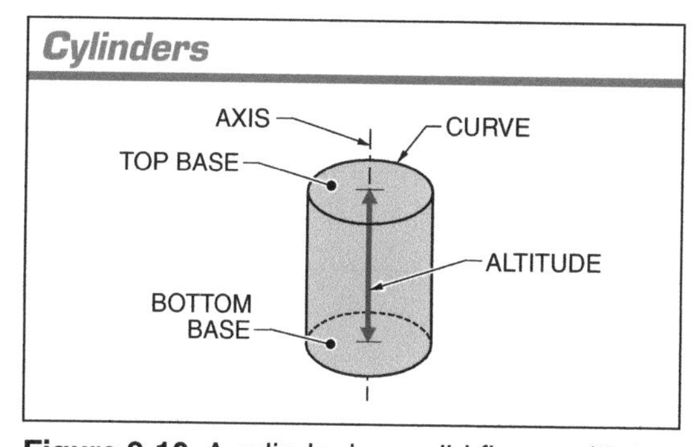

Figure 9-10. A cylinder is a solid figure with two circular bases.

When the radius and altitude are known, the surface area of a cylinder is found by applying the following formula:

$A = 2\pi r(r + h)$

where

A = total surface area

π = 3.1416

r = radius

h = altitude

Examples — Cylinders

1. Find the volume of a cylinder with an 8″ radius and a 60″ altitude.

ANS: **12,063.74 cu in.**

❶ Determine the radius (8) and altitude (60) of the cylinder.

❷ Find the value of 8^2 ($8 \times 8 = 64$).

❸ Multiply π by 64 by 60.

QUICK REFERENCE

- *The formula for the volume of a cylinder is $V = \pi r^2 \times h$.*

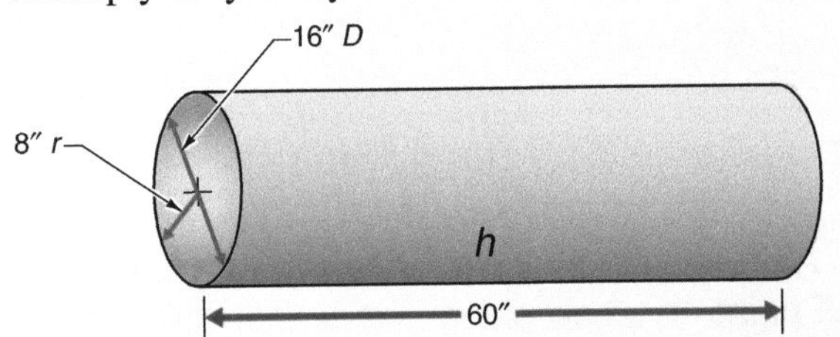

$V = \pi r^2 \times h$

❶ $\{V = 3.1416 \times (8 \times 8) \times 60$

❷ $\{V = 3.1416 \times 64 \times 60$

❸ $\{V = 12{,}063.74$ **cu in.**

2. Find the surface area of a cylinder with an 8″ radius and an altitude of 60″.

ANS: **3418.06 sq in.**

❶ Determine the radius (8) and altitude (60) of the cylinder.

❷ Add 8 and 60 ($8 + 60 = 68$).

❸ Multiply 2 by π by 8 by 68.

QUICK REFERENCE

- *The formula for the surface area of a cylinder is $A = 2\pi r(r + h)$.*

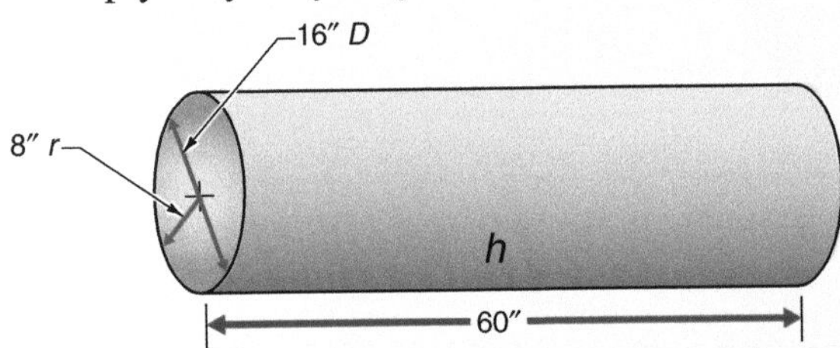

$A = 2\pi r(r + h)$

❶ $\{A = 2 \times 3.1416 \times 8(8 + 60)$

❷ $\{A = 2 \times 3.1416 \times 8(68)$

❸ $\{A = 3418.06$ **sq in.**

MATH EXERCISES — Cylinders

________________ **1.** What is the volume of a cylinder that has a 1.75′ radius and a 6′ altitude?

________________ **2.** What is the surface area of a cylinder that has a 1.75′ radius and a 6′ altitude?

________________ **3.** What is the volume in cubic feet of a cylinder that has a 10″ radius and a 4′ altitude?

________________ **4.** What is the surface area in square feet of a cylinder that has a 10″ radius and a 4′ altitude?

________________ **5.** How many cubic feet of liquid will the storage drum hold? (*Round the answer to the nearest cubic foot.*)

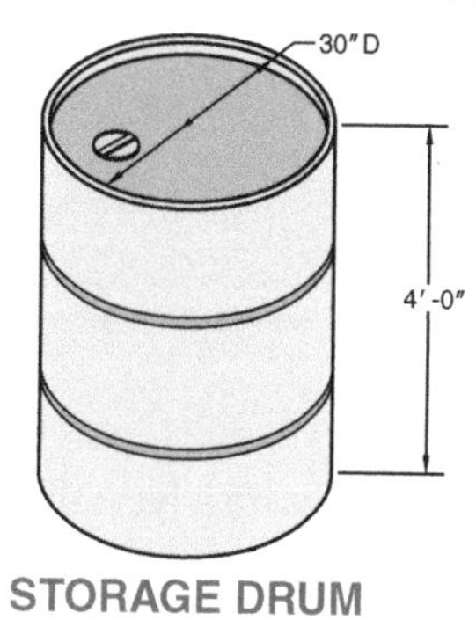

STORAGE DRUM

________________ **6.** Which of the three pipes will hold the least liquid?

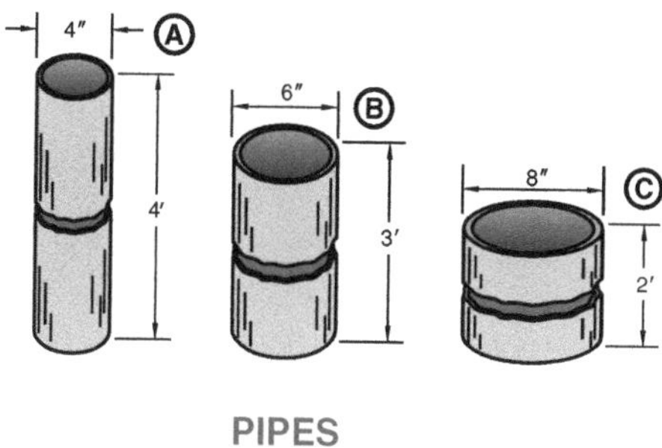

PIPES

PRACTICAL APPLICATIONS—Cylinders

________________ **7.** **Alternative Energy:** A geothermal system includes a cylindrical heat exchanger that is 18″ in diameter and 42″ tall. What is the volume of the heat exchanger? *(Round the answer to the nearest cubic inch.)*

________________ **8.** **Maintenance:** A waste removal contractor uses 55 gal. steel drums to remove solid waste. The drums measure a standard 23½″ in diameter by 36¼″ in height. What is the volume of each 55 gal. steel drum in cubic feet?

9. Construction: How much sheet metal is needed to construct a cylindrical tank with a length of 20′ and a radius of 3½′? *(Round the answer to the nearest hundredth.)*

Pyramids

A *pyramid* is a solid figure with a base that is a polygon and sides that are triangles. **See Figure 9-11.** The *base* of a pyramid is the bottom. The *apex* is the point opposite the base. The *altitude* is the perpendicular distance from the base to the apex. The *slant height* is the distance from the side of the base to the apex.

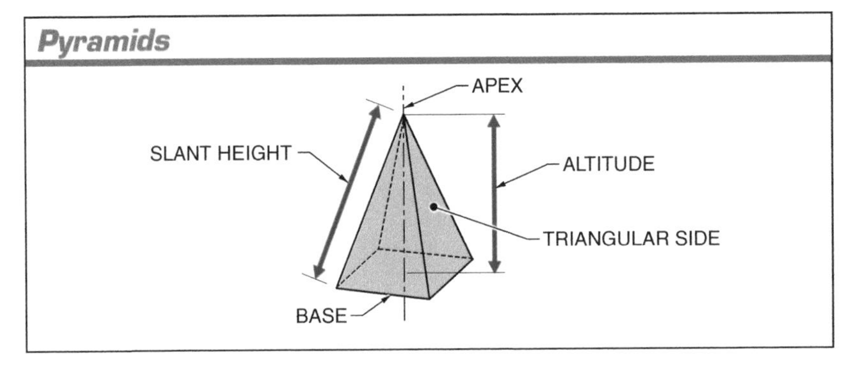

Figure 9-11. The square pyramid is named for the shape of its base.

Pyramids are named according to the kind of polygon forming the base. The square pyramid, which has a square base, is the most commonly seen pyramid.

The volume of a square pyramid is found by applying the following formula:

$V = \frac{1}{3}Ah$

where

V = volume

A = area of square base (or l^2)

h = altitude

When the slant height and the length of a side are known, the surface area of a square pyramid is found by applying the following formula:

$A = \frac{1}{2}ps + l^2$

where

A = total surface area

p = perimeter of base

s = slant height

l = base length

$\frac{1}{2}ps$ = surface area of triangular faces

l^2 = surface area of square base

Examples—Square Pyramids

1. Find the volume of a square pyramid with a base length of 14″ and an altitude of 10″.
ANS: **653.34 cu in.**

❶ Determine the values for the base length (14) and altitude (10) of the square pyramid.
❷ Find the value of 14^2 ($14 \times 14 = 196$).
❸ Multiply ⅓ by 196 by 10.

QUICK REFERENCE

- *The formula for the volume of a square pyramid is $V = ⅓Ah$.*

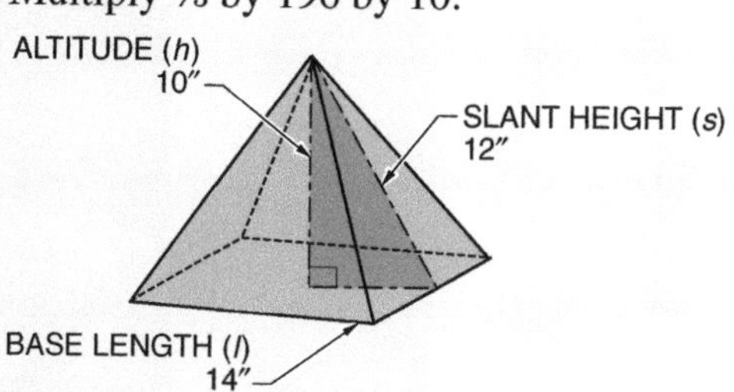

$$V = \frac{1}{3}Ah \quad \text{or} \quad V = \frac{1}{3}l^2h$$

❶ $V = \frac{1}{3} \times 14^2 \times 10$

❷ $V = \frac{1}{3} \times 196 \times 10$

❸ $V = 653.34$ **cu in.**

2. Find the surface area of a square pyramid with a base length of 14″ and a slant height of 12″.
ANS: **532 sq in.**

❶ Determine the perimeter of the base ($p = 14 + 14 + 14 + 14 = 56$).
❷ Determine the values for the slant height (12) and base length (14).
❸ Multiply ½ by 56 by 12 to find the surface area of the triangular faces (½ × 56 × 12 = 336).
❹ Find the value of 14^2 to find the surface area of the square base.
❺ Add the surface area of the triangular faces to the surface area of the square base (336 + 196 = 532).

QUICK REFERENCE

- *The formula for the surface area of a square pyramid is $A = ½ps + l^2$.*

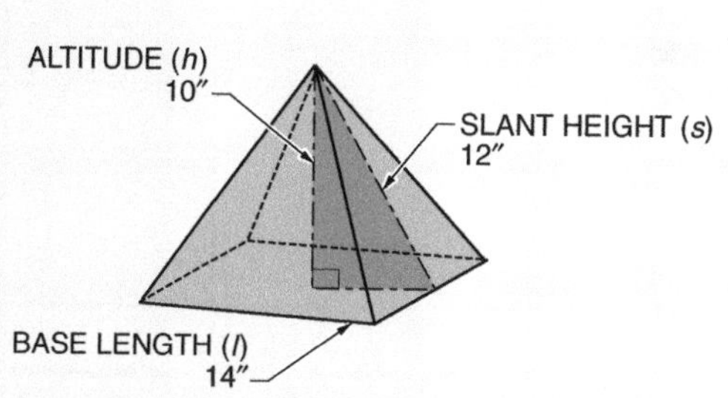

$$A = \frac{1}{2}ps + l^2$$

❶ $A = (\frac{1}{2} \times 4l \times s) + l^2$

$A = (\frac{1}{2} \times 56 \times s) + l^2$

❷ $A = (\frac{1}{2} \times 56 \times 12) + 14^2$

❸ $A = 336 + 14^2$

❹ $A = 336 + 196$

❺ $A = 532$ **sq in.**

MATH EXERCISES—Square Pyramids

__________ **1.** What is the volume of a square pyramid with a base length of 163 mm and an altitude of 97 mm?

__________ **2.** What is the surface area of a square pyramid with a base length of 20′ and a slant height of 22′?

PRACTICAL APPLICATIONS—Square Pyramids

________________ **3. Construction:** A pyramid-shaped roof needs to be covered with membrane material for waterproofing. The pyramid has a slant height of 20′ and a base length of 18′. What is the surface area to be covered? (*Exclude the base.*)

Cones

A *cone* is a solid figure with a circular base and a curved surface that tapers from the base to the apex. **See Figure 9-12.** The *altitude* is the perpendicular distance from the base to the apex. The *slant height* is the distance from the side of the base to the apex.

When the radius of the base and altitude are known, the volume of a cone is found by applying the following formula:

$V = \frac{1}{3}\pi r^2 h$

where

V = volume

π = 3.1416

r = radius

h = altitude

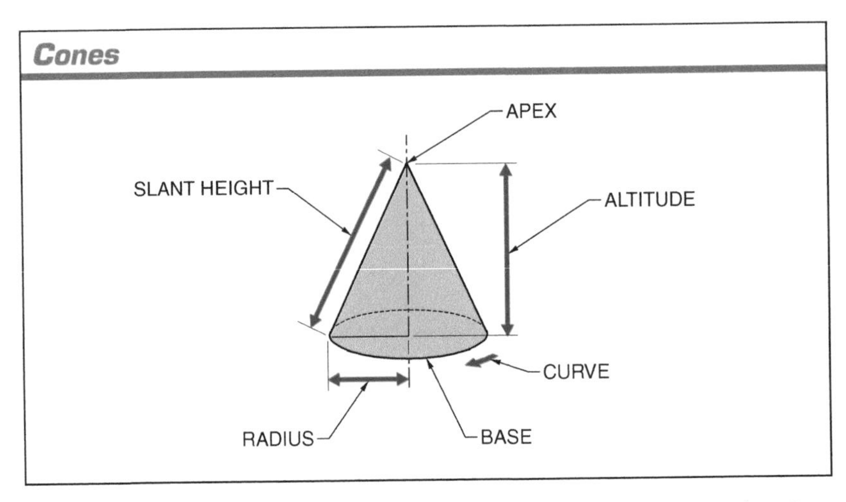

Figure 9-12. A cone is a solid figure with a circular base and a curved surface that tapers to an apex.

When the radius of the base, the altitude, and the slant height are known, the surface area of a cone is found by applying the following formula:

$A = \pi r(r + s)$

where

A = total surface area

$\pi = 3.1416$

r = radius

s = slant height

Examples—Cones

1. Find the volume of a cone with a 1.625′ radius and a 5′ altitude.

ANS: **13.83 cu ft**

❶ Determine the values for the radius (1.625) and altitude (5) of the cone.

❷ Find the value of 1.625^2.

❸ Multiply ⅓ by π by 2.640625 by 5.

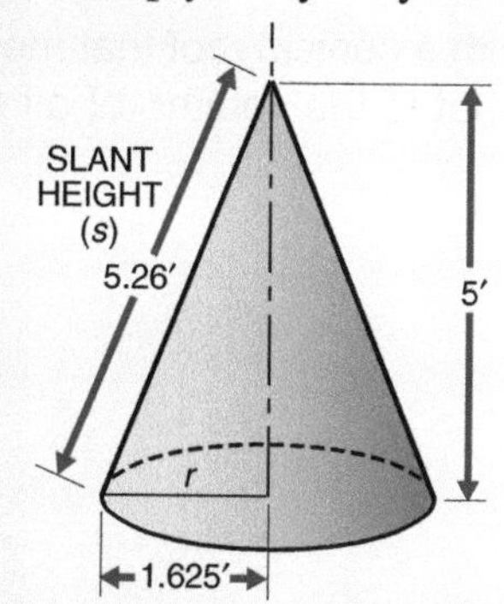

$$V = \frac{1}{3}\pi r^2 h$$

❶ $V = \frac{1}{3} \times 3.1416 \times 1.625^2 \times 5$

❷ $V = \frac{1}{3} \times 3.1416 \times 2.64025 \times 5$

❸ $V = 13.83$ cu ft

QUICK REFERENCE

- *The formula for the volume of a cone is $V = ⅓\pi r^2 h$.*

2. Find the surface area of a cone with a 1.625′ radius and a 5.26′ slant height.

ANS: **35.15 sq ft**

❶ Determine the values of the radius (1.625) and slant height (5.26) of the cone.

❷ Add 1.625 and 5.26 (1.625 + 5.26 = 6.885).

❸ Multiply π by 1.625 by 6.885.

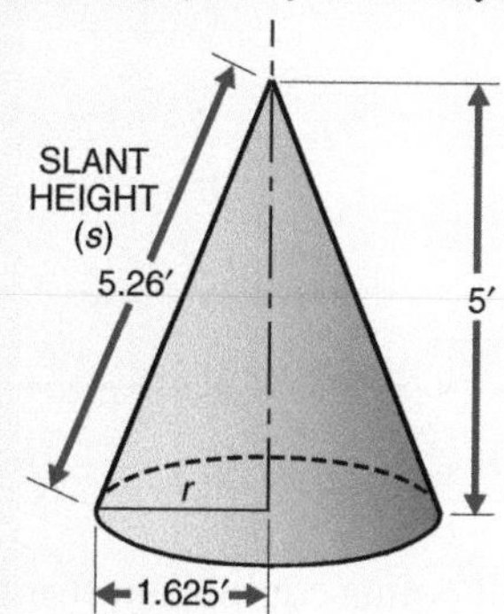

$$A = \pi r(r + s)$$

❶ $A = \pi 1.625\ (1.625 + 5.26)$

❷ $A = 3.1416 \times 1.625\ (6.885)$

❸ $A = 35.15$ sq ft

QUICK REFERENCE

- *The formula for the surface area of a cone is $A = \pi r(r + s)$.*

MATH EXERCISES—Cones

______________ **1.** What is the volume of a cone that has a base radius of 1″ and an altitude of 9″?

______________ **2.** What is the volume of a cone that has a base radius of 7 mm and an altitude of 21 mm?

______________ **3.** What is the surface area of a cone that has a base radius of 10″ and a slant height of 16″?

PRACTICAL APPLICATIONS—Cones

______________ **4. Construction:** A grain silo is topped with a conical roof that measures 12′ tall and has a base diameter of 40′-6″. What is the volume of air present in the roof section?

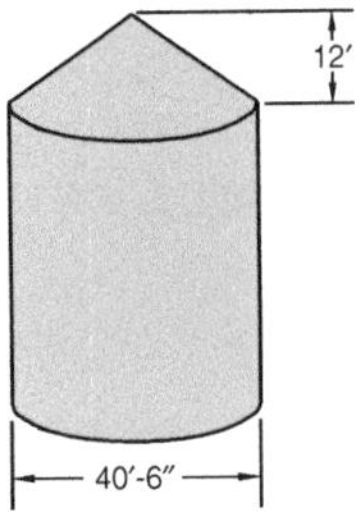

12′

40′-6″

______________ **5. Manufacturing:** A custom drill bit for a mining operation has a conical tip that measures 18′ across at the base and 6′ in height. What is the average volume of earth the conical tip can displace?

Spheres

A *sphere* is a solid figure with all points equidistant from its centerpoint. A circle is formed by passing a cutting plane through any point of a sphere. **See Figure 9-13.**

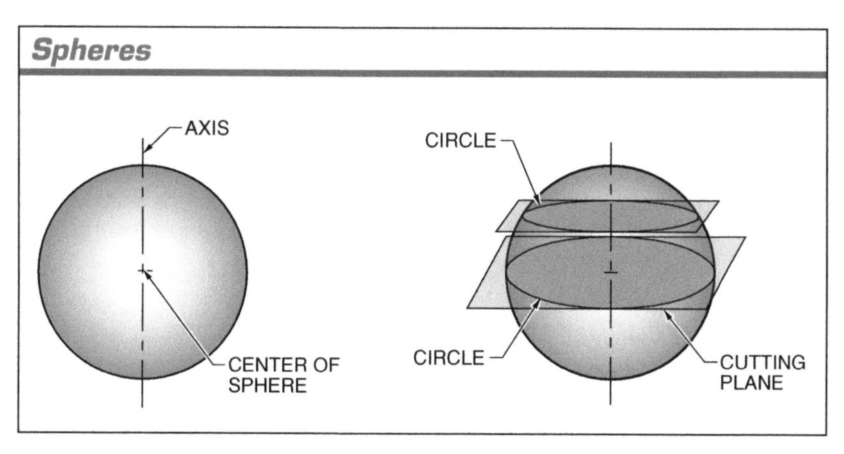

Figure 9-13. A sphere is a solid figure with all points equidistant from its centerpoint.

When the radius is known, the volume of a sphere is found by applying the following formula:

$V = \frac{4}{3}\pi r^3$

where

V = volume

π = 3.1416

r = radius

When the radius is known, the surface area of a sphere is found by applying the following formula:

$A = 4\pi r^2$

where

A = total surface area

π = 3.1416

r = radius

Examples — Spheres

1. Find the volume of a sphere with a 3.5′ radius.

   ANS: **179.59 cu ft**

   ❶ Determine the radius (3.5) of the sphere.

   ❷ Find the value of 3.5^3 ($3.5 \times 3.5 \times 3.5 = 42.875$).

   ❸ Multiply 4/3 by π by 42.875.

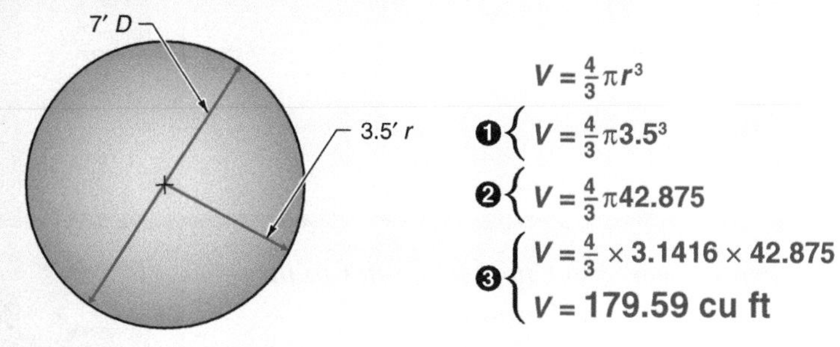

$V = \frac{4}{3}\pi r^3$

❶ $V = \frac{4}{3}\pi 3.5^3$

❷ $V = \frac{4}{3}\pi 42.875$

❸ $V = \frac{4}{3} \times 3.1416 \times 42.875$

V = **179.59 cu ft**

QUICK REFERENCE

- *The formula for the volume of a sphere is $V = \frac{4}{3}\pi r^3$.*

2. Find the surface area of a sphere with a 3.5′ radius.

ANS: **153.94 sq ft**

❶ Determine the radius (3.5) of the sphere.
❷ Find the value of 3.5^2 ($3.5 \times 3.5 = 12.25$).
❸ Multiply 4 by π by 12.25.

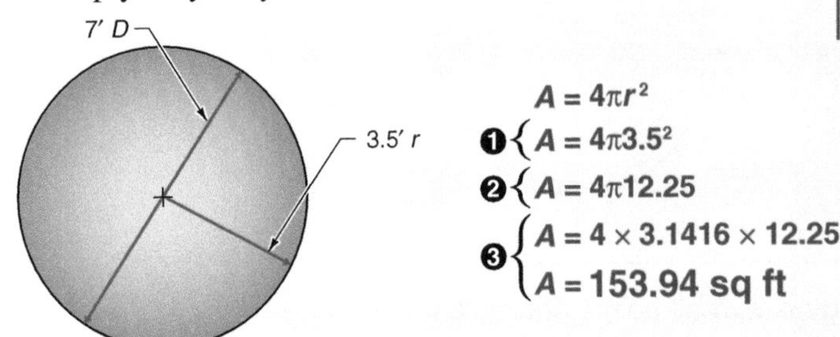

QUICK REFERENCE

- *The formula for the surface area of a sphere is $A = 4\pi r^2$.*

MATH EXERCISES—Spheres

______________ **1.** What is the volume of a sphere that has an 11″ radius?

______________ **2.** What is the volume of a sphere that has an 12″ diameter?

PRACTICAL APPLICATIONS—Spheres

______________ **3. Plumbing:** A water storage tower has a spherical tank with a radius of 10′. What is the capacity of the tank?

______________ **4. Manufacturing:** A ball mill with a 10 cu ft capacity is used to grind paint pigment. The balls used to grind the pigment must take up 6 cu ft of the ball mill to grind the pigment properly. Each ball is 2″ in diameter. How many balls are needed to fill the ball mill?

For an interactive review of the concepts covered in Chapter 9, refer to the corresponding Quick Quiz® included on the Digital Resources.

Working with Plane and Solid Figures 9

Review

Name ____________________ Date ____________

Math Problems

Round decimals to nearest hundredth.

________ **1.** What is the area of a rectangle that measures 125 m by 300 m?

________ **2.** What is the area of a triangle with an 8″ base and a 12″ height?

________ **3.** What is the length of the hypotenuse of a triangle with sides of 9 cm and 12 cm?

________ **4.** What is the radius of a circle with a 22″ diameter?

________ **5.** What is the circumference of a circle with an 8″ radius?

________ **6.** What is the area of a circle with a 15 cm radius?

________ **7.** What is the volume of a rectangular prism that measures 3′ × 6′ × 8′?

________ **8.** What is the volume of a cylinder that has a 2′ radius and is 12′ long?

________ **9.** What is the surface area of a cylinder that has a 3.5 cm radius and is 20 cm long?

9 Review (continued)

______________ **10.** What is the volume of a square pyramid with a base length of 4.5″ and an altitude of 12″?

______________ **11.** What is the volume of a cone with a radius of 9″ and an altitude of 35″?

______________ **12.** What is the surface area of a sphere with a 3′ radius?

Practical Applications

______________ **13.** **Electrical:** An electrician is routing conductors through a piece of rigid metal conduit (RMC) that has a 4″ diameter. In order to figure out how many conductors can be safely routed through the RMC, the electrician must determine the cross-sectional area of the conduit. What is the cross-sectional area?

______________ **14.** **Construction:** A technician for a concrete engineering firm must perform a compression test on a cylindrical piece of concrete that is 6″ in diameter and 12″ high. What is the volume of the concrete?

______________ **15.** **Construction:** How many square feet of sheet metal are needed to fabricate a 3′ diameter, 10′ long cylindrical tank?

Working with Plane and Solid Figures 9

Test

Name ______________________________ Date ______________

Math Problems

Round decimals to nearest hundredth.

__________ **1.** What is the area of a square with 12.25″ sides?

__________ **2.** What is the area of a triangle with a base of 12″ and a height of 18″?

__________ **3.** What is the length of the hypotenuse of a right triangle with a base of 7″ and a height of 16″?

__________ **4.** What is the length of the third leg of a right triangle if the hypotenuse is 14″ and the second leg is 6″?

__________ **5.** What is the radius of a circle with a 4.5″ diameter?

__________ **6.** What is the circumference of a circle with a 36″ radius?

__________ **7.** What is the area of the circle with a 1 m radius?

__________ **8.** What is the volume of a 5″ × 5″ × 5″ cube?

__________ **9.** What is the volume of a 2′ × 3′ × 4′ rectangular prism?

9 Test (continued)

_______________ **10.** What is the surface area of a $2' \times 3' \times 4'$ rectangular prism?

_______________ **11.** What is the surface area of a cylinder that has a 0.45 m radius and is 2.8 m long?

_______________ **12.** What is the volume of a square pyramid with a base length of 30 cm and an altitude of 12 cm?

_______________ **13.** What is the surface area of a square pyramid with a base length of 98″ and a slant height of 210″?

_______________ **14.** What is the volume of a sphere with a radius of 0.5′?

_______________ **15.** What is the surface area of a sphere with a diameter of 24″?

Practical Applications

_______________ **16.** **HVAC:** What is the internal surface area of a 4 ft long section of ductwork that measures 8 in. × 10 in.? (*Exclude the ends.*)

9 Test (continued)

__________ **17. Boiler Operation:** The total force on a boiler safety valve is determined by the area of the valve disk. What is the area of a safety valve disk with a radius of 1.75 in.?

__________ **18. Construction:** What is the length of *c* on the concrete ramp? (*Round decimal to nearest hundredth.*)

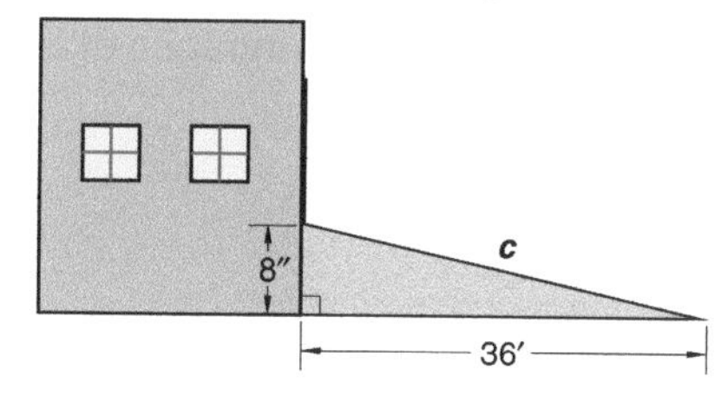

__________ **19. Maintenance:** A building that is a square pyramid, with a base length of 120′ and a slant height of 100′, needs to be painted. What is the surface area of the building?

__________ **20. Construction:** What is the total circumference of a drain with a radius of 9″?

__________ **21. Boiler Operation:** What is the volume of a cylindrical tank that is 4′ in diameter and 12′ long?

9 Test (continued)

______ **22. Electrical:** An electrician needs to know the volume of an outlet box to determine the box's fill rate. The dimensions of the outlet box are $4\frac{1}{2}'' \times 4\frac{1}{2}'' \times 1\frac{1}{2}''$. What is the volume?

______ **23. Maintenance:** A metal sphere with a radius of 5′ is to be treated with a weather-resistant material and displayed at the entrance of a museum. What is the surface area of the sphere?

______ **24. Agriculture:** A 24 ft × 36 ft rectangle must be excavated to a depth of 8 ft to accommodate a storm shelter for a large dairy operation. How many cubic feet of earth must be removed?

______ **25. Culinary Arts:** A walk-in freezer measures 18′ × 10′ wide. What is the area of the floor of the freezer?

10 Working with Graphs

Graphs are used to visually represent relationships between variables. Very little text is used in a graph. Line graphs, bar graphs, and pie graphs are commonly used to convey important data simply.

OBJECTIVES

1. Explain the purpose of a graph.
2. Describe how variables are represented on a line graph.
3. Create straight-line, curved-line, and broken-line graphs.
4. Describe how variables are represented on a bar graph.
5. Create simple, stacked-bar, and grouped-bar graphs.
6. Describe how variables are represented on a pie graph.
7. Create pie graphs.

KEY TERMS

- graph
- line graph
- x-axis
- y-axis
- origin
- coordinates
- bar graph
- stacked-bar graph
- grouped-bar graph
- pie graph

Digital Resources
ATPeResources.com/QuickLinks
Access Code: 764460

SECTION 10-1 USING LINE GRAPHS

A *graph* is a diagram that shows the relationship between two or more variables. Graphs present information visually, using very little text. Three common types of graphs are the line graph, the bar graph, and the pie graph (often called a pie chart).

Line Graphs

A *line graph* is a graph in which points representing variables are connected by a line. The axes are the lines serving as references for the values of the variables. The *x-axis* is the horizontal line, and the *y-axis* is the vertical line. The *origin* is the point where the two axes intersect. **See Figure 10-1.**

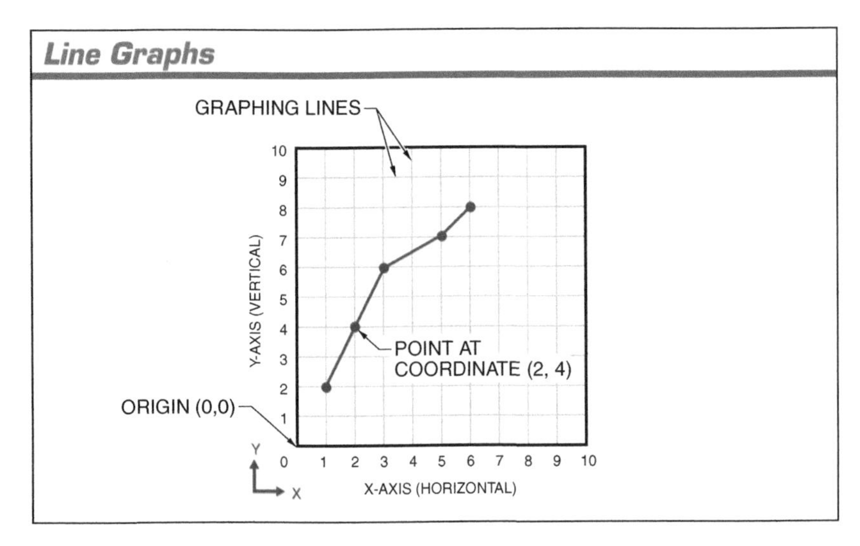

Figure 10-1. A line graph is a graph in which points representing variables are connected by a line.

In a line graph, *coordinates* are two numbers written in parentheses used to indicate points on a graph. The first number represents the location of the point on the x-axis. The second number represents the location of the point on the y-axis. For example, the point (2, 4) is located two units from the origin on the x-axis and four units from the origin on the y-axis.

To interpret a line graph, note what items the graph is comparing and the value of each interval on the axis. Then note the changes in the comparison from coordinate to coordinate.

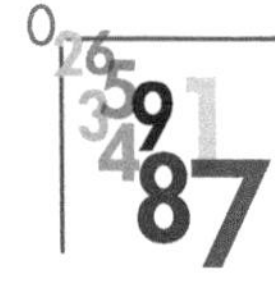

When time is one of the variables to be graphed, it is typically placed on the x-axis.

The type of line produced on a line graph depends on the relationship between the variables. Variables can have a causal relationship or no causal relationship. Variables with a causal relationship are related to each other by a mathematical formula or rule, producing straight-line or curved-line graphs. Variables with no causal relationship are related to each other only through the line graph, producing broken-line graphs.

Straight-Line Graphs. Straight-line graphs reveal a linear, or proportional relationship between variables. **See Figure 10-2.**

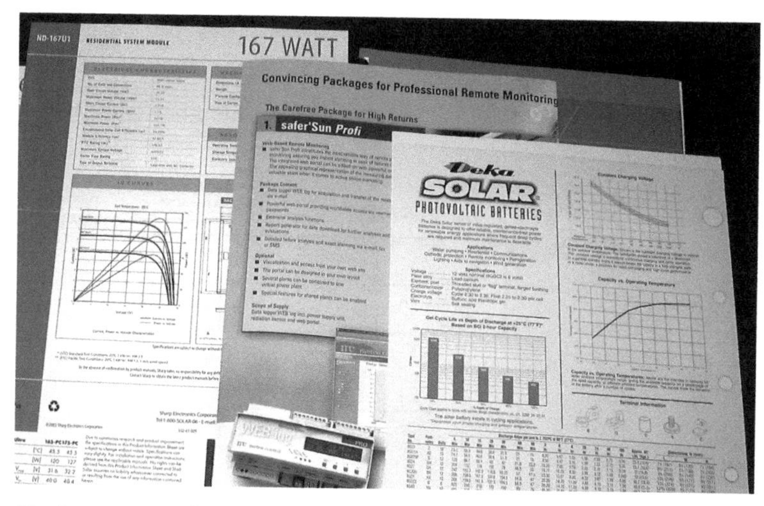

Equipment specifications often use line graphs to display electrical output.

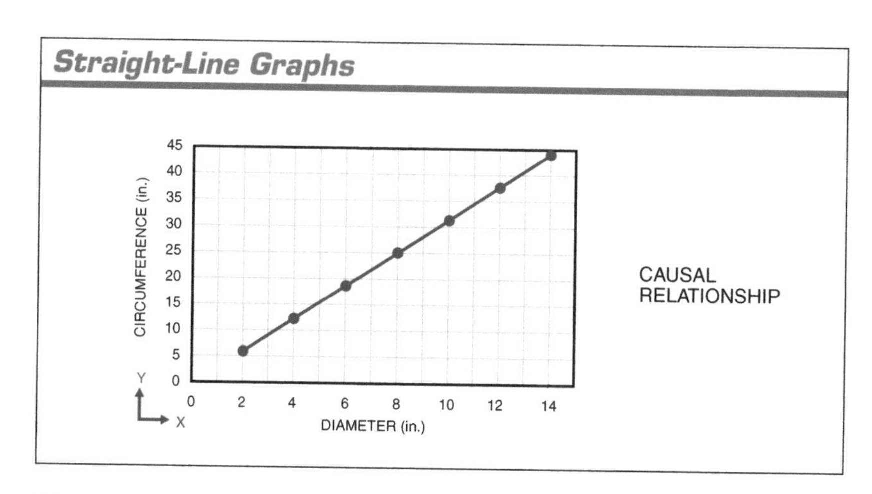

Figure 10-2. A straight-line graph reveals a linear relationship between variables.

For example, if the diameters of circles are 2″, 4″, 6″, 8″, 10″, 12″, and 14″, then the circumferences are 6.28″, 12.56″, 18.84″, 25.12″, 31.42″, 37.70″, and 43.98″. When plotted with the x-axis as the diameter and the y-axis as the circumference, these points (2, 6.28), (4, 12.56), etc., produce a straight line.

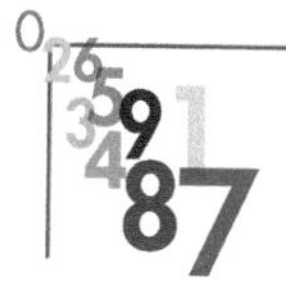

A table is often useful for organizing data in a logical manner and can be used to create any type of graph. It is also useful when interpreting a graph.

Example—Straight-Line Graphs

1. Create a line graph for the perimeter of square glass tiles with side lengths of 5 cm, 2 cm, 1.5 cm, 7 cm, 6 cm, and 9 cm.

ANS:

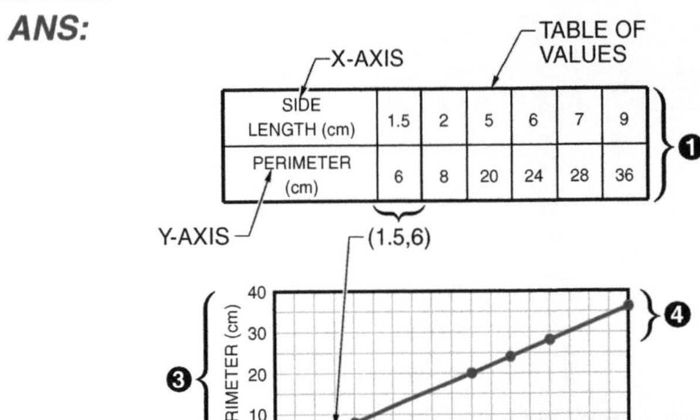

❶ Arrange the values from smallest to largest in a table of values.

❷ Label the x-axis as the value of the side length and use the range of the variables to determine the side length intervals.

❸ Label the y-axis as the value of the perimeter and use the range of the variables to determine the perimeter intervals.

❹ Pair the x and y coordinates and plot (1.5, 6), (2, 8), (5, 20), (6, 24), (7, 28), and (9, 36). Connect the points to create a straight line.

Curved-Line Graphs. Curved-line graphs reveal a non-linear, causal relationship between variables. **See Figure 10-3.** For example, if the length of the sides of a number of squares is plotted on the x-axis in centimeters and the areas of the squares are plotted on the y-axis, a curved-line graph is produced.

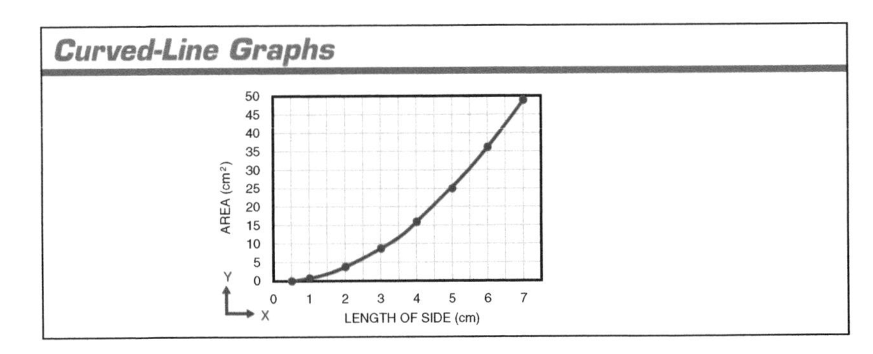

Figure 10-3. A curved-line graph reveals a non-linear, causal relationship between variables.

Example—Curved-Line Graphs

1. Create a line graph for the volume of cubic storage containers with side lengths of 1′, 1.25′, 1.75′, 2′, and 2.5′. (*Hint: The volume is the length cubed.*)

ANS:

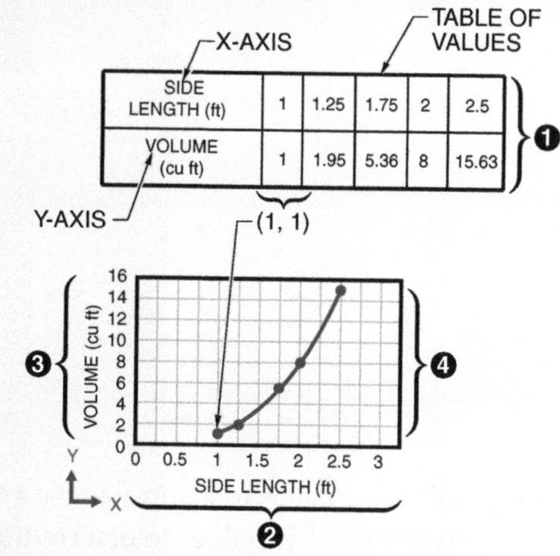

❶ Arrange the values from smallest to largest in a table of values.
❷ Label the x-axis as the value of the side length and use the range of the variables to determine the side length intervals.
❸ Label the y-axis as the value of the volume and use the range of the variables to determine the volume intervals.
❹ Pair the x and y coordinates and plot (1, 1), (1.25, 1.95), (1.75, 5.36), (2, 8), and (2.5, 15.63). Connect the points to create a curved line.

Broken-Line Graphs. Broken-line graphs do not reveal a predictable causal relationship between variables. However, they do reveal trends. For example, if a graph is plotted that shows the hardness measurements of a certain copper alloy after it is annealed at various temperatures for an hour, no distinct relationship will be revealed. What will be evident is the trend between the temperature and the hardness, that is, the lower the temperature, the less hard the copper alloy. **See Figure 10-4.**

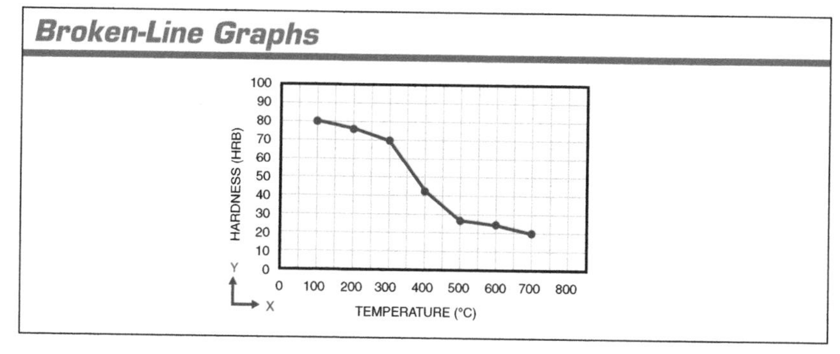

Figure 10-4. A broken-line graph reveals trends, not casual relationships.

Example—Broken-Line Graphs

1. Create a line graph for the changing price per watt for photovoltaic modules if the price was \$3.15 in 2003, \$3.00 in 2004, \$3.25 in 2005, \$3.50 in 2006, and \$3.40 in 2007.

ANS:

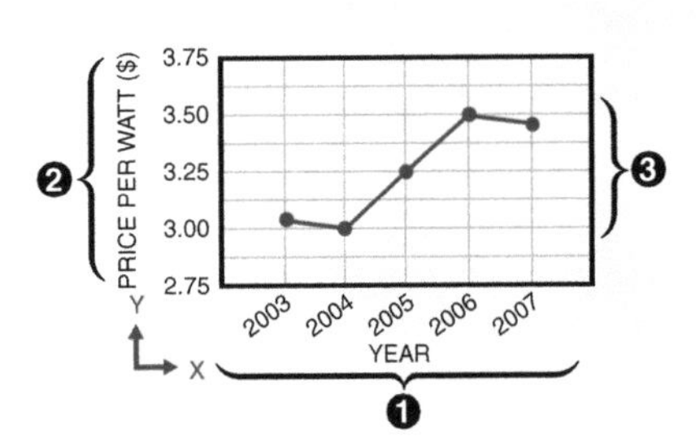

❶ Label the x-axis as the year and use the range of the variables to determine the year intervals.
❷ Label the y-axis as the price per watt and use the range of variables to determine the price-per-watt intervals.
❸ Plot the coordinates and connect the points to create a broken line.

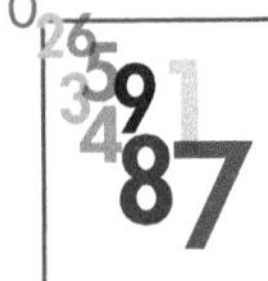

Guidelines to follow when creating a line graph include the following:

1. *Title all graphs.*
2. *Give each axis 5 to 10 intervals with a corresponding value.*
3. *Give each axis a physical quantity.*
4. *Include a table of values if necessary.*
5. *Include a legend to define abbreviations or symbols if necessary.*
6. *When drawing graphs by hand, use a ruler for straight lines and a french curve for curved lines.*

MATH EXERCISES—Line Graphs

________________ 1. Use the Current-Voltage graph to find the current in the circuit if the voltage is 180 V.

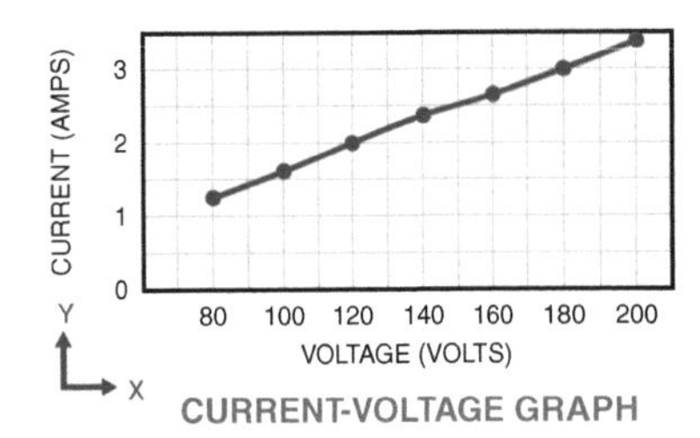

CURRENT-VOLTAGE GRAPH

2. Create a line graph using the following double-pulley force-load information. *Note:* Force equals load divided by 2 ($F = w/2$).

Force in pounds (x-axis) 4, 10, 11, 15, 16, 17.5, 20.5
Load in pounds (y-axis) 8, 20, 22, 30, 32, 35, 41

Use the Metro Population Growth graph to answer questions 3–6.

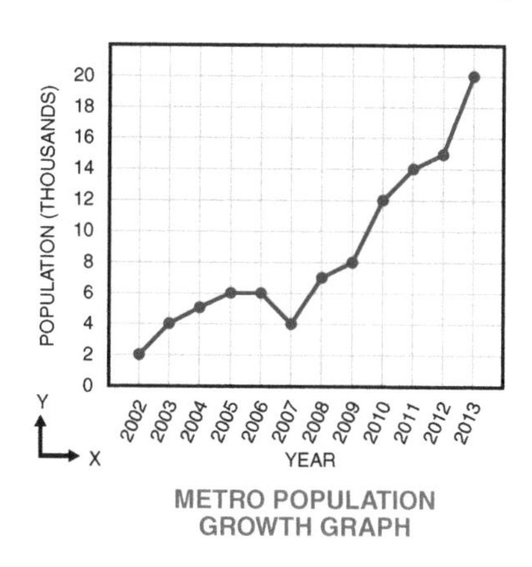

METRO POPULATION GROWTH GRAPH

____________________ **3.** In which year did the population grow by 4000?

____________________ **4.** What was the net population growth from 2003 to 2010?

____________________ **5.** Which year had the largest population increase?

____________________ **6.** In which year was there a decrease in population?

Use the Monthly Rainfall graph to answer questions 7–8.

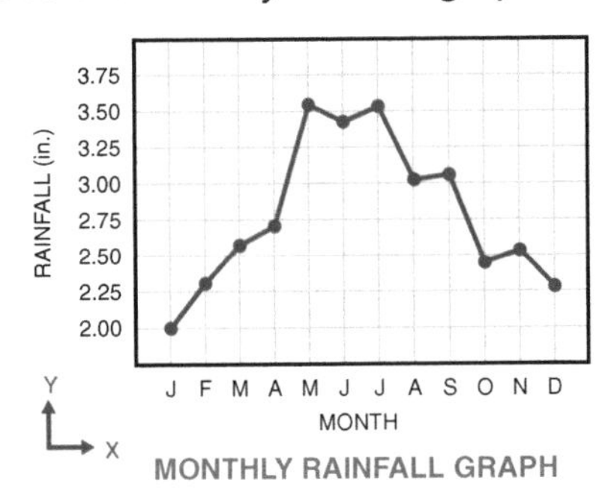

MONTHLY RAINFALL GRAPH

________________ **7.** Which month had the least rainfall?

________________ **8.** Which three-month period had the most rainfall?

PRACTICAL APPLICATIONS—Line Graphs

9. Boiler Operation: Boiler operators use a daily water treatment log to track the condition of boiler water. Create a line graph of the daily measurement of product 8570.

Day	Boiler Water					Feedwater	Condensate			Products		
	P	M	OH	TDS	Na_2SO_3	TDS	pH	TDS	Hard	938	8570	960
1	320	384	256	2700	45	45	8.8	15	0	44	4	4
2	376	440	312	2900	60	39	8.8	16	0	32	8	0
3	384	400	296	3000	55	36	8.9	16	0	32	6	0
4	324	380	268	2700	40	39	8.5	14	0	32	6	4
5	340	392	288	2900	45	34	8.7	14	0	32	6	4
6	272	328	216	2300	45	36	8.6	13	0	32	8	4
7	290	364	228	2600	35	34	8.8	13	0	36	4	4

________________ **10. Alternative Energy:** Use the Collector Comparison graph to determine which type of collector drops the most in efficiency as conditions change.

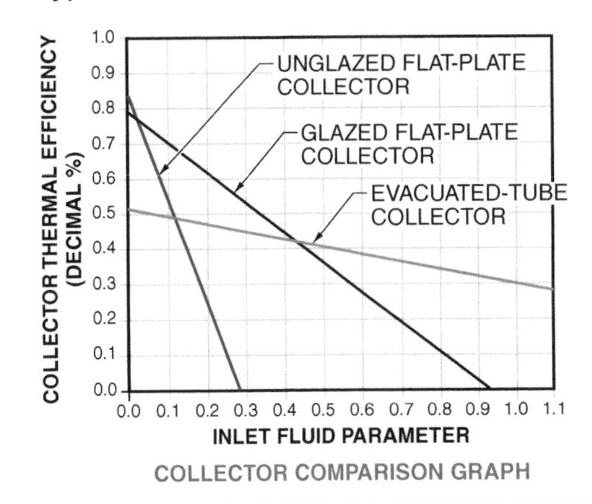

COLLECTOR COMPARISON GRAPH

SECTION 10-2 USING BAR GRAPHS

A *bar graph* is a graph in which the values of variables are represented by bars. The bars may be drawn horizontally or vertically. The length or height of a bar indicates a value. **See Figure 10-5.** Bar graphs can also be used to compare two sets of data, each with different values for a variable.

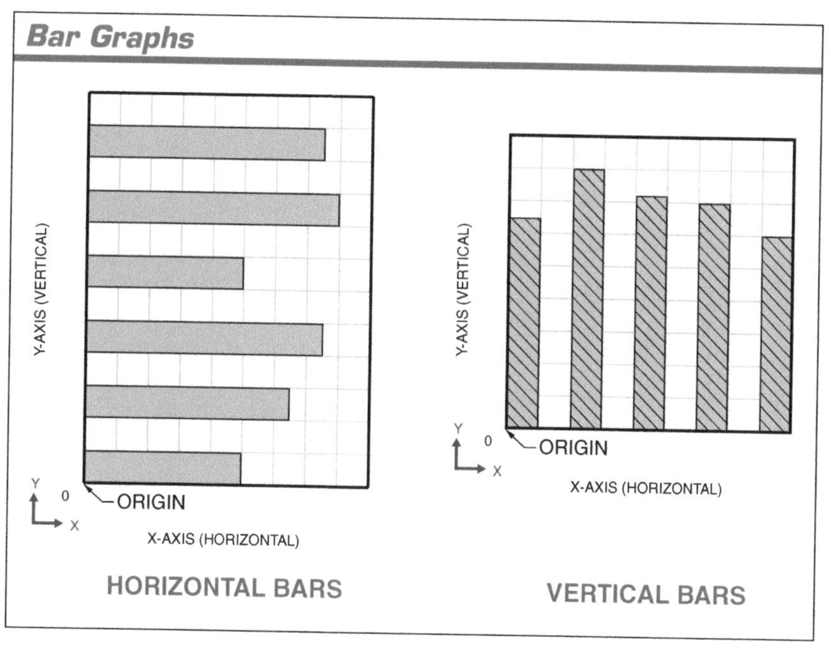

Figure 10-5. A bar graph is a graph in which values of variables are represented by bars.

To interpret a bar graph, determine the value of each space on the axis. If the bars are horizontal, determine the space value on the horizontal scale. If the bars are vertical, determine the space value on the vertical scale. If the end of a bar is not on a graph line, the value must be approximated. To find the value of a variable, determine the size of the bar.

To create a bar graph, determine which variable to locate on the x-axis and which to locate on the y-axis. Label the x-axis and y-axis and assign the values to the spaces on the scales. The assigned values on the graph should conveniently represent the values to be plotted. Draw each bar to the length or height corresponding to its value.

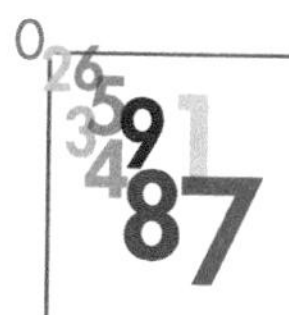

A histogram is a type of bar graph where each bar represents a range of values instead of a single value. For example, the number of gallons of paint in a paint factory that must be reworked over the course of a month can range on a weekly basis from 0–10 gal., 6–20 gal., 21–30 gal., etc. A histogram will reflect the number of times a value falls within each of these ranges.

Example—Bar Graphs

1. Create a bar graph to show the density of common metals where aluminum (Al) is 2.7 g/cm³, iron (Fe) is 7.86 g/cm³, lead (Pb) is 11.34 g/cm³, copper (Cu) is 8.89 g/cm³, gold (Au) is 19.3 g/cm³, and tin (Sn) is 7.29 g/cm³.

ANS:

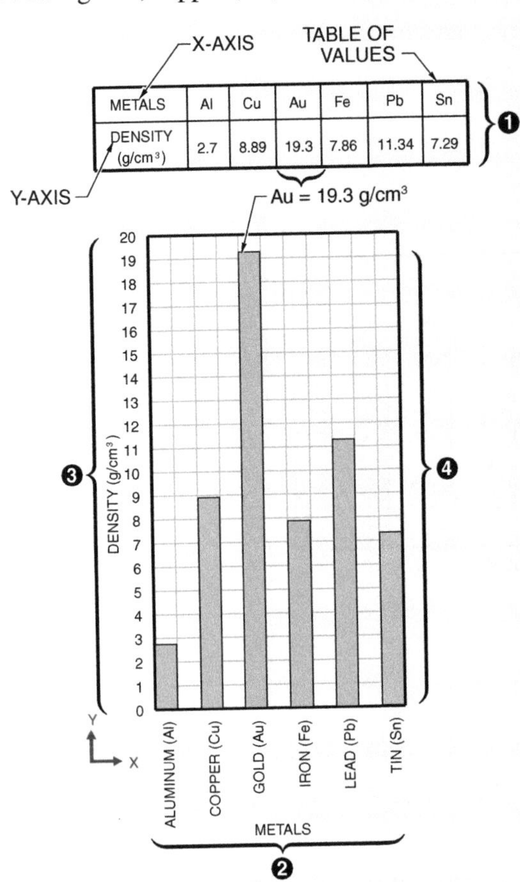

❶ Arrange the metals in alphabetical order in a table to help with the plotting process.
❷ Label the x-axis as metals.
❸ Label the y-axis as density.
❹ Draw each bar to a length corresponding to its value.

Stacked-Bar Graphs. A *stacked-bar graph* is a bar graph where each bar is divided into more than one variable. Each bar is divided into different sections that are either shaded, colored, or cross-hatched. Each section represents a value of one of the variables. **See Figure 10-6.**

For example, a stacked-bar graph is created to show the number of male and female students enrolled in different classes. The total enrollment is divided into the number of males (solid section) and the number of females (cross-hatched section). To find the number of students in each class for each gender, determine the total value of the bar representing a class. Next, find the value of each section of the bar on the y-axis. Then subtract each value from the total value of the bar. For example, the total number of students in the industrial technology class is 240, 40 who are female and 200 who are male (240 – 40 = 200).

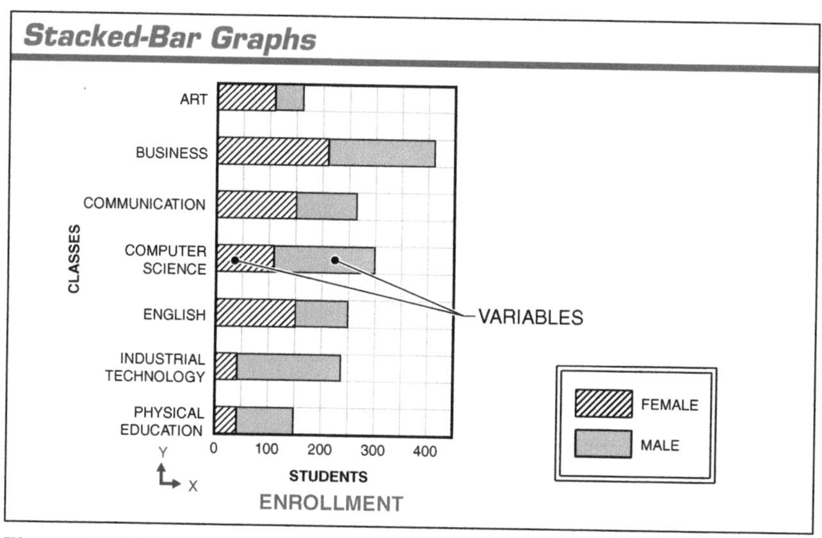

Figure 10-6. A stacked-bar graph is a graph where each bar is divided into more than one variable.

Example—Stacked-Bar Graphs

1. Create a graph for the total peak kilowatts of photovoltaic cells and modules installed commercially in 2003 through 2007 using the following information:

YEAR	PHOTOVOLTAIC CELL	PHOTOVOLTAIC MODULE	TOTAL
2003	30,000	80,000	110,000
2004	35,000	145,000	180,000
2005	25,000	200,000	225,000
2006	20,000	315,000	335,000
2007	25,000	495,000	520,000

ANS:

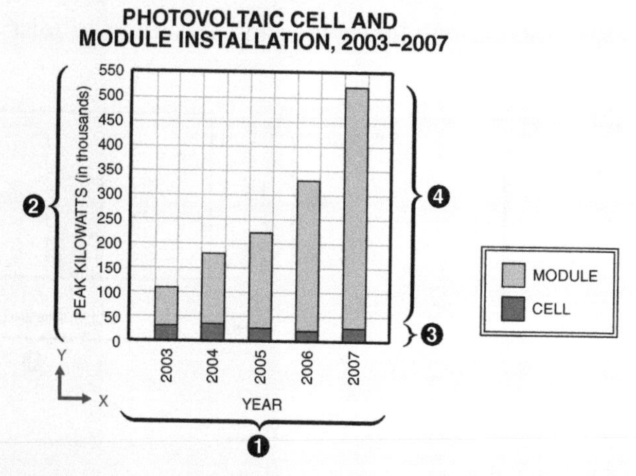

❶ Label the x-axis as the year.

❷ Label the y-axis as the peak kilowatts.

❸ Decide which variable to draw on the bottom of the stack. Draw each bar according to its value and color.

❹ Draw each bar for the values of the other variable on top of the first bars and shade.

Grouped-Bar Graphs. A *grouped-bar graph* is a bar graph where more than one bar is used to illustrate the variables in a group. **See Figure 10-7.** For example, a grouped-bar graph shows the number of male and female students in different departments. The total enrollment for each department is divided into two bars that represent the number of males (solid section) and the number of females (cross-hatched section). It can be seen from the bar graph that the total number of students in the business department is the sum of the two bars, or 220 (male students) plus 210 (female students). Therefore, the total number of students in the business department is 430.

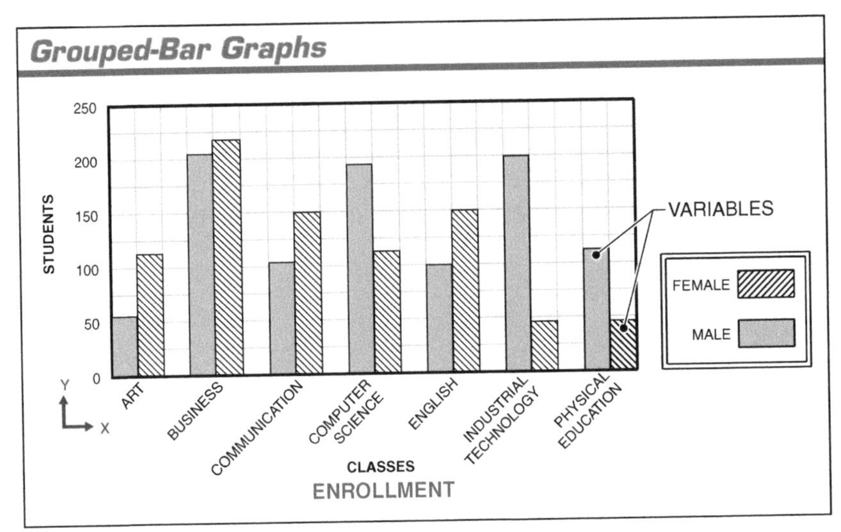

Figure 10-7. A grouped-bar graph is a graph where more than one bar is used to illustrate the variables in a group.

Example—Grouped-Bar Graphs

1. Create a graph from the following information on allowable spans according to joist spacing for joint designation 550S162-33.

ALLOWABLE SPANS OF 550S162-33

SPACING	30 PSF	40 PSF
12″	11′-7″	10′-7″
16″	10′-7″	9′-7″
19.2″	9′-11″	9′-0″
24″	9′-1″	8′-1″

ANS:

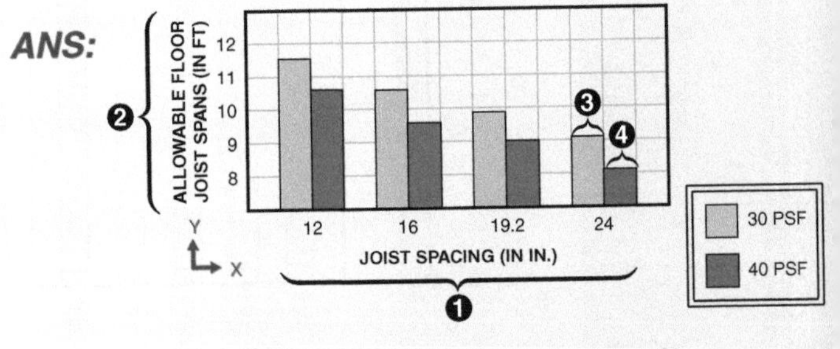

❶ Label the x-axis as the joist spacing.
❷ Label the y-axis as the allowable span.
❸ Draw each bar for the 30 psf live load joist spacing corresponding to the value of its allowable span. Color the bars as needed.
❹ Directly beside the 30 psf bars, draw each bar for the 40 psf live load joist spacing corresponding to the value of its allowable span. Shade the bars.

MATH EXERCISES—Bar Graphs

See the Motor Failure graph to answer questions 1–5.

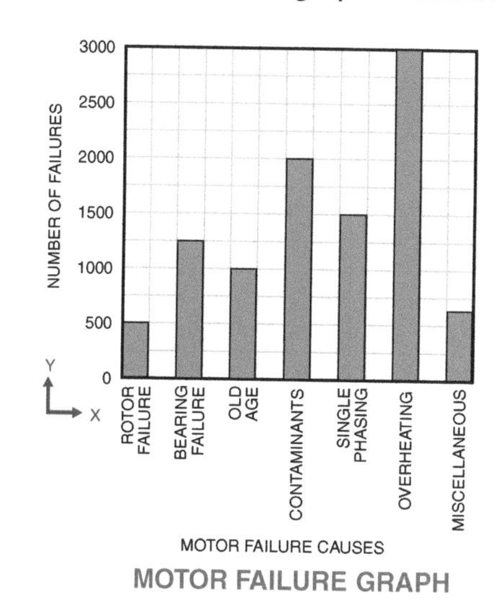

MOTOR FAILURE GRAPH

________________ **1.** More motors failed because of ___ than for any other reason.

________________ **2.** ___ was the least-occurring cause of motor failure.

________________ **3.** Contaminants and bearing failure accounted for ___ motor failures.

________________ **4.** ___ was the third largest cause of motor failure.

________________ **5.** ___ accounted for 1000 motor failures.

6. Create a bar graph entitled Automobile Production Data from the following data.

YEAR	AUTOMOBILES
1	7,300,000
2	8,000,000
3	8,250,000
4	9,800,000
5	9,500,000
6	8,400,000
7	10,400,000

7. Create a horizontal grouped-bar graph entitled Sales Data from the following data.

MONTH	SALESPERSON	UNITS SOLD
JAN	JAY	6
	KRIS	3
	KEVIN	4
FEB	JAY	2
	KRIS	6
	KEVIN	3
MAR	JAY	4
	KRIS	6
	KEVIN	4

See the Course Enrollments graph to answer questions 8–10.

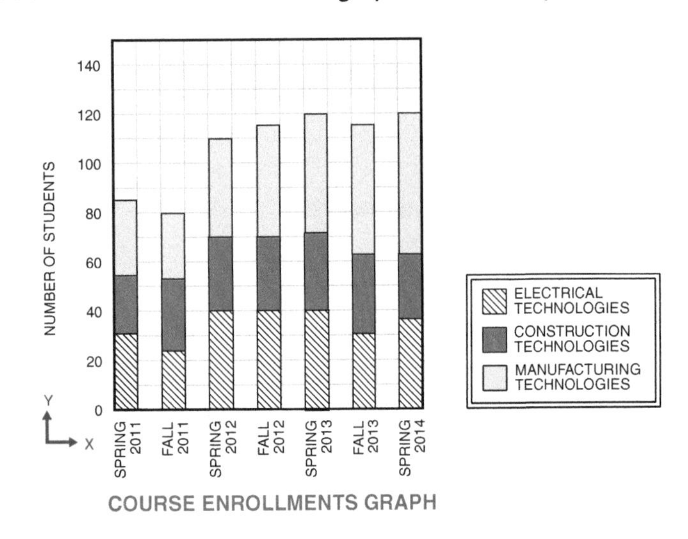

COURSE ENROLLMENTS GRAPH

____________ **8.** Electrical technologies enrollment increased by ___ from the fall of 2013 to the spring of 2014.

____________ **9.** The total enrollment for the spring of 2012 was ___ students.

____________ **10.** The largest growth in student enrollment from the spring of 2011 through the spring of 2014 occurred in the area of ___.

PRACTICAL APPLICATIONS—Bar Graphs

______________ 11. **Boiler Operation:** Soot buildup on heating surfaces increases the fuel consumption in boilers. What soot thickness results in fuel consumption increases above 9%?

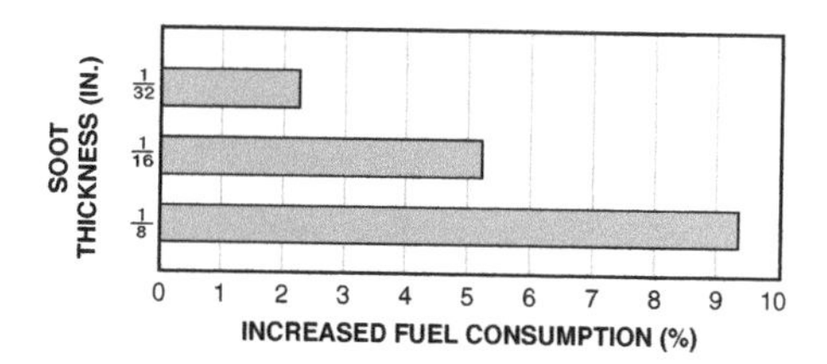

______________ 12. **Culinary Arts:** Use the Sources of Soluble and Insoluble Fiber graph to determine which vegetable and which fruit offers the highest amount of insoluble fiber.

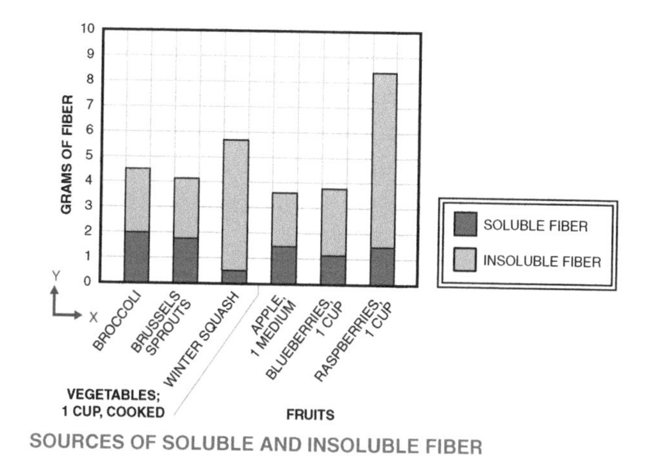

SOURCES OF SOLUBLE AND INSOLUBLE FIBER

SECTION 10-3 USING PIE GRAPHS

A *pie graph* is a graph in which a circle represents 100% of a variable and the sectors of the circle represent parts of the total. **See Figure 10-8.** To interpret a pie graph, determine the values of 100% by the sizes of the sectors. A pie graph usually has a percentage of a total in each sector. The percentage can be used to find the sector value if the total is known.

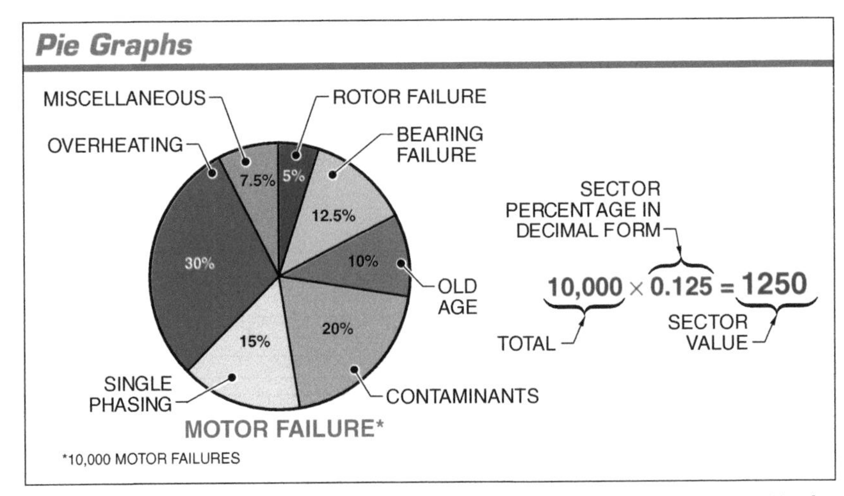

Figure 10-8. A pie graph is a graph in which a circle represents 100% of a variable and the sectors represent parts of the total.

To find the sector value when the total is known, multiply the total by the sector percentage in decimal form. For example, to find the number of motor failures caused by bearing failure, multiply total motor failures (10,000) by 12.5% bearing failures (10,000 × 0.125 = 1250). Of the 10,000 motor failures, 1250 were caused by bearing failure.

Pie graphs can be created on a computer or by hand. First, each value should be divided by the total to convert the percentages to decimal form. Then each decimal number should be multiplied by 360° to determine each sector angle. The sum of all sector angles must be equal to 360°. When creating a pie graph by hand, a protractor can be used to draw the proper angle for each sector. The sectors can then be shaded, colored, or hatched as needed.

Example—Pie Graphs

1. Create a pie graph for a parts inspection of 200 total parts, where 175 passed inspection, 15 had faulty assemblies, 8 had electronic failures, and 2 had miscellaneous damage.

ANS:

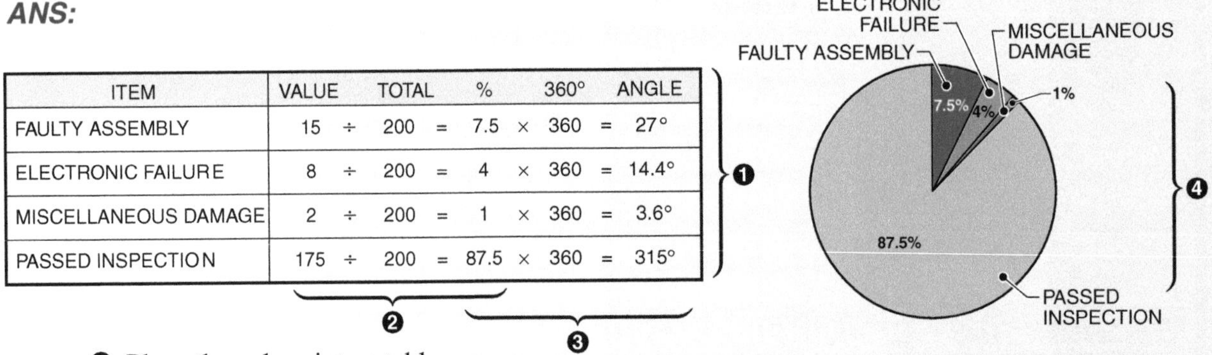

ITEM	VALUE		TOTAL		%		360°		ANGLE
FAULTY ASSEMBLY	15	÷	200	=	7.5	×	360	=	27°
ELECTRONIC FAILURE	8	÷	200	=	4	×	360	=	14.4°
MISCELLANEOUS DAMAGE	2	÷	200	=	1	×	360	=	3.6°
PASSED INSPECTION	175	÷	200	=	87.5	×	360	=	315°

❶ Place the values into a table.

❷ Find the percentage of each sector by dividing the number of parts for each section by the total number of parts inspected. Then find the angle of each sector by multiplying the percentage by 360°.

❸ Draw each angle. Shade, color, or cross-hatch the sectors as needed.

MATH EXERCISES—Pie Graphs

Complete the shaded areas of the table, develop a circle graph entitled PC-Tech Manufacturing Co., and answer questions 1–7.

PC-TECH VALUES TABLE									
GROUP	**VALUE**		**TOTAL**		**%**		**360°**		**ANGLE**
MANAGEMENT	5	÷		=		×	360	=	
MARKETING	12	÷		=		×	360	=	
CLERICAL	7	÷		=		×	360	=	
ACCOUNTING	6	÷		=		×	360	=	
CUSTOMER SERVICE	3	÷		=		×	360	=	
MAINTENANCE	1	÷		=		×	360	=	
PRODUCTION	46	÷		=		×	360	=	

__________ **1.** Marketing represents a(n) ___° sector on the circle graph.

__________ **2.** Production represents a(n) ___° sector on the circle graph.

__________ **3.** The clerical workers comprise ___% of the total workforce.

__________ **4.** The second-largest department at PC-Tech Manufacturing Co. is ___.

__________ **5.** Accounting represents a(n) ___° sector on the circle graph.

__________ **6.** The total percentage of employees in the three smallest departments is ___%.

__________ **7.** The total percentage of all workers, excluding management, not in the production department is ___%.

PRACTICAL APPLICATIONS—Pie Graphs

______________ 8. **Mechanics:** Use the Fuel Combustion Efficiency graph to determine the percentage of total heat loss.

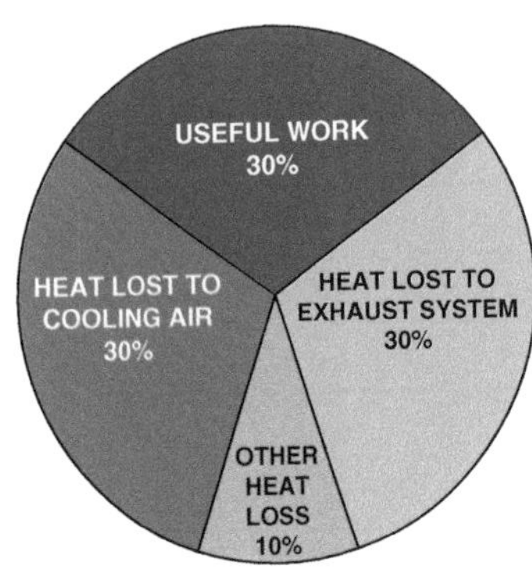

FUEL COMBUSTION EFFICIENCY

______________ 9. **Mechanics:** Determine the percentage of heat lost to cooling air as well as other heat loss as indicated by the Fuel Combustion Efficiency graph.

10. **Culinary Arts:** Create a pie graph using the following data from the restaurant profit and loss statement below.

Payroll Expenses	37.3%
Cost of Goods Sold	29.9%
Net Profit	10.6%
Rent	8.0%
Utilities	3.2%
Taxes	6.0%
Supplies	2.3%
Insurance	1.0%
Miscellaneous	1.7%

For an interactive review of the concepts covered in Chapter 10, refer to the corresponding Quick Quiz® included on the Digital Resources.

QUICK QUIZ®
Working with Graphs

Working with Graphs 10
Review

Name ______________________ Date ____________

Math Problems

Use the Insulation Step Voltage Test graph to answer questions 1–5.

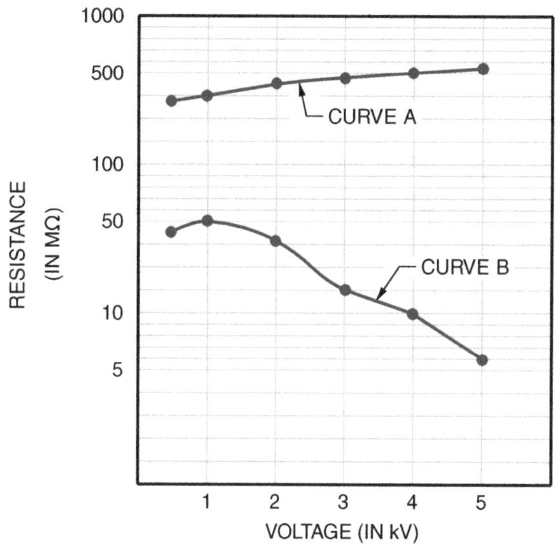

__________ **1.** Curve B shows a resistance of ___ MΩ (megohms) at 1 kV (kilovolt).

__________ **2.** Curve A shows a continual ___ in resistance.

__________ **3.** Curve A shows ___ MΩ when Curve B shows 10 MΩ.

__________ **4.** The maximum resistance shown on Curve B is ___ MΩ.

__________ **5.** The lowest reading on Curve A is approximately ___ MΩ.

INSULATION STEP VOLTAGE TEST GRAPH

Use the Public Opinion Poll graph to answer questions 6–10.

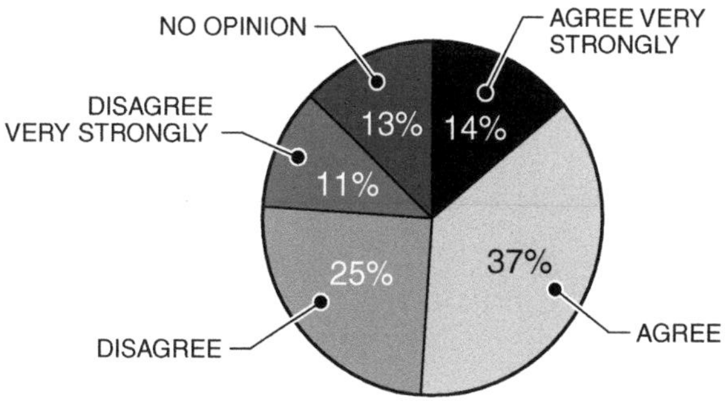

__________ **6.** The majority interviewed ___ with the question.

PUBLIC OPINION POLL

10 Review (continued)

_______________ 7. No opinion is expressed by ___% of the people interviewed.

_______________ 8. Of those interviewed, ___% agree very strongly.

_______________ 9. Of those interviewed, ___% disagree very strongly.

_______________ 10. A total of ___% disagree or disagree very strongly.

Practical Applications

Use the Electrical Device Energy Consumption graph to answer questions 11 and 12.

_______________ 11. **Electrical:** ___ represents the greatest percentage of total energy consumption.

_______________ 12. **Electrical:** Audible outputs account for ___% of total energy consumption.

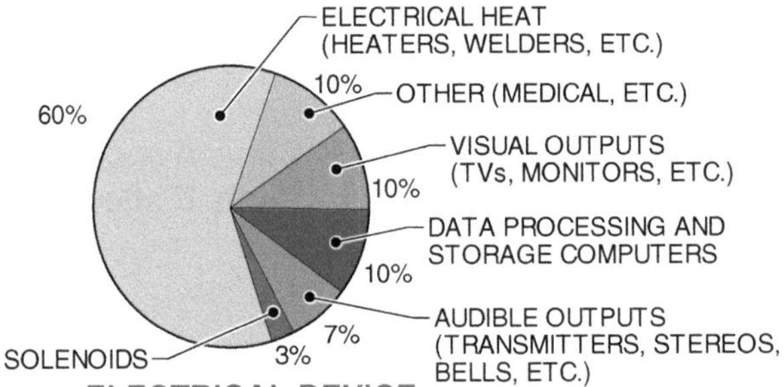

ELECTRICAL DEVICE ENERGY CONSUMPTION

13. **Mechanics:** Create a line graph representing the horsepower of a small engine if there is a decrease of 1% for each 10°F above 60°F, starting at 60°F and ending at 130°F. At 60°F, the engine has 10.0 horsepower.

_______________ 14. **Mechanics:** Use the line graph created in question 13 to determine the engine HP at 80°F.

_______________ 15. **Mechanics:** Use the line graph created in question 13 to determine the engine HP at 130°F.

Working with Graphs 10

Test

Name ______________________ Date ____________

Math Problems

1. Complete the Main Panelboard Table and then answer questions 2–7.

MAIN PANELBOARD TABLE										
ITEM	USE	VALUE		TOTAL		%		360°		ANGLE
15 A SPCB	GENERAL LIGHTING AND RECEPTACLES	6	÷		=		×	360	=	
20 A SPCB	SPECIAL APPLIANCES	8	÷		=		×	360	=	
20 A SPCB	SMALL APPLIANCES AND LAUNDRY	3	÷		=		×	360	=	
20 A SPCB	SPARES	7	÷		=		×	360	=	
30 A DPCB	OVENS AND COOKTOPS	2	÷		=		×	360	=	
40 A DPCB	RANGES	1	÷		=		×	360	=	
30 A DPCB	AC	1	÷		=		×	360	=	
30 A DPCB	WATER HEATERS	1	÷		=		×	360	=	
125 A DPCB	HEATING	1	÷		=		×	360	=	
30 A DPCB	SPARES	2	÷		=		×	360	=	

NOTE: SINGLE POLE CIRCUIT BREAKERS (SPCB) USED FOR 120 V CIRCUITS
DOUBLE POLE CIRCUIT BREAKERS (DPCB) USED FOR 240 V CIRCUITS

10 Test (continued)

_______________ **2.** Special appliances constitute ___% of the CBs in the main panelboard.

_______________ **3.** The main panelboard has a total of ___ SPCBs.

_______________ **4.** Spares for 20 A SPCBs account for ___% of the total CBs. (*Round to 2 places.*)

_______________ **5.** A total of ___ CBs each account for angles of 11.25° of the circle graph.

_______________ **6.** Ovens and cooktops have the same number of circuits as ___.

_______________ **7.** Ovens and cooktops and ___ constitute the greatest use for DPCBs.

Use the Monthly Snowfall graph to answer questions 8–10.

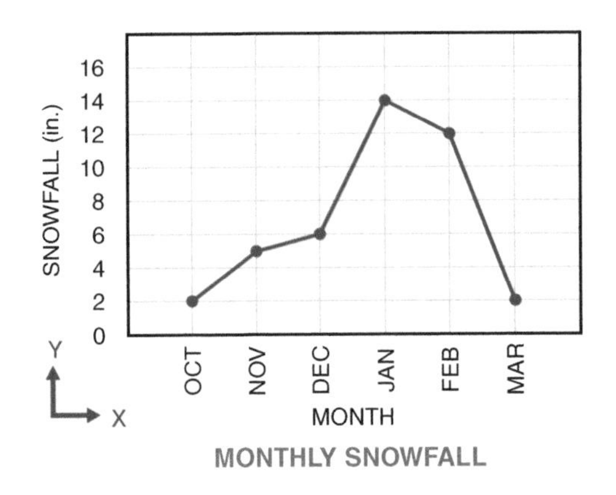

MONTHLY SNOWFALL

_______________ **8.** More snow fell during ___ than during any other month.

_______________ **9.** A total of ___ inches of snow fell from October through March.

_______________ **10.** The snowfall during December was ___ inches.

10 Test (continued)

Use the Monthly Sales of Outdoor Power Equipment graph to answer questions 11–20.

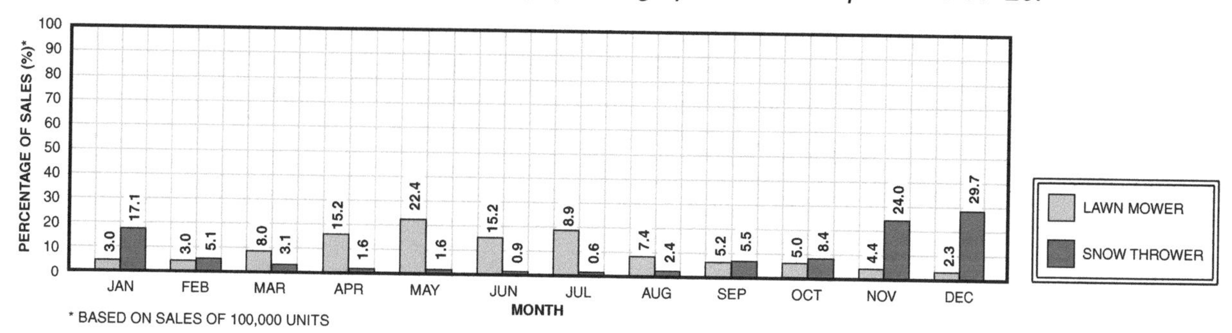

MONTHLY SALES OF OUTDOOR POWER EQUIPMENT

__________ **11.** ___ is the month when the most snow throwers were sold.

__________ **12.** How many more snow throwers were sold in January than lawn mowers were sold in April?

__________ **13.** ___ is the month when the least amount of lawn mowers were sold.

__________ **14.** ___ is the month when 600 snow throwers were sold.

__________ **15.** If 85,000 total units were sold the previous year and the percentage of total sales for November was the same for both years, how many more snow throwers were sold in November of this year than November of last year?

__________ **16.** The last two months of the year accounted for ___% of total sales of snow throwers.

__________ **17.** April, May, and June accounted for ___% of total sales for lawn mowers.

__________ **18.** How many snow throwers were sold between April and August?

__________ **19.** How many lawn mowers were sold in January, February, November, and December?

__________ **20.** If total cost to produce a snow thrower is $325.00 per unit and each unit sold for $565.00, what was the profit/loss on snow throwers for December?

10 Test (continued)

Practical Applications

21. HVAC: Create a grouped-bar graph entitled Winter/Summer Indoor Air Temperature for Educational Facilities using the data table below.

SCHOOL SPACE	WINTER TEMP*	SUMMER TEMP*
Classrooms, auditoriums, libraries, and office areas	72	78
Corridors	68	80
Laboratories	72	78
Locker rooms and shower areas	75	—
Mechanical rooms	60	—
Industrial technologies	72	78
Storage	65	—
Restrooms	72	—

* in degrees Fahrenheit

Use the Lamp Energy Consumption graph to answer questions 22–24.

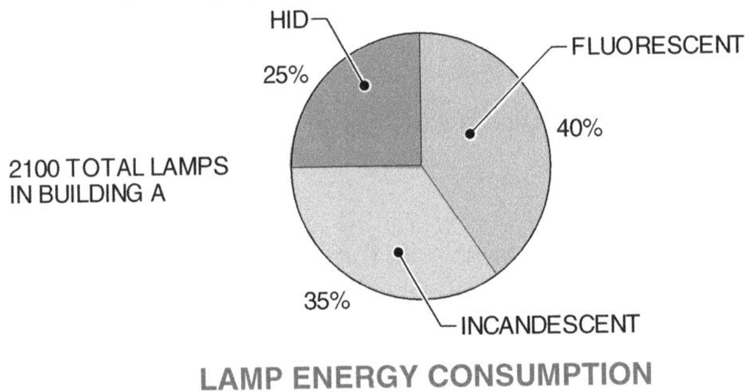

LAMP ENERGY CONSUMPTION

__________ **22. Alternative Energy:** How many fluorescent lamps are in Building A?

__________ **23. Alternative Energy:** How many incandescent lamps are in Building A?

__________ **24. Alternative Energy:** How many HID lamps are in Building A?

10 Test (continued)

Use the Battery Self-Discharge Rates graph to answer questions 25–28.

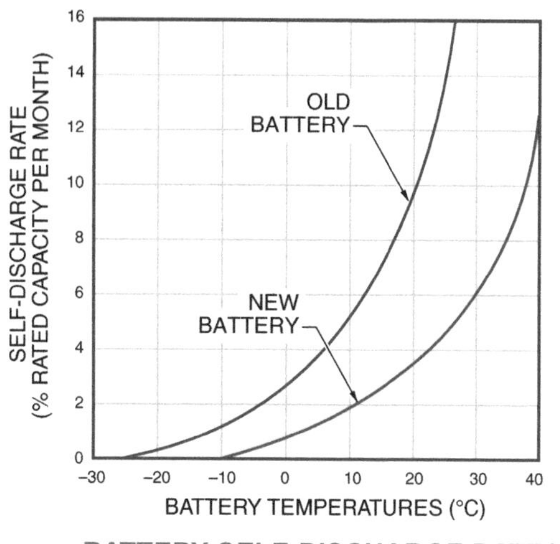

BATTERY SELF-DISCHARGE RATES

______________ **25. Electrical:** What is the self-discharge rate of a new electrical storage battery at 30°C?

______________ **26. Electrical:** What is the self-discharge rate of an old electrical storage battery at 10°C.

______________ **27. Electrical:** Approximately what is the difference between the discharge rates of an old electrical storage battery and a new storage battery when both are at 0°C?

______________ **28. Electrical:** If the discharge rate of the old electrical storage battery is 14%, what is its temperature?

10 Test (continued)

_______________ **29. Culinary Arts:** Determine the total percentage of expenses using the Restaurant Profit and Loss graph shown below.

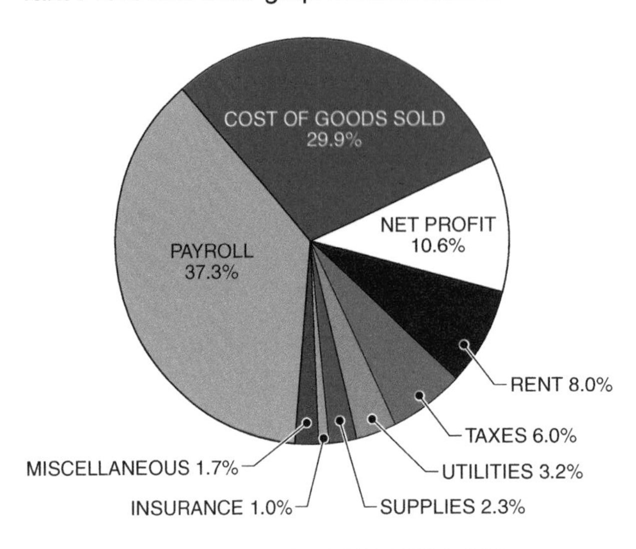

RESTAURANT PROFIT AND LOSS

30. Agriculture: Create a bar graph using the following data from the Power, Structural, and Technical (PST) Systems Areas of Study table below.

PST Area of Study	Percentage
Power and machine mechanics	30%
Agricultural structures	20%
Welding and metalwork	15%
Electrical power and processes	15%
Environmental systems	10%
Tool and equipment safety	10%

POWER, STRUCTURAL, AND TECHNICAL SYSTEMS AREAS OF STUDY

Answers*

**Provided for selected Math Exercises and Practical Applications*

CHAPTER 1—USING WHOLE NUMBERS

Expressing Whole Numbers ________ 4

1. One hundred twenty-three thousand, five hundred seven
3. Six thousand, three hundred
5. 506,925
7. XXI
9. MCMXCIII

Adding Whole Numbers ________ 7

1. 179
3. 32,447
5. 174,292
7. $5555.00

Subtracting Whole Numbers ________ 10

1. 383
3. 1218
5. 530,793
7. 16°
9. 85 cm

Multiplying Whole Numbers ________ 15

1. 1728
3. 19,692
5. 140 cm
7. $291.00

Dividing Whole Numbers ________ 19

1. 72
3. 23
5. 41 hrs
7. 15 spools
9. $2.25/sq ft

CHAPTER 2—WORKING WITH SIMPLE FRACTIONS

Proper Fractions, Improper Fractions, and Mixed Numbers ________ 32

1. $\frac{59}{7}$
3. $\frac{221}{11}$
5. $\frac{250}{3}$
7. $\frac{467}{10}$
9. $42\frac{7}{18}$

Finding Prime Factors ________ 34

1. 2, 3, 13
3. 311
5. 2, 3, 5, 7, 11

Finding Common Factors ________ 36

1. 1, 2
3. 16
5. 36

Reducing Fractions to Lowest Terms ________ 37

1. $\frac{1}{2}$
3. $\frac{3}{4}$
5. $\frac{11}{16}''$
7. $\frac{7}{8}$

Reducing Fractions to Lowest Terms with a Given Denominator ________ 39

1. $\frac{3}{4}$
3. $\frac{4}{6}$
5. $\frac{3}{9}$
7. $\frac{3}{8}''$

Changing Fractions to Higher Terms with a Given Denominator ____ 41

1. 10/12
3. 32/40
5. 24/64″

Finding the Lowest Common Denominator ____ 43

1. 24
3. 60
5. 8

Changing Fractions to Their Lowest Common Denominator ____ 45

1. 24/168, 63/168, 42/168, 112/168
3. 24/72, 16/72, 30/72, 27/72
5. 77/231, 66/231, 84/231
7. 6/8″, 8 4/8″, 15 7/8″, 21 6/8″

CHAPTER 3—WORKING WITH COMPLEX FRACTIONS

Adding Fractions with Common Denominators ____ 55

1. 1 1/5
3. 1 7/12
5. 2 3/4″

Adding Fractions with Uncommon Denominators ____ 57

1. 37/60
3. 2 17/18
5. 2 3/4
7. 1 1/8″

Adding Mixed Numbers ____ 58

1. 11 2/3
3. 13 23/105
5. 12 11/12
7. 76″

Subtracting Fractions with Common Denominators ____ 60

1. 1/3
3. 14/27
5. 1 1/2″

Subtracting Fractions with Uncommon Denominators ____ 62

1. 43/200
3. 2/35
5. 5/8 full

Subtracting Mixed Numbers ____ 64

1. 3 1/4
3. 3 11/40
5. 1′-9 3/4″
7. 11/16″

Multiplying Two or More Fractions ____ 66

1. 1/10
3. 8/15
5. 3/40
7. 9/20
9. 3/20″
11. 15/32″

Multiplying Improper Fractions ____ 68

1. 27
3. 1 3/8
5. 7
7. 72 1/4

Multiplying a Fraction by a Whole Number ____ 70

1. 4/9
3. 7/20
5. 9 3/5
7. 6 2/7
9. 1 4/5 gal.

Multiplying a Mixed Number by a Whole Number ______ 71

1. 4011
3. $3100\frac{1}{2}$
5. $130\frac{15}{19}$
7. 54′
9. 120″

Multiplying Mixed Numbers ______ 73

1. $476\frac{4}{9}$
3. $3769\frac{8}{9}$
5. $348\frac{2}{21}$
7. $680\frac{3}{5}$
9. $214\frac{7}{12}$ lb

Dividing Two Fractions ______ 75

1. $1\frac{3}{10}$
3. $1\frac{7}{15}$
5. $3\frac{3}{4}$
7. $2\frac{1}{2}$
9. $3\frac{1}{2}$ doors

Dividing a Fraction by a Whole Number ______ 77

1. $\frac{3}{16}$
3. $\frac{31}{96}$
5. $\frac{7}{80}$
7. $\frac{3}{128}$
9. $\frac{1}{4}$ tank

Dividing a Mixed Number by a Whole Number ______ 79

1. $50\frac{1}{10}$
3. $197\frac{7}{20}$
5. $90\frac{19}{36}$
7. $2\frac{2}{9}$
9. $1\frac{9}{64}$″

Dividing a Whole Number by a Fraction ______ 81

1. 35
3. $58\frac{1}{2}$
5. 99
7. 12 spacers

Dividing Mixed Numbers ______ 83

1. $1\frac{1}{3}$
3. $1\frac{11}{46}$
5. $48\frac{4}{15}$
7. 2
9. 6 small pieces

Dividing Complex Fractions ______ 87

1. $\frac{1}{30}$
3. $\frac{1}{50}$
5. 225
7. $\frac{13}{32}$
9. $11\frac{7}{11}$ cripples
11. 2 brace posts, 66 line posts

CHAPTER 4—WORKING WITH DECIMALS

Rounding Decimals ______ 100

1. 0.8
3. 6.31830

Adding and Subtracting Decimals ______ 102

1. 130.96297
3. 55.3759
5. 94.08825
7. 442.950 sq ft

Multiplying Decimals ______ 105

1. 1.4
3. 0.102172
5. 2.5115102
7. 56,980
9. 56.25
11. 118″

Dividing Decimals ______ 109

1. 50.6
3. 36
5. 0.000015625
7. 1.62
9. 4.905
11. 1.375″

Changing Fractions to Decimals ___ 111

1. 0.15625
3. 0.31
5. 0.375
7. 0.75
9. 3.75″

Changing Decimals to Fractions ___ 114

1. 13/40
3. 41/2000
5. 193/1000
7. 8 3/8
9. ±3/50

Changing Decimals to Fractions with a Given Denominator ______ 116

1. 9/12
3. 23/32
5. 5/8
7. 3/64″
9. 16/32″
11. 6273 7/8 cu yd

CHAPTER 5—WORKING WITH PERCENTAGES

Percentages and Their Fraction and Decimal Equivalents ______ 131

1. 25%
3. 7/50
5. 2.75

Calculating Percentage ______ 133

1. 7
3. 255
5. 3885 kWh

Calculating Percent Rate ______ 135

1. 15%
3. 50%
5. 15%

Calculating Base Amount ______ 136

1. 10
3. 1900
5. 400′

Calculating Costs ______ 141

1. $1100.00
3. single discount of 15%
5. $43.75
7. 17%

Calculating Tax and Total Cost ___ 144

1. $6.01
3. $142,040.00
5. $2.70
7. $460.95
9. $8.65

CHAPTER 6—WORKING WITH MEASUREMENTS

Converting to Lower Units of Measure ______ 163

1. 316,224 sq in.
3. 5550 mm
5. 78″

Converting to Higher Units of Measure ______ 165

1. 188 gal., 1 qt, 1 pt
3. 3.2 mi
5. 750 patch cords

Converting Between Measurement Systems ______ 168

1. 6.3 qt
3. 1.4 L

5. 196.9″
7. 487.7 cm

Converting Temperature ______ 170

1. 374°F
3. 84.2°F
5. 2.8°C
7. 20°C to 22.2°C

Adding and Subtracting Units of Measure ______ 172

1. 10 yd, 2′, 9″
3. 33 gal., 1 qt, 1 pt

Multiplying and Dividing Units of Measure ______ 174

1. 24 yd, 1′, 4″
3. 36 cu ft, 18 cu in.
5. 22½ cu ft
7. 1 gal., 2 qt

CHAPTER 7—WORKING WITH EXPONENTS

Exponents ______ 192

1. 64
3. 8⁄729
5. 1⁄729
7. 1⁄16
9. 0.000064
11. 562,500
13. 625 sq ft
15. 484 W

Square Roots ______ 195

1. 6
3. 7
5. 0.43
7. 4.75
9. 1.133
11. 65′

CHAPTER 8—FINDING RATIOS AND PROPORTIONS

Ratios ______ 207

1. 9:1
3. 3:1
5. 33:2
7. 1:5
9. 4:1

Proportions ______ 210

1. 18
3. 2
5. 455
7. 60
9. 144
11. 10
13. 1.2
15. 2.85

Proportions ______ 214

1. 750 lb
3. 5 hr
5. 91 spot welds
7. 29 hangers

CHAPTER 9—WORKING WITH PLANE AND SOLID FIGURES

Squares and Rectangles ______ 227

1. 232 sq ft
3. 7.56 sq in.
5. 5200 sq ft
7. 178.5 sq ft

Triangles ______ 230

1. 176 sq in.
3. 36 m^2
5. 35.875 sq in.
7. 3326.4 sq in.

Circles — 233

1. 25.13″
3. 53.41″
5. 113.10 cm
7. 7.0686 sq in.

Rectangular Solids — 236

1. 1152 cu in.
3. 393.75 cu in.
5. 400 cu ft

Cylinders — 238

1. 57.73 cu ft
3. 8.04 cu ft
5. 20 cu ft
7. 10,688 cu in.
9. 516.79 sq ft

Square Pyramids — 241

1. 859,064.3 mm^3
3. 720 sq ft

Cones — 244

1. 9.42 cu in.
3. 816.81 sq in.
5. 508.94 cu ft

Spheres — 246

1. 5575.28 cu in.
3. 4188.8 cu ft

CHAPTER 10—WORKING WITH GRAPHS

Line Graphs — 258

1. 3 A
3. 2009
5. 2012
7. January
9.

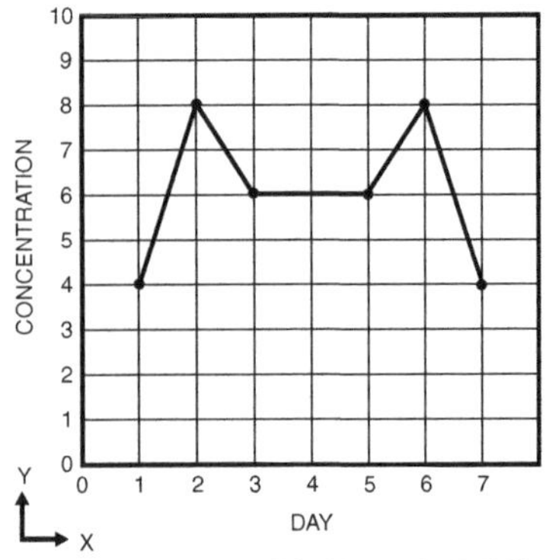

DAILY 8570 CONCENTRATION

Bar Graphs — 265

1. overheating
3. 3250
5. Old age
7.

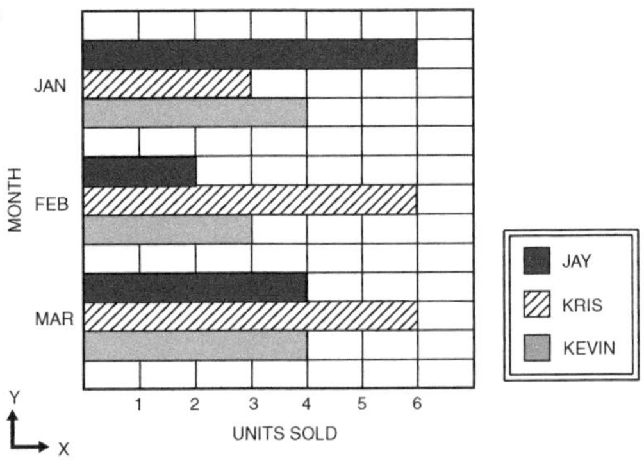

SALES DATA

9. 110
11. 9.25%

Circle Graphs — 269

1. 54
3. 8.75
5. 27
7. 36.25
9. 40%

Appendix

Common Math Symbols

Symbol	Term
+	addition
∠	angle
¢	cent
$\sqrt[3]{x}$	cube root
°	degree
ϕ	diameter
÷, ⌊, or ⌉	division
$	dollar
= or ::	equal
′	foot
>	greater than
″	inch
<	less than
′	minutes
×	multiplication
#	number
‖	parallel
%	percent
⊥	perpendicular
π	pi (3.14)
±	plus or minus
$\bar{x}$	repetend
∟	right angle
″	seconds
$\sqrt{x}$	square root
–	subtraction
Δ	triangle

Units of Power

Power	W	ft lb/s	HP	kW
Watt	1	0.7376	0.341×10^{-3}	0.001
Foot-pound/sec	1.356	1	0.818×10^{-3}	1.356×10^{-3}
Horsepower	745.7	550	1	0.7457
Kilowatt	1000	736.6	1.341	1

Motor Torque Formulas

Torque

$T = \frac{HP \times 5252}{rpm}$ where

T = torque

HP = horsepower

5252 = constant $\frac{33{,}000 \text{ lb-ft}}{\pi \times 2} = 5252$

rpm = revolutions per minute

Starting Torque

$T = \frac{HP \times 5252}{rpm} \times \%$ where

HP = horsepower

5252 = constant $\frac{33{,}000 \text{ lb-ft}}{\pi \times 2} = 5252$

rpm = revolutions per minute

$\%$ = motor class percentage

Normal Torque Rating

$T = \frac{HP \times 63{,}000}{rpm}$ where

T = nominal torque rating (in lb-in)

63,000 = constant

HP = horsepower

rpm = revolutions per minute

Exponents Formulas

Products	$c_a \times c_b = c^{a+b}$
Quotients	$\frac{c^a}{c^b} = c^{a-b}$ $(c \neq 0)$
Powers	$(c^a)^b = c^{ab} = (c^b)^a$
Products Raised to a Power	$(cd)^n = c^n \times d^n$
Quotients Raised to a Power	$\left(\frac{c}{d}\right)^n = \frac{c^n}{d^n}$ $(d \neq 0)$
Zero Exponents	$c^0 = 1$ $(c \neq 0)$
Negative Exponents	$c^{-a} = \frac{1}{c^a}$ $(c \neq 0)$

Common Units of Measure

Length and Area

1 inch (in.)	= 2.54 centimeters (cm)
1 foot (ft)	= 12 inches (in.)
	= 30.48 centimeters (cm)
1 yard (yd)	= 3 feet (ft)
	= 0.914 meter (m)
1 rod (rd)	= 16.5 feet (ft)
	= 5.5 yards (yd)
1 furlong (fur)	= 660 feet (ft)
	= 40 rods (rd)
1 mile (mi)	= 5280 feet (ft)
	= 1760 yards (yd)
	= 320 rods (rd)
	= 8 furlongs (fur)
	= 1.609 kilometers (km)
1 meter (m)	= 100 centimeters (cm)
	= 1000 millimeters (mm)
	= 1.094 yards (yd)
	= 3.28 feet (ft)
	= 39.27 inches (in.)
1 kilometer (km)	= 1000 meters (m)
	= 0.621 mile (mi)
1 millimeter (mm)	= 1000 microns (or micrometers, µm)
1 square inch (sq in.)	= 6.45 square centimeters (cm^2)
1 square foot (sq ft)	= 144 square inches (sq in.)
	= 0.0929 square meter (m^2)
1 square yard (sq yd)	= 9 square feet (sq ft)
	= 0.836 square meter (m^2)
1 square mile (sq mi)	= 640 acres (A)
	= 1 section
1 acre (A)	= 43,560 square feet (sq ft)
	= 4840 square yards (sq yd)
1 square meter (m^2)	= 10,000 square centimeters (cm^2)
	= 11.196 square yards (sq yd)
	= 10.76 square feet (sq ft)
1 square centimeter (cm^2)	= 100 square millimeters (mm^2)
	= 0.155 square inch (sq in.)
1 nautical mile (nmi)	= 6080 feet (ft)
	= 1.853 kilometers (km)
1 nautical mile per hour (nmi/h)	= 1 knot (kn)

Power and Heat

1 British thermal unit (Btu)	= 778 foot-pounds (ft-lb)
	= 0.252 calorie (cal)
1 calorie (cal)	= 3088 foot-pounds (ft-lb)
	= 3.968 Btu
1 kilowatt (kW)	= 1000 watts (W)
	= 738 foot-pounds per second (ft-lb/sec)
	= 1.341 horsepower (HP)
1 horsepower (HP)	= 33,000 foot-pounds per minute (ft-lb/min)
	= 0.746 kilowatt (kW)
1 kilowatt-hour (kWh)	= 3414 Btu
	= 860 calories (cal)
1 horsepower-hour (HP-h)	= 2544 Btu

Weight

1 pound (lb)	= 16 ounces (oz)
	= 7000 grains (gr)
	= 454 grams (g)
	= 0.454 kilogram (kg)
1 grain (gr)	= 64.8 milligrams (mg)
	= 0.0648 gram (g)
	= 0.0023 ounce (oz)
1 gram (g)	= 1000 milligrams mg)
	= 0.03527 ounce (oz)
	= 15.43 grains (gr)
1 kilogram (kg)	= 1000 grams (g)
	= 2.205 pounds (lb)
1 U.S. short ton (sh tn)	= 2000 pounds (lb)
	= 907 kilograms (kg)
1 U.S. long ton (long tn)	= 2240 pounds (lb)
	= 1016 kilograms (kg)
1 metric ton (t)	= 1000 kilograms (kg)
	= 0.984 U.S. long ton (long tn)
	= 1.102 U.S. short ton (sh tn)
	= 22,205 pounds (lb)

Volume

1 cubic inch (cu in.)	= 16.39 cubic centimeters (cm^3)
1 cubic foot (cu ft)	= 1728 cubic inches (cu in.)
	= 28.32 liters (L)
1 cubic yard (cu yd)	= 27 cubic feet (cu ft)
	= 0.765 cubic meter (m^3)
1 cubic meter (cm^3)	= 1000 liters (L)
	= 1.308 cubic yards (cu yd)
	= 35.31 cubic feet (cu ft)
1 imperial gallon (gal. imp)	= 277.4 cubic inches (cu in.)
	= 4.55 liters (L)
1 U.S. gallon (gal. U.S.)	= 0.833 imperial gallon (gal. imp)
	= 3.785 liters (L)
	= 231 cubic inches (cu in.)
1 U.S. barrel (petroleum)	= 42 U.S. gallons (gal. U.S.)
	= 35 imperial gallons (gal. imp)
1 liter (L)	= 1000 cubic centimeters (cm^3)
	= 0.22 imperial gallon (gal. imp)
	= 0.2642 U.S. gallon (gal. U.S.)
	= 61 cubic inches (cu in.)

Density of Water at 62°F

1 cubic foot (cu ft)	= 62.35 pounds (lb)
	= 7.48 gallons (gal.)
1 pound (lb)	= 0.01604 cubic foot (cu ft)
1 gallon (gal.)	= 8.33 pounds (lb)

Pressure

1 atmosphere (atm)	= 14.696 pounds per square inch (psi)
1 inch of water at 62°F	= 0.0361 psi
	= 5.20 pounds per square foot (psf)
1 foot head of water at 62°F	= 0.433 psi
1 inch of mercury	= 0.491 psi

Fractional and Decimal Inch Equivalents*

Fraction	Decimal	Fraction	Decimal	Fraction	Decimal	Fraction	Decimal
1/64	0.015625	17/64	0.265625	33/64	0.515625	47/64	0.765625
1/32	0.03125	9/32	0.28125	17/32	0.53125	25/32	0.78125
3/64	0.046875	19/64	0.296875	35/64	0.546875	51/64	0.796875
1/16	0.0625	5/16	0.3125	9/16	0.5625	13/16	0.8125
5/64	0.078125	21/64	0.328125	37/64	0.578125	53/64	0.828125
3/32	0.09375	11/32	0.34375	19/32	0.59375	27/32	0.84375
7/64	0.109375	23/64	0.359375	39/64	0.609375	55/64	0.859375
1/8	0.125	3/8	0.375	5/8	0.625	7/8	0.875
9/64	0.140625	25/64	0.390625	41/64	0.640625	57/64	0.890625
5/32	0.15625	13/32	0.40625	21/32	0.65625	29/32	0.90625
11/64	0.171875	27/64	0.421875	43/64	0.671875	59/64	0.921875
3/16	0.1875	7/16	0.4375	11/16	0.6875	15/16	0.9375
13/64	0.203125	29/64	0.453125	45/64	0.703125	61/64	0.953125
7/32	0.21875	15/32	0.46875	23/32	0.71875	31/32	0.96875
15/64	0.234375	31/64	0.484375	47/64	0.734375	63/64	0.984375
1/4	0.250	1/2	0.500	3/4	0.750	1	1.000

*based on 1/100 mm = 0.003973″ 10 mm = 1 centimeter = 0.3937″ 25.4 mm = 1″

Millimeter and Decimal Inch Equivalents*

mm in.	mm in.	mm in.	mm in.	mm in.	mm in.
1/50 = 0.00079	25/50 = 0.01969	1 = 0.03937	26 = 1.02362	51 = 2.00787	76 = 2.99212
2/50 = 0.00157	26/50 = 0.02047	2 = 0.07874	27 = 1.06299	52 = 2.04724	77 = 3.03149
3/50 = 0.00236	27/50 = 0.02126	3 = 0.11811	28 = 1.10236	53 = 2.08661	78 = 3.07086
4/50 = 0.00315	28/50 = 0.02205	4 = 0.15748	29 = 1.14173	54 = 2.12598	79 = 3.11023
	29/50 = 0.02283				
5/50 = 0.00394	30/50 = 0.02362	5 = 0.19685	30 = 1.18110	55 = 2.16535	80 = 3.14960
6/50 = 0.00472	31/50 = 0.02441	6 = 0.23622	31 = 1.22047	56 = 2.20472	81 = 3.18897
7/50 = 0.00551	32/50 = 0.02520	7 = 0.27559	32 = 1.25984	57 = 2.24409	82 = 3.22834
8/50 = 0.00630	33/50 = 0.02598	8 = 0.31496	33 = 1.29921	58 = 2.28346	83 = 3.26771
9/50 = 0.00709	34/50 = 0.02677	9 = 0.35433	34 = 1.33858	59 = 2.32283	84 = 3.30708
10/50 = 0.00787	35/50 = 0.02756	10 = 0.39370	35 = 1.37795	60 = 2.36220	85 = 3.34645
11/50 = 0.00866	36/50 = 0.02835	11 = 0.43307	36 = 1.41732	61 = 2.40157	86 = 3.38582
12/50 = 0.00945	37/50 = 0.02913	12 = 0.47244	37 = 1.45669	62 = 2.44094	87 = 3.42519
13/50 = 0.01024	38/50 = 0.02992	13 = 0.51181	38 = 1.49606	63 = 2.48031	88 = 3.46456
14/50 = 0.01102	39/50 = 0.03071	14 = 0.55118	39 = 1.53543	64 = 2.51968	89 = 3.50393
15/50 = 0.01181	40/50 = 0.03150	15 = 0.59055	40 = 1.57480	65 = 2.55905	90 = 3.54330
16/50 = 0.01260	41/50 = 0.03228	16 = 0.62992	41 = 1.61417	66 = 2.59842	91 = 3.58267
17/50 = 0.01339	42/50 = 0.03307	17 = 0.66929	42 = 1.65354	67 = 2.63779	92 = 3.62204
18/50 = 0.01417	43/50 = 0.03386	18 = 0.70866	43 = 1.69291	68 = 2.67716	93 = 3.66141
19/50 = 0.01496	44/50 = 0.03465	19 = 0.74803	44 = 1.73228	69 = 2.71653	94 = 3.70078
20/50 = 0.01575	45/50 = 0.03543	20 = 0.78740	45 = 1.77165	70 = 2.75590	95 = 3.74015
21/50 = 0.01654	46/50 = 0.03622	21 = 0.82677	46 = 1.81102	71 = 2.79527	96 = 3.77952
22/50 = 0.01732	47/50 = 0.03701	22 = 0.86614	47 = 1.85039	72 = 2.83464	97 = 3.81889
23/50 = 0.01811	48/50 = 0.03780	23 = 0.90551	48 = 1.88976	73 = 2.87401	98 = 3.85826
24/50 = 0.01890	49/50 = 0.03858	24 = 0.94488	49 = 1.92913	74 = 2.91338	99 = 3.89763
		↓	↓	↓	↓
		25 = 0.98425	50 = 1.96850	75 = 2.95275	100 = 0.93700

*based on 1/100 mm = 0.003973″ 10 mm = 1 centimeter = 0.3937″ 25.4 mm = 1″

Sizes of Boards, Dimensional Lumber, and Timbers

	THICKNESS					WIDTH				
	Nominal Inch	Minimum Dressed				Nominal Inch	Minimum Dressed			
		Dry		Green			Dry		Green	
		inch	mm	inch	mm		inch	mm	inch	mm
Boards	¾	⅝	16	11/16	17	2	1½	38	1 9/16	40
	1	¾	19	25/32	20	3	2½	64	2 9/16	65
	1¼	1	25	1 1/32	26	4	3½	89	3 9/16	90
	1½	1¼	32	1 9/32	33	5	4½	114	4⅝	117
						6	5½	140	5⅝	143
						7	6½	165	6⅝	168
						8	7¼	184	7½	190
						9	8¼	210	8½	216
						10	9¼	235	9½	241
						11	10¼	260	10½	267
						12	11¼	286	11½	292
						14	13¼	337	13½	343
						16	15¼	387	15½	394
Dimensional Lumber	2	1½	38	1 9/16	40	2	1½	38	1 9/16	40
	2½	2	51	2 1/16	52	2½	2	51	2 1/16	52
	3	2½	64	2 9/16	65	3	2½	64	2 9/16	65
	3½	3	76	3 1/16	78	3½	3	76	3 1/16	78
	4	3½	89	3 9/16	90	4	3½	89	3 9/16	90
	4½	4	102	4 1/16	103	4½	4	102	4 1/16	103
						5	4½	114	4⅝	117
						6	5½	140	5⅝	143
						8	7¼	184	7½	190
						10	9¼	235	9½	241
						12	11¼	286	11½	292
						14	13¼	337	13½	343
						16	15¼	387	15½	394
Timbers	5 and thicker			½ off	13 off	5 and wider			½ off	13 off

Standard Lumber Sizes

Type	Nominal Size in Inches		Actual Size in Inches	
	Thickness	Width	Thickness	Width
Common Boards	1	2	¾	1½
Nominal: 1″ × 4″; Actual: ¾″ × 3½″	1	4	¾	3½
	1	6	¾	5½
	1	8	¾	7¼
	1	10	¾	9¼
	1	12	¾	11¼
Dimensional Lumber	2	2	1½	1½
Nominal: 2″ × 4″; Actual: 1½″ × 3½″	2	4	1½	3½
	2	6	1½	5½
	2	8	1½	7¼
	2	10	1½	9¼
	2	12	1½	11¼
Timbers	5	5	4½	4½
Nominal: 6″ × 6″; Actual: 5½″ × 5½″	6	6	5½	5½
	6	8	5½	7½
	6	10	5½	9½
	8	8	7½	7½
	8	10	7½	9½

Concrete Coverage Estimates

Thickness*	Coverage†	Thickness*	Coverage†	Thickness*	Coverage†
1	324	5	65	9	36
1¼	259	5¼	62	9¼	35
1½	216	5½	59	9½	34
1¾	185	5¾	56	9¾	33
2	162	6	54	10	32.5
2¼	144	6¼	52	10¼	31.5
2½	130	6½	50	10½	31
2¾	118	6¾	48	10¾	30
3	108	7	46	11	29.5
3¼	100	7¼	45	11¼	29
3½	93	7½	43	11½	28
3¾	86	7¾	42	11¾	27.5
4	81	8	40	12	27
4¼	76	8¼	39	12¼	21.5
4½	72	8½	38	12½	18
4¾	68	8¾	37	12¾	13.5

Note: Coverage and thickness are based on 1 cu yd of concrete.
* in in.
† in sq ft

Concrete Masonry Units

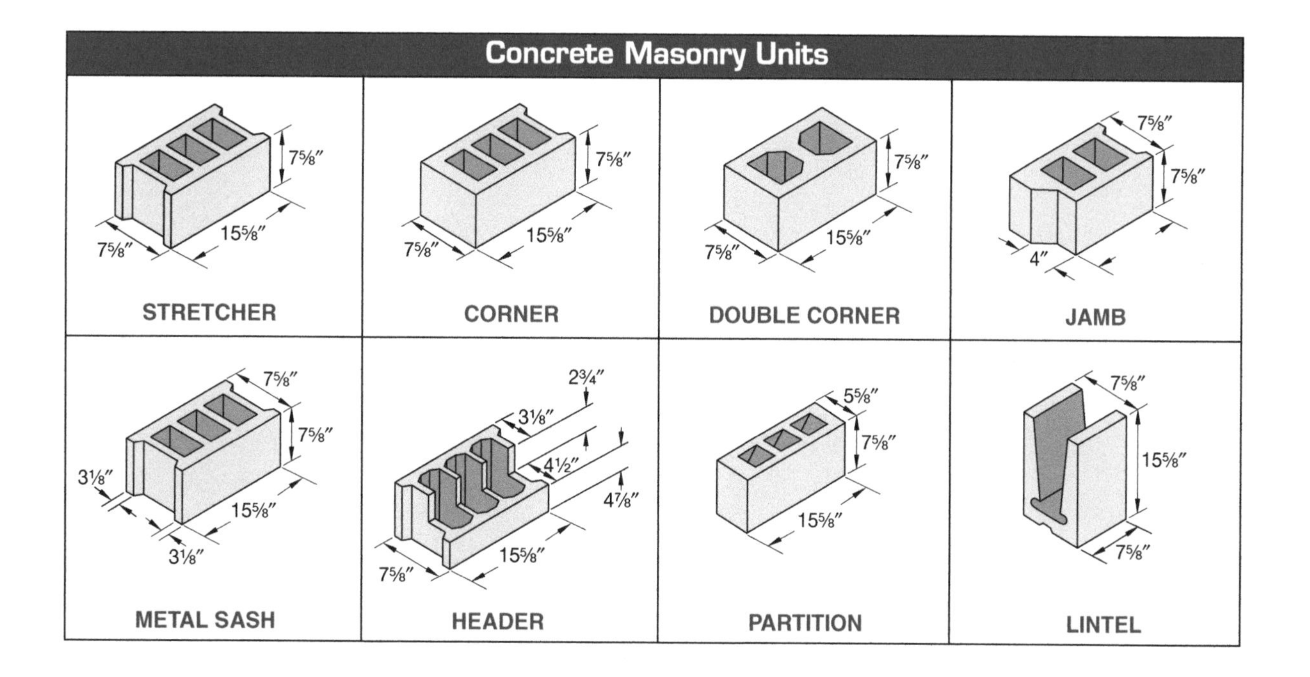

Pipe Wall Thickness							
Nominal ID*	OD (BW Guage)	Nominal Wall Thickness			Nominal Wall Thickness		
		Standard	Extra-Heavy	Double Extra-Heavy	Schedule 40	Schedule 60	Schedule 80
⅛	0.405	0.269	0.215	—	0.068	0.095	—
¼	0.540	0.364	0.302	—	0.088	0.119	—
⅜	0.675	0.493	0.423	—	0.091	0.126	—
½	0.840	0.622	0.546	0.252	0.109	0.147	0.294
¾	1.050	0.824	0.742	0.434	0.113	0.154	0.308
1	1.315	1.049	0.957	0.599	0.133	0.179	0.358
1¼	1.660	1.380	1.278	0.896	0.140	0.191	0.382
1½	1.900	1.610	1.500	1.100	0.145	0.200	0.400
2	2.375	2.067	1.939	1.503	0.154	0.218	0.436
2½	2.875	2.469	2.323	1.771	0.203	0.276	0.552
3	3.500	3.068	2.900	2.300	0.216	0.300	0.600
3½	4.000	3.548	3.364	2.728	0.226	0.318	—
4	4.500	4.026	3.826	3.152	0.237	0.337	0.674
5	5.563	5.047	4.813	4.063	0.258	0.375	0.750
6	6.625	6.065	5.761	4.897	0.280	0.432	0.864
8	8.625	7.981	7.625	6.875	0.322	0.500	0.875
10	10.750	10.020	9.750	8.750	0.365	0.500	—
12	12.750	12.000	11.750	10.750	0.406	0.500	—

* in in.

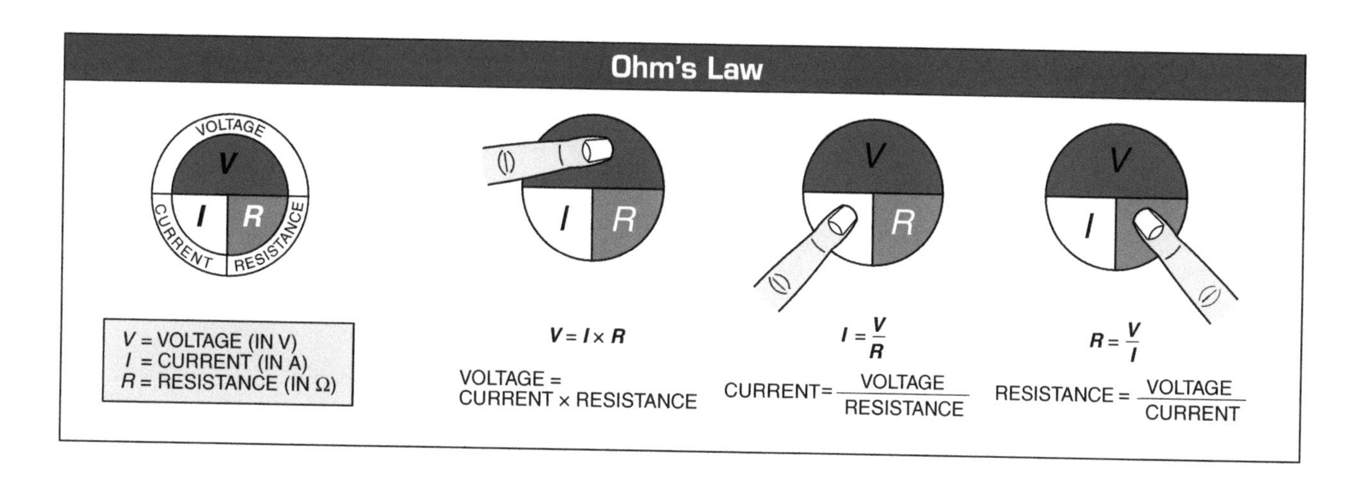
Ohm's Law
VOLTAGE
CURRENT
RESISTANCE
V
I
R
V = VOLTAGE (IN V)
I = CURRENT (IN A)
R = RESISTANCE (IN Ω)
V = I × R
VOLTAGE = CURRENT × RESISTANCE
I = V/R
CURRENT = VOLTAGE/RESISTANCE
R = V/I
RESISTANCE = VOLTAGE/CURRENT

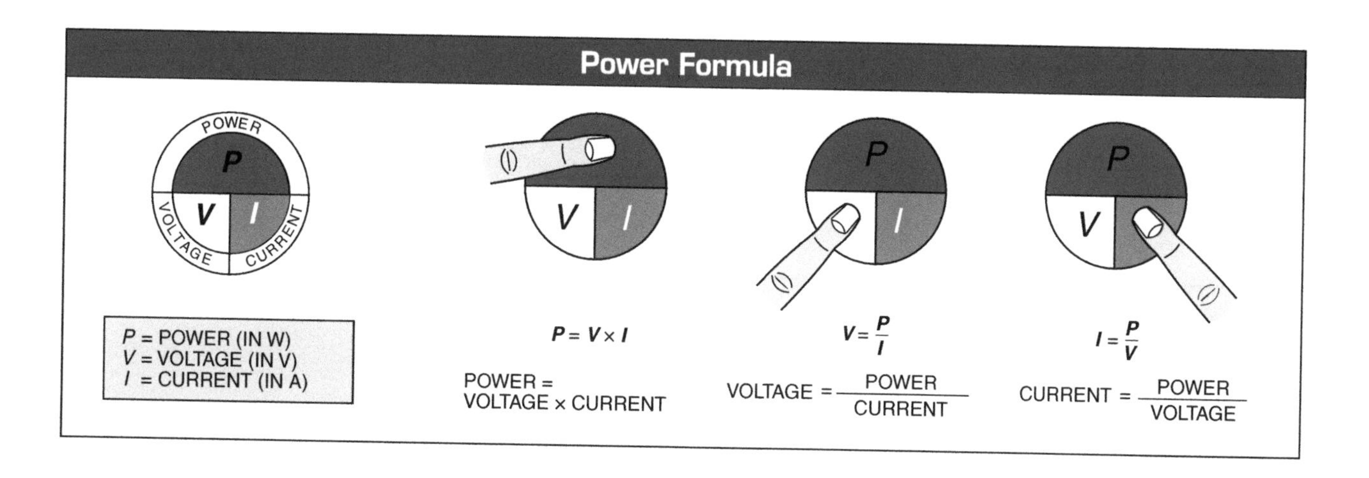
Power Formula
POWER
VOLTAGE
CURRENT
P
V
I
P = POWER (IN W)
V = VOLTAGE (IN V)
I = CURRENT (IN A)
P = V × I
POWER = VOLTAGE × CURRENT
V = P/I
VOLTAGE = POWER/CURRENT
I = P/V
CURRENT = POWER/VOLTAGE

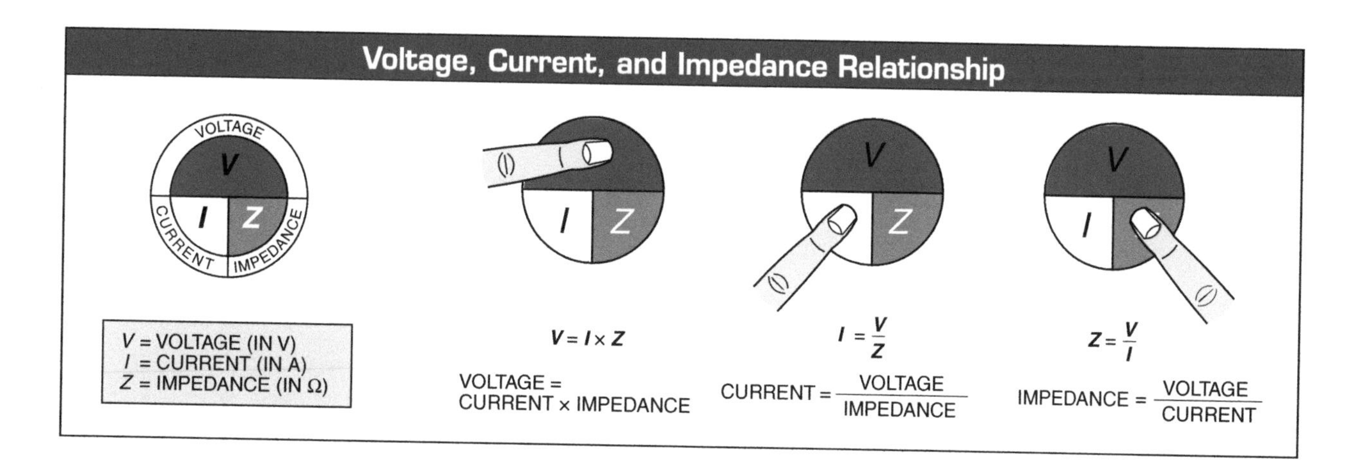
Voltage, Current, and Impedance Relationship
VOLTAGE
CURRENT
IMPEDANCE
V
I
Z
V = VOLTAGE (IN V)
I = CURRENT (IN A)
Z = IMPEDANCE (IN Ω)
V = I × Z
VOLTAGE = CURRENT × IMPEDANCE
I = V/Z
CURRENT = VOLTAGE/IMPEDANCE
Z = V/I
IMPEDANCE = VOLTAGE/CURRENT

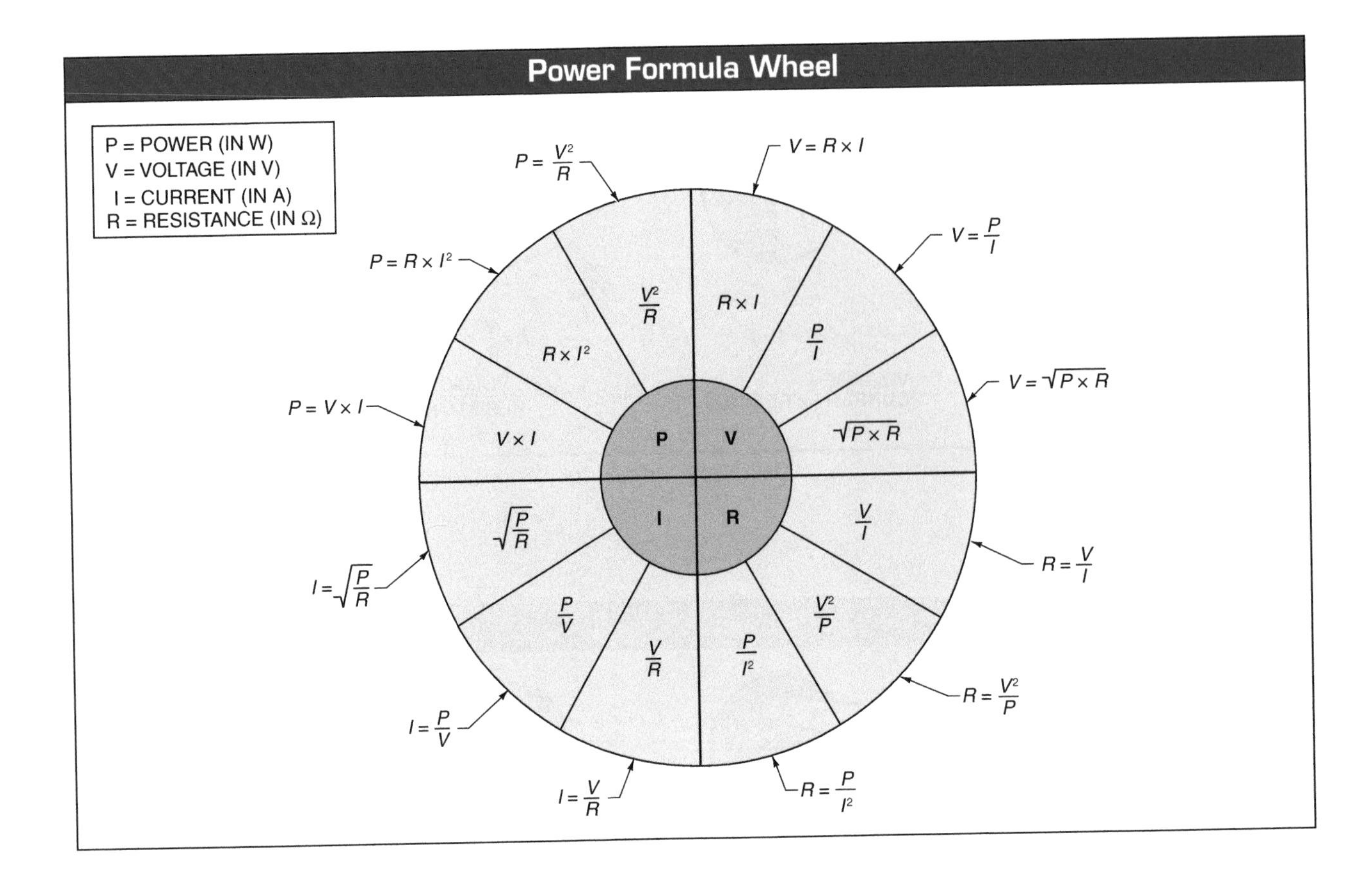

Power Formula Wheel
P = POWER (IN W)
V = VOLTAGE (IN V)
I = CURRENT (IN A)
R = RESISTANCE (IN Ω)
P
V
I
R
$P = \frac{V^2}{R}$
$V = R \times I$
$V = \frac{P}{I}$
$V = \sqrt{P \times R}$
$R = \frac{V}{I}$
$R = \frac{V^2}{P}$
$R = \frac{P}{I^2}$
$I = \frac{V}{R}$
$I = \frac{P}{V}$
$I = \sqrt{\frac{P}{R}}$
$P = V \times I$
$P = R \times I^2$
$\frac{V^2}{R}$
$R \times I$
$\frac{P}{I}$
$\sqrt{P \times R}$
$\frac{V}{I}$
$\frac{V^2}{P}$
$\frac{P}{I^2}$
$\frac{V}{R}$
$\frac{P}{V}$
$\sqrt{\frac{P}{R}}$
$V \times I$
$R \times I^2$

Stock Material Weight*

Material	Weight	Material	Weight	Material	Weight
Metals		Chestnut	30	Granite	172
Aluminum, bronze	481	Cypress, southern	32	Greenstone, trap	187
Aluminum, cast-hammered	165	Douglas fir	34	Gypsum, alabaster	159
Antimony	416	Elm, American	35	Limestone	160
Arsenic	358	Hemlock, eastern, western	28	Magnesite	187
Bismuth	608	Hickory	53	Marble	168
Brass, cast-rolled	534	Larch, western	36	Phosphate rock, apatite	200
Chromium	428	Maple, red, black	38–40	Pumice, natural	40
Cobalt	552	Oak	51	Quartz, flint	165
Copper, cast-rolled	556	Pine, white, yellow, western	27–28	Sandstone, bluestone	147
Gold, cast-hammered	1205	Poplar, yellow	28	Slate, shale	172
Iron, cast, pig	450	Redwood	30	Soapstone, talc	169
Iron, slag	172	Spruce	28	**Bituminous Substances**	
Iron, wrought	485	Tamarack	37	Asphaltum	81
Lead	706	Walnut	39–40	Coal, anthracite	97
Magnesium	109	**Liquids**		Coal, bituminous	84
Manganese	456	Acids, muriatic, 40%	75	Coal, coke	75
Mercury	848	Acids, nitric, 91%	94	Coal, lignite	78
Molybdenum	562	Acids, sulphuric, 87%	112	Graphite	131
Nickel	545	Alcohol, 100%	49	Paraffin	56
Platinum, cast-hammered	1330	Gasoline	42	Petroleum, crude	55
Silver, cast-hammered	656	Lye, soda, 66%	106	Petroleum, refined	50
Steel	490	Oils	58	Pitch	69
Tin, cast-hammered	459	Petroleum	55	Tar, bituminous	75
Tungsten	1180	Water, 4°C	62	**Brick Masonry**	
Vanadium	350	Water, seawater	64	Common brick	120
Zinc, cast-rolled	440	Water, ice	56	Pressed brick	140
Solids		Water, snow, fresh fallen	8	Soft brick	100
Carbon, amorphous, graphitic	129	**Gases**		**Concrete**	
Cork	15	Air, 0°C	0.08071	Cement, cinder, etc.	100
Ebony	76	Ammonia	0.0478	Cement, slag, etc.	130
Fats	58	Carbon dioxide	0.1234	Cement, stone, sand	144
Glass, common, plate	160	Carbon monoxide	0.0781	**Building Material**	
Glass, crystal	184	Gas, natural	0.038–0.039	Ashes, cinders	40–45
Phosphorous, white	114	Hydrogen	0.00559	Cement, Portland, loose	90
Resins, rosin, amber	67	Nitrogen	0.0784	Cement, Portland, set	183
Rubber	58	Oxygen	0.0892	Lime, gypsum, loose	65–75
Silicon	155	**Minerals**		Mortar, set	103
Sulphur, Amorphous	128	Asbestos	153	Slags, bank screenings	98–117
Wax	60	Basalt	184	Slags, bank slag	67–72
Timber, U.S. Seasoned		Bauxite	159	Slags, machine slag	96
Ash, white	41	Borax	109	**Earth**	
Beech	44	Chalk	137	Clay, damp, plastic	110
Birch, yellow	43	Clay	137	Dry, packed	95
Cedar, white, red	22–23	Dolomite	181	Mud, packed	115

* in lb/cu ft

U.S. Customary System of Measurement

		Unit	Abbr	Equivalents	Metric Equivalent
Length		mile	mi	5280′, 320 rd, 1760 yd	1.609 km
		rod	rd	5.50 yd, 16.5′	5.029 m
		yard	yd	3′, 36″	0.9144 m
		foot	ft *or* ′	12″, 0.333 yd	30.48 cm
		inch	in. *or* ″	0.083′, 0.028 yd	2.54 cm
Area		square mile	sq mi *or* mi^2	640 A, 102,400 sq rd	2.590 km^2
		acre	A	4840 sq yd, 43,560 sq ft	0.405 hectare, 4047 m^2
		square rod	sq rd *or* rd^2	30.25 sq yd, 0.00625 A	25.293 m^2
		square yard	sq yd *or* yd^2	1296 sq in., 9 sq ft	0.836 m^2
		square foot	sq ft *or* ft^2	144 sq in., 0.111 sq yd	0.093 m^2
		square inch	sq in. *or* in^2	0.0069 sq ft, 0.00077 sq yd	6.452 cm^2
Volume		cubic yard	cu yd *or* yd^3	27 cu ft, 46,656 cu in.	0.765 m^3
		cubic foot	cu ft *or* ft^3	1728 cu in., 0.0370 cu yd	0.028 m^3
		cubic inch	cu in. *or* in^3	0.00058 cu ft, 0.000021 cu yd	16.387 cm^3
Capacity	*U.S. liquid measure*	gallon	gal.	4 qt (231 cu in.)	3.785 L
		quart	qt	2 pt (57.75 cu in.)	0.946 L
		pint	pt	4 gi (28.875 cu in.)	0.473 L
		gill	gi	4 fl oz (7.219 cu in.)	118.294 mL
		fluid ounce	fl oz	8 fl dr (1.805 cu in.)	29.573 mL
		fluid dram	fl dr	60 min (0.226 cu in.)	3.697 mL
		minim	min	⅙ fl dr (0.003760 cu in.)	0.061610 mL
	U.S. dry measure	bushel	bu	4 pk (2150.42 cu in.)	35.239 L
		peck	pk	8 qt (537.605 cu in.)	8.810 L
		quart	qt	2 pt (67.201 cu in.)	1.101 L
		pint	pt	½ qt (33.600 cu in.)	0.551 L
	British imperial liquid and dry measure	bushel	bu	4 pk (2219.36 cu in.)	0.036 m^3
		peck	pk	2 gal. (554.84 cu in.)	0.091 m^3
		gallon	gal.	4 qt (277.420 cu in.)	4.546 L
		quart	qt	2 pt (69.355 cu in.)	1.136 L
		pint	pt	4 gi (34.678 cu in.)	568.26 cm^3
		gill	gi	5 fl oz (8.669 cu in.)	142.066 cm^3
		fluid ounce	fl oz	8 fl dr (1.7339 cu in.)	28.412 cm^3
		fluid dram	fl dr	60 min (0.216734 cu in.)	3.5516 cm^3
		minim	min	1/60 fl dr (0.003612 cu in.)	0.059194 cm^3
Mass/Weight	*avoirdupois*	ton	t	2000 lb	
		short ton		2000 lb	0.907 t
		long ton		2240 lb	1.016 t
		pound	lb *or* #	16 oz, 7000 gr	0.454 kg
		ounce	oz	0.0625 lb, 16 dr, 437.5 gr	28.350 g
		dram	dr	27.344 gr, 0.0625 oz	1.772 g
		grain	gr	0.037 dr, 0.002286 oz	0.0648 g
	troy	pound	lb	12 oz, 240 dwt, 5760 gr	0.373 kg
		ounce	oz	20 dwt, 480 gr	31.103 g
		pennyweight	dwt *or* pwt	24 gr, 0.05 oz	1.555 g
		grain	gr	0.042 dwt, 0.002083 oz	0.0648 g

Metric System of Measurement

	Unit	Abbr	Number of Base Units	U.S. Customary Equivalent		
Length	kilometer	km	1000	0.62 mi		
	hectometer	hm	100	109.36 yd		
	dekameter	dam	10	32.81″		
	meter*	m	1	39.37″, 3.281′		
	decimeter	dm	0.1	3.94″		
	centimeter	cm	0.01	0.39″		
	millimeter	mm	0.001	0.039″		
Area	square kilometer	sq km *or* km^2	1,000,000	003861 sq mi		
	hectare	ha	10,000	2.47 A		
	are	a	100	119.60 sq yd		
	square centimeter	sq cm *or* cm^2	0.0001	0.155 sq in.		
Volume	cubic centimeter	cu cm, cm^3, *or* cc	0.000001	0.061 cu in.		
	cubic decimeter	dm^3	0.001	61.023 cu in.		
	cubic meter*	m^3	1	1.307 cu yd		
Capacity				**cubic**	**dry**	**liquid**
	kiloliter	kL	1000	1.31 cu yd		
	hectoliter	hL	100	3.53 cu ft	2.84 bu	
	dekaliter	daL	10	0.35 cu ft	1.14 pk	2.64 gal.
	liter*	L	1	61.02 cu in.	0.908 qt	1.057 qt
	cubic deciliter	dL^3	1	61.02 cu in.	0.908 qt	1.057 qt
	deciliter	dL	0.10	6.1 cu in.	0.18 pt	0.21 pt
	centiliter	cL	0.01	0.61 cu in.		338 fl oz
	milliliter	mL	0.001	0.061 cu in.		0.27 fl dr
Mass/Weight	metric ton	t	1,000,000	1.102 t		
	kilogram*	kg	1000	2.2046 lb		
	hectogram	hg	100	3.527 oz		
	dekagram	dag	10	0.353 oz		
	gram	g	1	0.035 oz		
	decigram	dg	0.10	1.543 gr		
	centigram	cg	0.01	0.154 gr		
	milligram	mg	0.001	0.015 gr		

* base units

Units of Energy

Energy	Btu	ft lb	J	kcal	kWh
British thermal unit (Btu)	1	777.9	1.056	0.252	2.930×10^{-4}
Foot-pound (ft lb)	1.285×10^{-3}	1	1.356	3.240×10^{-4}	3.766×10^{-7}
Joule (J)	9.481×10^{-4}	0.7376	1	2.390×10^{-4}	2.778×10^{-7}
Kilocalorie (kcal)	3.968	3.086	4.184	1	1.163×10^{-3}
Kilowatt-hour (kWh)	3.413	2.655×10^{6}	3.6×10^{6}	860.2	1

Metric Prefixes			
Multiples and Submultiples	**Prefixes**	**Symbols**	**Meanings**
$1{,}000{,}000{,}000{,}000 = 10^{12}$	tera-	T	trillion
$1{,}000{,}000{,}000 = 10^{9}$	giga-	G	billion
$1{,}000{,}000 = 10^{6}$	mega-	M	million
$1000 = 10^{3}$	kilo-	k	thousand
$100 = 10^{2}$	hecto-	h	hundred
$10 = 10^{1}$	deka-	d	ten
Unit $1 = 10^{0}$			
$0.1 = 10^{-1}$	deci-	d	tenth
$0.01 = 10^{-2}$	centi-	c	hundredth
$0.001 = 10^{-3}$	milli-	m	thousandth
$0.000001 = 10^{-6}$	mirco-	μ	millionth
$0.000000001 = 10^{-9}$	nano-	n	billionth
$0.000000000001 = 10^{-12}$	pico-	p	trillionth

Metric Conversions												
Initial Units	**Final Units**											
	giga-	**mega-**	**kilo-**	**hecto-**	**deka-**	**base unit**	**deci-**	**centi-**	**milli-**	**micro-**	**nano-**	**pico-**
giga-		3R	6R	7R	8R	9R	10R	11R	12R	15R	18R	21R
mega-	3L		3R	4R	5R	6R	7R	8R	9R	12R	15R	18R
kilo-	6L	3L		1R	2R	3R	4R	5R	6R	9R	12R	15R
hecto-	7L	4L	1L		1R	2R	3R	4R	5R	8R	11R	14R
deka-	8L	5L	2L	1L		1R	2R	3R	4R	7R	10R	13R
base unit	9L	6L	3L	2L	1L		1R	2R	3R	6R	9R	12R
deci-	10L	7L	4L	3L	2L	1L		1R	2R	5R	8R	11R
centi-	11L	8L	5L	4L	3L	2L	1L		1R	4R	7R	10R
milli-	12L	9L	6L	5L	4L	3L	2L	1L		3R	6R	9R
micro-	15L	12L	9L	8L	7L	6L	5L	4L	3L		3R	8R
nano-	18L	15L	12L	11L	10L	9L	8L	7L	6L	3L		3R
pico-	21L	18L	15L	14L	13L	12L	11L	10L	9L	6L	3L	

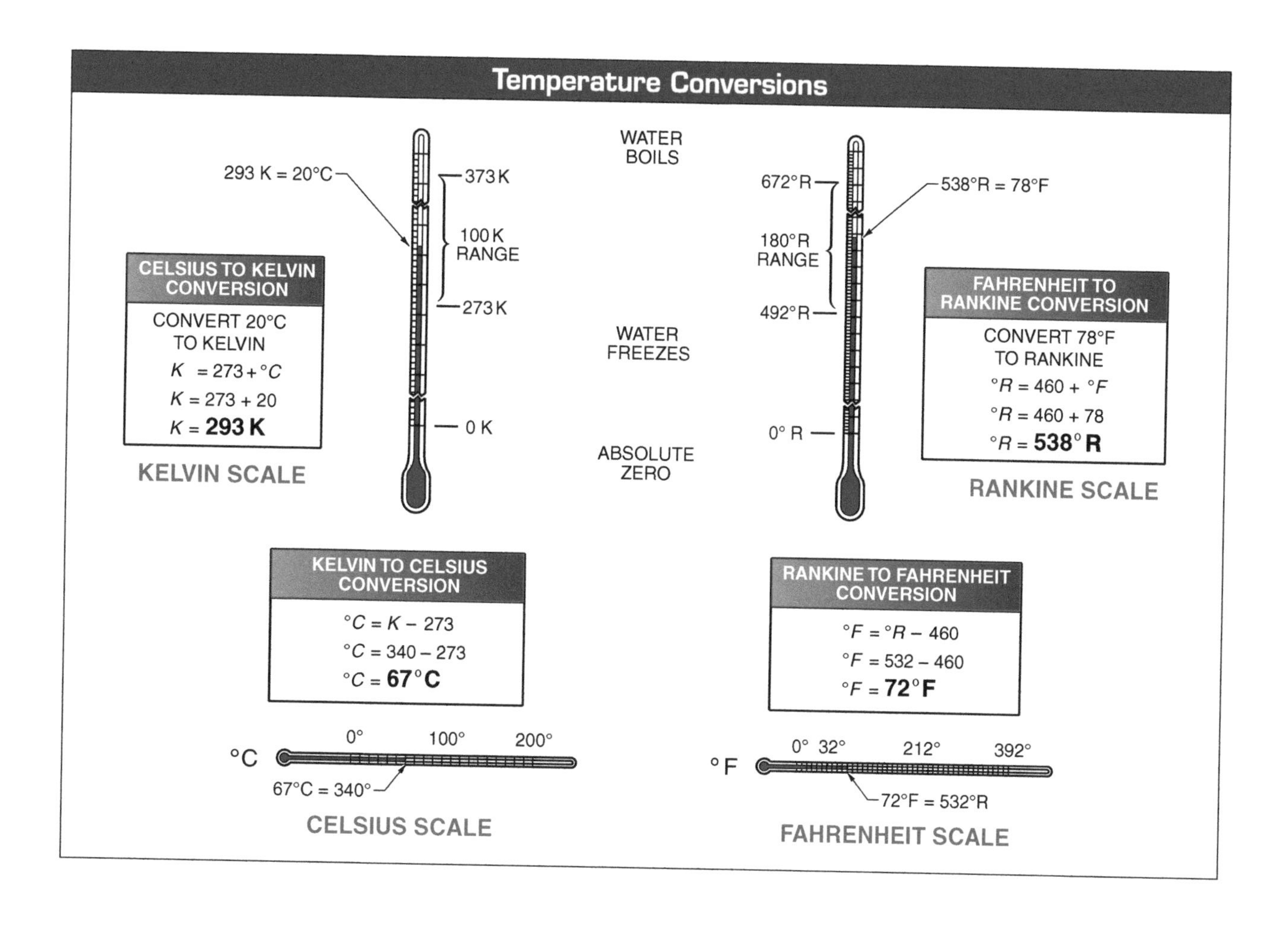

Volume and Capacity Conversion Factors

Factors	Cubic Inches	Cubic Feet	Cubic Centimeters	Liters	U.S. Gallons	Imperial Gallons	Water at Max Density: Pounds of Water	Water at Max Density: Kilograms of Water
cu in.	1	0.0005787	16.384	0.016384	0.004329	0.0036065	0.361275	0.0163872
cu ft	1728	1	0.037037	28.317	7.48052	6.23210	62.4283	28.3170
cu cm	0.0610	0.0000353	1	0.001	0.000264	0.000220	0.002205	0.0001
L	61.0234	0.0353145	0.001308	1	0.264170	0.220083	2.20462	1
gal. (U.S.)	231	0.133681	0.004951	3.78543	1	0.833111	8.34545	3.78543
gal. (imperial)	277.274	0.160459	0.0059429	4.54374	1.20032	1	10.0172	4.54373
water lb	27.6798	0.0160184	0.0005929	0.453592	0.119825	0.0998281	1	0.453593

Temperature and Refrigerant Pressure

Temp*	R-11†	R-12†	R-22†	R-113†	R-114†	R-500†	R-502†	R-134a†	R-123†
–50	28.9‡	15.4‡	6.2‡		27.1‡		0.0	18.7‡	
–45	28.7‡	13.3‡	2.7‡		26.6‡		1.9	16.9‡	
–40	28.4‡	11.0‡	0.5		26.0‡	7.6‡	4.1	14.8‡	
–35	28.1‡	8.4‡	2.6		25.4‡	4.6‡	6.5	12.5‡	
–30	27.8‡	5.5‡	4.9	29.3‡	24.6‡	1.2‡	9.2	9.5‡	
–25	27.4‡	2.3‡	7.4	29.2‡	23.8‡	1.2	12.1	6.9‡	
–20	27.0‡	0.6	10.1	29.1‡	22.9‡	3.2	15.3	3.7‡	27.8‡
–15	26.5‡	2.4	13.2	26.9‡	21.9‡	5.4	18.8	0.6	27.4‡
–10	26.0‡	4.5	16.5	28.7‡	20.5‡	7.8	22.6	1.9	26.9‡
–5	25.4‡	6.7	20.1	28.5‡	19.3‡	10.4	26.7	4.0	26.4‡
0	24.7‡	9.2	24.0	28.2‡	17.8‡	13.3	31.1	6.5	25.9‡
5	23.9‡	11.8	28.2	27.9‡	16.2‡	16.4	35.9	9.1	25.2‡
10	23.1‡	14.5	32.8	27.6‡	14.4‡	19.7	41.0	11.9	24.5‡
15	22.1‡	17.7	37.7	27.2‡	12.4‡	23.4	46.5	15.0	23.8‡
20	21.1‡	21.0	43.0	26.8‡	10.2‡	27.3	52.4	18.4	22.8‡
25	19.9‡	24.5	48.8	26.3‡	7.8‡	31.5	58.8	22.1	21.8‡
30	18.6‡	28.5	54.9	25.8‡	5.2‡	36.0	65.6	26.1	20.7‡
35	17.2‡	32.6	31.5	25.2‡	2.3‡	40.9	72.8	30.4	19.5‡
40	15.5‡	37.0	68.5	25.5‡	0.4	46.1	80.5	34.1	18.1‡
50	13.9‡	41.7	76.0	25.8‡	2.0	51.4	88.7	40.1	16.6‡
50	12.0‡	46.7	84.0	22.9‡	3.8	57.6	97.4	45.5	14.9‡
55	10.0‡	52.0	92.6	22.2‡	5.8	63.9	106.6	51.3	13.0‡
60	7.8‡	57.7	101.6	21.0‡	7.9	70.6	116.4	57.5	11.2‡
65	5.4‡	63.8	111.2	19.9‡	10.1	77.8	126.7	54.1	8.9‡
70	2.8	70.2	121.4	18.7‡	12.6	85.4	137.6	71.2	6.5‡
75	0.0	77.0	132.2	17.3‡	15.2	93.3	149.1	78.8	4.1‡
80	1.5	84.2	143.6	15.9‡	18.0	102.0	161.2	86.8	1.2‡
85	3.2	91.8	155.7	14.3‡	20.9	111.0	174.0	95.4	0.9
90	4.9	99.8	168.4	12.5‡	24.1	120.6	187.2	104.4	2.5
95	6.8	108.3	181.8	10.6‡	27.56	130.6	201.4	114.1	4.3
100	8.8	117.2	195.9	8.6‡	31.2	141.2	216.2	124.3	6.1
105	10.9	126.6	210.8	6.4‡	35.0	162.4	231.7	135.1	8.1
110	13.2	136.4	226.4	4.0‡	39.1	165.1	247.9	146.5	10.3
115	15.6	146.8	242.7	1.4‡	43.4	176.5	264.9	158.6	12.6
120	18.2	157.7	259.9	0.7	48.0	189.4	282.7	171.3	15.1
125	21.0	169.1	277.9	2.2	52.8	203.0	301.4	184.7	17.8
130	24.0	181.0	296.8	3.7	58.0	217.2	320.8	196.9	20.6
135	27.1	192.5	316.6	5.4	63.4	232.1	341.2	213.7	23.6
140	30.4	206.6	337.3	7.2	68.1	247.7	362.6	229.4	26.8
145	34.0	220.3	358.9	9.2	75.1			245.8	30.2
150	37.7	234.5	381.5	11.2	81.4			263.0	33.9

* in °F
† in psi
‡ in inches of mercury

Fluid Weights and Temperature

Fluid	Weight*	Temperature†
Air	4.33×10^{-5}	20°C/68°F @ 29.92 in. Hg
Gasoline	0.0237–0.0249	20°C/68°F
Kerosene	0.0296	20°C/68°F
Mercury	0.49116	0°C/32°F
Oil, lubricating	0.0307–0.0318	15°C/59°F
Oil, fuel	0.0036–0.0353	15°C/59°F
Water	0.0361	4°C/39°F
Seawater	0.0370	15°C/59°F

* in lb/cu in.
† laboratory temperature under which numerical values are defined

Weight of Water

Temperature in °F	Weight per Cu Ft	Weight per Gal.	Temperature in °F	Weight per Cu Ft	Weight per Gal.
32	62.418	8.344	**130**	61.563	8.230
35	62.422	8.345	**135**	61.472	8.218
39.2	62.425	8.346	**140**	61.381	8.206
40	62.425	8.346	**145**	61.291	8.193
45	62.422	8.345	**150**	61.201	8.181
50	62.409	8.343	**155**	61.096	8.167
55	62.394	8.341	**160**	60.991	8.153
60	62.372	8.338	**165**	60.843	8.134
65	62.344	8.334	**170**	60.783	8.126
70	62.313	8.331	**175**	60.665	8.110
75	62.275	8.325	**180**	60.548	8.094
80	62.232	8.321	**185**	60.430	8.078
85	62.182	8.313	**190**	60.314	8.063
90	62.133	8.306	**195**	60.198	8.047
95	62.074	8.297	**200**	60.081	8.032
100	62.022	8.291	**205**	60.980	8.018
105	61.960	8.283	**210**	59.820	7.997
110	61.868	8.271	**212**	59.760	7.989
115	61.807	8.261	**250**	58.750	7.854
120	61.715	8.250	**300**	56.970	7.616
125	61.654	8.242	**400**	54.250	7.252

Properties of Metals

Metal	Weight per Cubic Inch, Pound	Weight per Cubic Foot, Pound	Melting Point, °F	Linear Expansion per Unit Length per °F	Temp., °F*
Brass: 80C., 20Z.	0.3105	536.6	1823	0.00001	76 to 212
70C., 30Z.	0.3048	526.7	1706		
60C., 40Z.	0.3018	521.7	1652		
50C., 50Z.	0.2961	511.7	1616		
Bronze: 90C., 10T.	0.3171	547.9	1841	0.00001	68
Carbon steel	0.283–0.284	489.0–490.8	2500	0.00000633	68
Cast iron	0.254–0.279	438.7–482.4	1990–2300	0.00000655	68
Wrought iron	0.282–0.285	486.7–493.0	2750	0.00000661	

* Temperature given for each metal is that at which expansion coefficient shown in previous column was determined.

Cooling Heat Transfer Factors			
Type of Construction	Cooling Factor		
	15°F	20°F	25°F
Walls			
Wood Framed with Sheeting, Siding, and Veneer or Other Finish			
No insulation, ½″ gypsum board	5.0	6.4	7.8
R-11 cavity insulation + ½″ gypsum board	1.7	2.1	2.6
R-13 cavity insulation + ½″ gypsum board	1.5	1.9	2.3
R-13 cavity insulation + ¾″ bead board (R-2.7)	1.3	1.7	2.0
R-19 cavity insulation + ½″ gypsum board	1.1	1.4	1.7
R-19 cavity insulation + ¾″ extruded board	0.9	1.2	1.4
Masonry			
Above grade, no insulation	5.8	8.3	10.9
Above grade + R-5	1.6	2.3	3.1
Above grade + R-11	0.9	1.3	1.6
Below grade, no insulation	0.0	0.0	0.0
Below grade + R-5	0.0	0.0	0.0
Below grade + R-11	0.0	0.0	0.0
Ceilings			
No insulation	17.0	19.2	21.4
2″–2½″ insulation R-7	4.4	4.9	5.5
3″–3½″ insulation R-11	3.2	3.7	4.1
5¼″–6½″ insulation R-19	2.1	2.3	2.6
6″–7″ insulation R-22	1.9	2.1	2.4
10″–12″ insulation R-38	1.0	1.1	1.3
12″–13″ insulation R-44	0.9	1.0	1.1
Catherdral Type (Roof/Ceiling Combination)			
No insulation	11.2	12.6	14.1
R-11	2.8	3.2	3.5
R-19	1.9	2.2	2.4
R-22	1.8	2.0	2.2
Floors			
Over Unconditioned Space			
Over basement or enclosed crawl space (not vented)	0.0	0.0	0.0
Over vented space or garage	3.9	5.8	7.7
Over vented space or garage + R-11 insulation	0.8	1.3	1.7
Over vented space or garage + R-19 insulation	0.5	0.8	1.1
Basement Concrete Slab Floor, Unheated			
No edge insulation	0.0	0.0	0.0
1″ edge insulation R-5	0.0	0.0	0.0
2″ edge insulation R-9	0.0	0.0	0.0
Basement Concrete Slab Floor, Duct in Slab			
No edge insulation	0.0	0.0	0.0
1″ edge insulation R-5	0.0	0.0	0.0
2″ edge insulation R-9	0.0	0.0	0.0

Note: R values on this chart refer to thermal resistance value.

Air Conditioning Contractors of America

Air Density Effects

Temperature

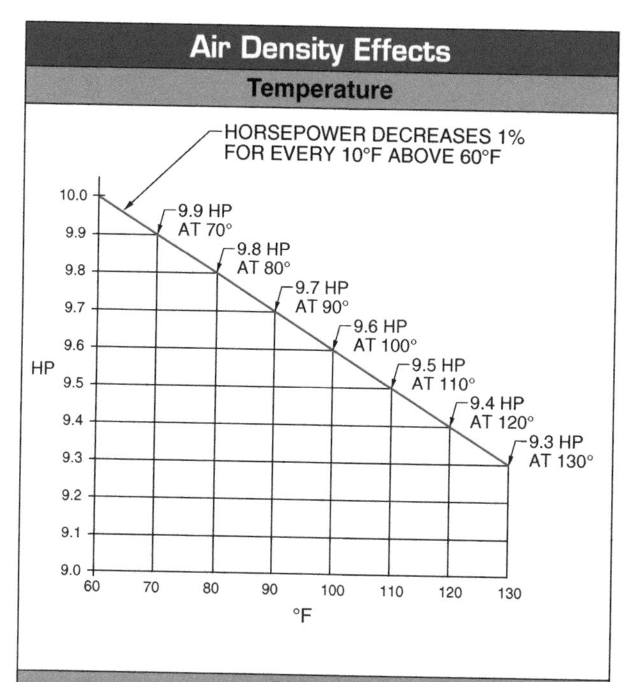

Altitude

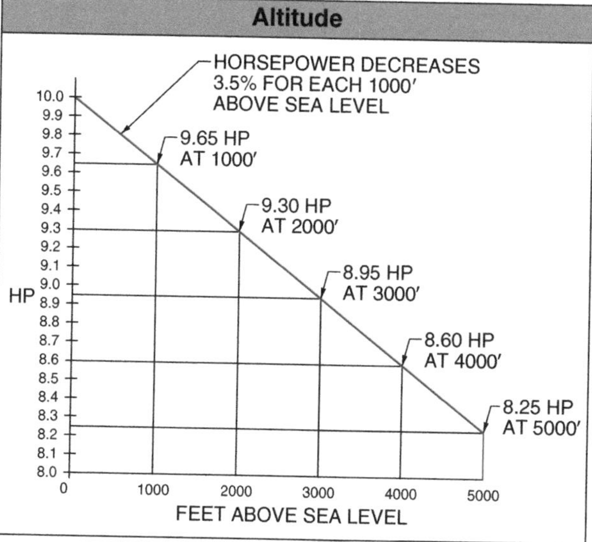

Fuel and Oil Mixes

	U.S. Gallons		Imperial Gallons		Metric	
	Fuel*	Oil†	Fuel*	Oil†	Petrol‡	Oil‡
16 : 1	1	8	1	10	4	0.250
	3	24	3	30	12	0.750
	5	40	5	50	20	1.250
	6	48	6	60	24	1.500
24 : 1	1	5.33	1	6.4	4	0.160
	3	16	3	19.2	12	0.470
	5	26.66	5	32.0	20	0.790
	6	32	6	38.4	24	0.940
32 : 1	1	4	1	5	4	0.125
	3	12	3	15	12	0.375
	5	20	5	25	20	0.625
	6	24	6	30	24	0.750
50 : 1	1	2.5	1	3	4	0.080
	3	8.0	3	9	12	0.240
	5	13.0	5	15	20	0.400
	6	15.5	6	18.5	24	0.480

* in gal.
† in oz
‡ in L

Boiler Conversion Factors

Atmosphere (standard)	=	29.92 inches of mercury
Atmosphere (standard)	=	14.7 pounds per square inch
1 horsepower	=	746 watts
1 horsepower	=	33,000 foot-pounds of work per minute
1 British thermal unit	=	778 foot-pounds
1 cubic foot	=	7.48 gallons
1 gallon	=	231 cubic inches
1 cubic foot of fresh water	=	62.5 pounds
1 cubic foot of salt water	=	64 pounds
1 foot of head of water	=	0.434 pounds per square inch
1 inch of head of mercury	=	0.491 pounds per square inch
1 gallon of fresh water	=	8.33 pounds
1 barrel (oil)	=	42 gallons
1 long ton of fresh water	=	36 cubic feet
1 long ton of salt water	=	35 cubic feet
1 ounce (avoirdupois)	=	437.5 grains

Motor Efficiencies*					
HP	Standard Motor (%)	Energy-Efficient Motor (%)	HP	Standard Motor (%)	Energy-Efficient Motor (%)
1	76.5	84.0	30	88.1	93.1
1½	78.5	85.5	40	89.3	93.6
2	79.9	86.5	50	90.4	93.7
3	80.8	88.5	75	90.8	95.0
5	83.1	88.6	100	91.6	95.4
7½	83.8	90.2	125	91.8	95.8
10	85.0	90.3	150	92.3	96.0
15	86.5	91.7	200	93.3	96.1
20	87.5	92.4	250	93.6	96.2
25	88.0	93.0	300	93.8	96.5

* Data represents typical motor efficiencies.

Full-Load Currents for DC Motors		
Motor Rating (HP)	Current (A)	
	120 V	240 V
¼	3.1	1.6
⅓	4.1	2
½	5.4	2.7
¾	7.6	3.8
1	9.5	4.7
1½	13.2	6.6
2	17	8.5
3	25	12.2
5	40	20
7½	48	29
10	76	38

Full-Load Currents for 1ϕ, AC Motors		
Motor Rating (HP)	Current (A)	
	115 V	230 V
⅙	4.4	2.2
¼	5.8	2.9
⅓	7.2	3.6
½	9.8	4.9
¾	13.8	6.9
1	16	8
1½	20	10
2	24	12
3	34	17
5	56	28
7½	80	40

Full-Load Currents for 3ϕ, AC Induction Motors				
Motor Rating (HP)	Current (A)			
	208 V	230 V	460 V	575 V
¼	1.11	0.96	0.48	0.38
⅓	1.34	1.18	0.59	0.47
½	2.2	2	1	0.8
¾	3.1	2.8	1.4	1.1
1	4	3.6	1.8	1.4
1½	5.7	5.2	2.6	2.1
2	7.5	6.8	3.4	2.7
3	10.6	9.6	4.8	3.9
5	16.7	15.2	7.6	6.1
7½	24	22	11	9
10	31	28	14	11
15	46	42	21	17
20	59	54	27	22
25	75	68	34	27
30	88	80	40	32
40	114	104	52	41
50	143	130	65	52
60	169	154	77	62
75	211	192	96	77
100	273	248	124	99
125	343	312	156	125
150	396	360	180	144
200	—	480	240	192
250	—	602	301	242
300	—	—	362	288
350	—	—	413	337
400	—	—	477	382
500	—	—	590	472

Motor Horsepower

Pump Flow*	Pump Pressure†										
	100	**250**	**500**	**750**	**1000**	**1250**	**1500**	**2000**	**3000**	**4000**	**5000**
1	0.07	0.18	0.36	0.54	0.72	0.91	1.09	1.45	2.18	2.91	3.64
2	0.14	0.36	0.72	1.09	1.45	1.82	2.18	2.91	4.37	5.83	7.29
3	0.21	0.54	1.09	1.64	2.18	2.73	3.28	4.37	6.56	8.75	10.93
4	0.29	0.72	1.45	2.18	2.91	3.64	4.37	5.83	8.75	11.66	14.58
5	0.36	0.91	1.82	2.73	3.64	4.55	5.46	7.29	10.93	14.58	18.23
8	0.58	1.45	2.91	4.37	5.83	7.29	8.75	11.66	17.50	23.33	29.17
10	0.72	1.82	3.64	5.64	7.29	9.11	10.93	14.58	21.87	29.17	36.46
12	0.87	2.18	4.37	6.56	8.75	10.93	13.12	17.50	26.25	35.00	43.75
15	1.09	2.73	5.46	8.20	10.93	13.67	16.40	21.87	32.81	43.75	54.69
20	1.45	3.64	7.29	10.93	14.58	18.23	21.87	29.17	43.75	58.34	72.92
25	1.82	4.55	9.11	13.67	18.23	22.79	27.34	36.46	54.69	72.92	91.16
30	2.18	5.46	10.93	16.40	21.87	27.34	32.81	43.75	65.63	87.51	109.39
35	2.55	6.38	12.76	19.14	25.52	31.90	38.28	51.05	76.57	102.10	127.62
40	2.91	7.29	14.58	21.87	29.17	36.46	43.75	58.34	87.51	116.68	145.85
45	3.28	8.20	16.40	24.61	32.81	41.02	49.22	65.63	98.45	131.27	164.08
50	3.64	9.11	18.23	27.34	36.46	45.58	54.69	72.92	109.39	145.85	182.32
55	4.01	10.20	20.05	30.08	40.11	50.13	60.16	80.22	120.33	160.44	200.55
60	4.37	10.93	21.87	32.81	43.75	54.69	65.63	87.51	131.27	175.02	218.78
65	4.74	11.85	23.70	35.55	47.40	59.25	71.10	94.80	142.21	189.61	237.01
70	5.10	12.76	25.52	38.28	51.05	63.81	76.57	102.10	153.13	204.20	255.25
75	5.46	13.67	27.36	41.02	54.69	68.37	82.04	109.39	164.08	218.78	273.48
80	5.83	14.58	29.17	43.75	58.34	72.92	87.51	116.68	175.02	233.37	291.71
90	6.56	16.40	32.81	49.22	65.63	82.04	98.45	131.27	196.90	262.54	328.17
100	7.29	18.23	36.46	54.69	72.92	91.16	109.39	145.85	218.78	291.71	364.64

* in gpm
† pump pressure in psi (efficiency assumed to be 80%)

Horsepower Formulas

To Find	Use Formula	Example		
		Given	**Find**	**Solution**
HP	$HP = \frac{I \times V \times Eff}{746}$	240 V, 20 A, 85% Eff	HP	$HP = \frac{I \times V \times Eff}{746}$ $HP = \frac{20\text{ A} \times 240\text{ V} \times 85\%}{746}$ $HP = \mathbf{5.5}$
I	$I = \frac{HP \times 746}{V \times Eff \times PF}$	10 HP, 240 V, 90% Eff, 88% PF	I	$I = \frac{HP \times 746}{V \times Eff \times PF}$ $I = \frac{10\text{ HP} \times 746}{240\text{ V} \times 90\% \times 88\%}$ $I = \mathbf{39\text{ A}}$

Note: Eff = efficiency

Percentage Formulas
Amount = base × rate, $A = BR$
$\text{Percent change} = \dfrac{\text{new value} - \text{original value}}{\text{original value}} \times 100$
$\text{Percent error} = \dfrac{\text{measured value} - \text{known value}}{\text{known value}} \times 100$
$\text{Percent concetration of ingredient } A = \dfrac{\text{amount of } A}{\text{amount of mixture}} \times 100$
$\text{Percent efficiency} = \dfrac{\text{output}}{\text{input}} \times 100$

Percentage Equivalents

Percent	Proper Fraction	Decimal	Fractional Hundredth
5%	1/20	0.05	5/100
10%	1/10	0.10	10/100
12½%	1/8	0.125	$\frac{12\frac{1}{2}}{100}$
15⅝%	5/32	0.15375	$\frac{15\frac{5}{8}}{100}$
16⅔%	1/6	0.1666	$\frac{16\frac{2}{3}}{100}$
18¾%	3/16	0.1875	$\frac{18\frac{3}{4}}{100}$
20%	1/5	0.20	20/100
21⅞%	7/32	0.21875	$\frac{21\frac{7}{8}}{100}$
25%	1/4	0.25	25/100
28⅛%	9/32	0.28125	$\frac{28\frac{1}{8}}{100}$
31¼%	5/16	0.3125	$\frac{31\frac{1}{4}}{100}$
33⅓%	1/3	0.3333	$\frac{33\frac{1}{3}}{100}$
37½%	3/8	0.375	$\frac{37\frac{1}{2}}{100}$
40%	2/5	0.40	40/100
43¾%	7/16	0.4375	$\frac{43\frac{3}{4}}{100}$
50%	1/2	0.50	50/100
60%	3/5	0.60	60/100
62½%	5/8	0.625	$\frac{62\frac{1}{2}}{100}$
66⅔%	2/3	0.6666	$\frac{66\frac{2}{3}}{100}$
70%	7/10	0.70	70/100
75%	3/4	0.75	75/100
80%	4/5	0.80	80/100
83⅓%	5/6	0.8333	$\frac{83\frac{1}{3}}{100}$
87½%	7/8	0.875	$\frac{87\frac{1}{2}}{100}$
90%	9/10	0.90	90/100

Squares, Cubes, Roots, and Reciprocals						
No.	**Square**	**Sq Root**	**Cube**	**Cube Root**	**Reciprocal**	**No.**
1	1	1.00000	1	1.00000	1.0000000	**1**
2	4	1.41421	8	1.25992	0.5000000	**2**
3	9	1.73205	27	1.44225	0.3333333	**3**
4	16	2.00000	64	1.58740	0.2500000	**4**
5	25	2.23607	125	1.70998	0.2000000	**5**
6	36	2.44949	216	1.81712	0.1666667	**6**
7	49	2.64575	343	1.91293	0.1428571	**7**
8	64	2.82843	512	2.00000	0.1250000	**8**
9	81	3.00000	729	2.08008	0.1111111	**9**
10	100	3.16228	1000	2.15443	0.1000000	**10**
11	121	3.31662	1331	2.22398	0.0909091	**11**
12	144	3.46410	1728	2.28943	0.0833333	**12**
13	169	3.60555	2197	2.35133	0.0769231	**13**
14	196	3.74166	2744	2.41014	0.0714286	**14**
15	225	3.87298	3375	2.46621	0.0666667	**15**
16	256	4.00000	4096	2.51984	0.0625000	**16**
17	289	4.12311	4913	2.57128	0.0588235	**17**
18	324	4.24264	5832	2.62074	0.0555556	**18**
19	361	4.35890	6859	2.66840	0.0526316	**19**
20	400	4.47214	8000	2.71442	0.0500000	**20**
21	441	4.58258	9261	2.75892	0.0476190	**21**
22	484	4.69042	10,648	2.80204	0.0454545	**22**
23	529	4.79583	12,167	2.84387	0.0434783	**23**
24	576	4.89898	13,824	2.88450	0.0416667	**24**
25	625	5.00000	15,625	2.92402	0.0400000	**25**
26	676	5.09902	17,576	2.96250	0.0384615	**26**
27	729	5.19615	19,683	3.00000	0.0370370	**27**
28	784	5.29150	21,952	3.03659	0.0357143	**28**
29	841	5.38516	24,389	3.07232	0.0344828	**29**
30	900	5.47723	27,000	3.10723	0.0333333	**30**
31	961	5.56776	29,791	3.14138	0.0322581	**31**
32	1024	5.65685	32,768	3.17480	0.0312500	**32**
33	1089	5.74456	35,937	3.20753	0.0303030	**33**
34	1156	5.83095	39,304	3.23961	0.0294118	**34**
35	1225	5.91608	42,875	3.27107	0.0285714	**35**
36	1296	6.00000	46,656	3.30193	0.0277778	**36**
37	1369	6.08276	50,653	3.33222	0.0270270	**37**
38	1444	6.16441	54,872	3.36198	0.0263158	**38**
39	1521	3.24500	59,319	3.39121	0.0256410	**39**
40	1600	6.32456	64,000	3.41995	0.0250000	**40**
41	1681	6.40312	68,921	3.44822	0.0243902	**41**
42	1764	6.48074	74,088	3.47603	0.0238095	**42**
43	1849	6.55744	79,507	3.50340	0.0232558	**43**
44	1936	6.63325	85,184	3.53035	0.0227273	**44**
45	2025	6.70820	91,125	3.55689	0.0222222	**45**
46	2116	6.78233	97,336	3.58305	0.0217391	**46**
47	2209	6.85565	103,823	3.60883	0.0212766	**47**
48	2304	6.92820	110,592	3.63424	0.0208333	**48**
49	2401	7.00000	117,649	3.65931	0.0204082	**49**
50	2500	7.07107	125,000	3.68403	0.0200000	**50**

(Continued on next page)

Squares, Cubes, Roots, and Reciprocals (continued)						
No.	Square	Sq Root	Cube	Cube Root	Reciprocal	No.
51	2601	7.14143	132,651	3.70843	0.0196078	51
52	2704	7.21110	140,608	3.73251	0.0192308	52
53	2809	7.28011	148,877	3.75629	0.0188679	53
54	2916	7.34847	157,464	3.77976	0.0185185	54
55	3025	7.41620	166,375	3.80295	0.0181818	55
56	3136	7.48331	175,616	3.82586	0.0178571	56
57	3249	7.54983	185,193	3.84850	0.0175439	57
58	3364	7.61577	195,112	3.87088	0.0172414	58
59	3481	7.68115	205,379	3.89300	0.0169492	59
60	3600	7.74597	216,000	3.91487	0.0166667	60
61	3721	7.81025	226,981	3.93650	0.0163934	61
62	3844	7.87401	238,328	3.95789	0.0161290	62
63	3969	7.93725	250,047	3.97906	0.0158730	63
64	4096	8.00000	262,144	4.00000	0.0156250	64
65	4225	8.06226	274,625	4.02073	0.0153846	65
66	4356	8.12404	287,496	4.04124	0.0151515	66
67	4489	8.18535	300,763	4.06155	0.0149254	67
68	4624	8.24621	314,432	4.08166	0.0147059	68
69	4761	8.30662	328,509	4.10157	0.0144928	69
70	4900	8.36660	343,000	4.12129	0.0142857	70
71	5041	8.42615	357,911	4.14082	0.0140845	71
72	5184	8.48528	373,248	4.16017	0.0138889	72
73	5329	8.54400	389,017	4.17934	0.0136986	73
74	5476	8.60233	405,224	4.19834	0.0135135	74
75	5625	8.66025	421,875	4.21716	0.0133333	75
76	5776	8.71780	438,976	4.23582	0.0131579	76
77	5929	8.77496	456,533	4.25432	0.0128970	77
78	6084	8.83176	474,552	4.27266	0.0128205	78
79	6241	8.88819	493,039	4.29084	0.0126582	79
80	6400	8.94427	512,000	4.30887	0.0125000	80
81	6561	9.00000	531,441	4.32675	0.0123457	81
82	6724	9.05539	551,368	4.34448	0.0121951	82
83	6889	9.11043	571,787	4.36207	0.0120482	83
84	7056	9.16515	592,704	4.37952	0.0119048	84
85	7225	9.21954	614,125	4.39683	0.0117647	85
86	7396	9.27362	636,056	4.41400	0.0116279	86
87	7569	9.32738	658,503	4.43105	0.0114943	87
88	7744	9.38083	681,472	4.44797	0.0113636	88
89	7921	9.43398	704,969	4.46475	0.0112360	89
90	8100	9.48683	729,000	4.48140	0.0111111	90
91	8281	9.53939	753,571	4.49794	0.0109890	91
92	8464	9.59166	778,688	4.51436	0.0108696	92
93	8649	9.64365	804,357	4.53065	0.0107527	93
94	8836	9.69536	830,584	4.54684	0.0106383	94
95	9025	9.74679	857,375	4.56290	0.0105263	95
96	9216	9.79796	884,736	4.57886	0.0104167	96
97	9409	9.84886	912,673	4.59470	0.0103093	97
98	9604	9.89949	941,192	4.61044	0.0102041	98
99	9801	9.94987	970,299	4.62607	0.0101010	99
100	10,000	10.00000	1,000,000	4.64159	0.0100000	100

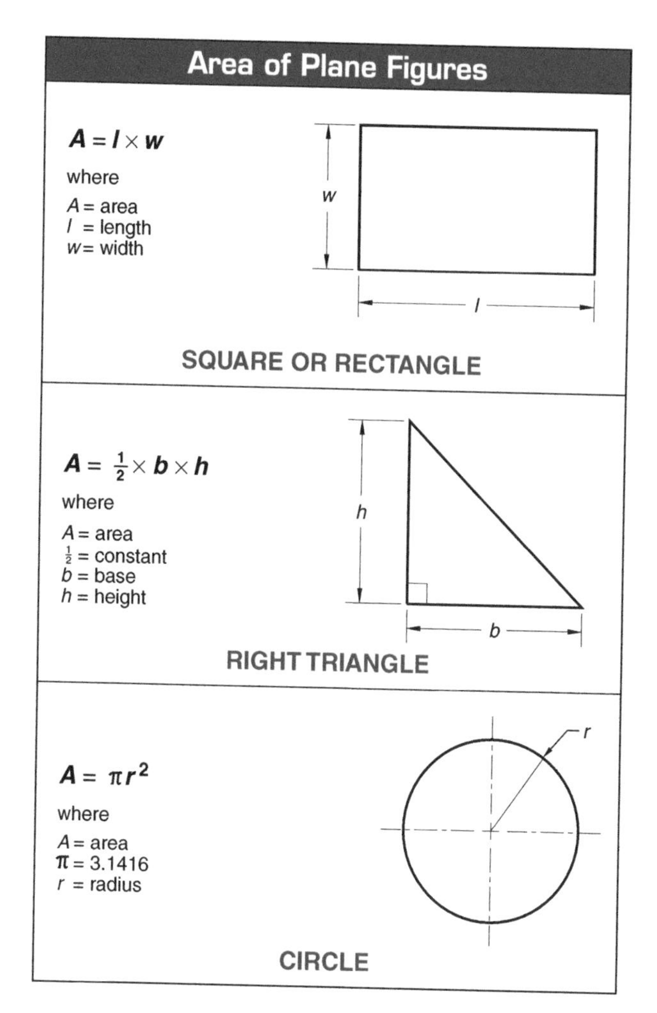

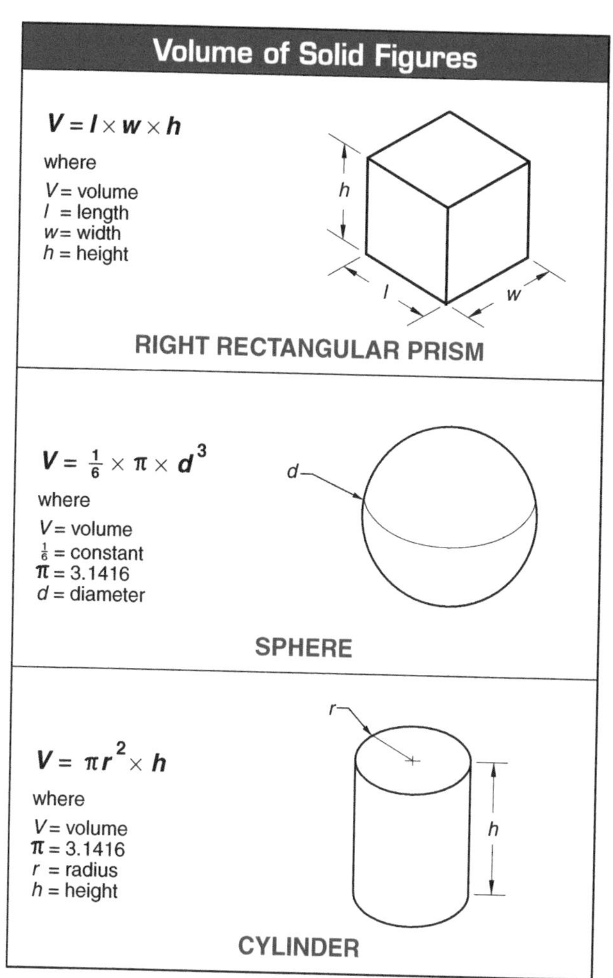

Straight Lines

Slope	$m = \frac{\text{rise}}{\text{run}} = \frac{\Delta y}{\Delta x} = \frac{y^2 - y^1}{x^2 - x^1}$	
	m = tan (angle of inclination) = tan θ	$0 \le \theta < 180°$
Equation of Straight Line	General Form	$Ax + By + C = 0$
	Parallel to *x-axis*	$y = b$
	Parallel to *y-axis*	$x = a$
	Slope-Intercept Form	$y = mx + b$
	Two-Point Form	$\frac{y - y^1}{x - x^1} = \frac{y^2 - y^1}{x^2 - x^1}$
	Point-Slope Form	$m = \frac{y - y^1}{x - x^1}$

y
x
0
d
(x_2, y_2)
Rise Δy
Run Δx
(x_1, y_1)

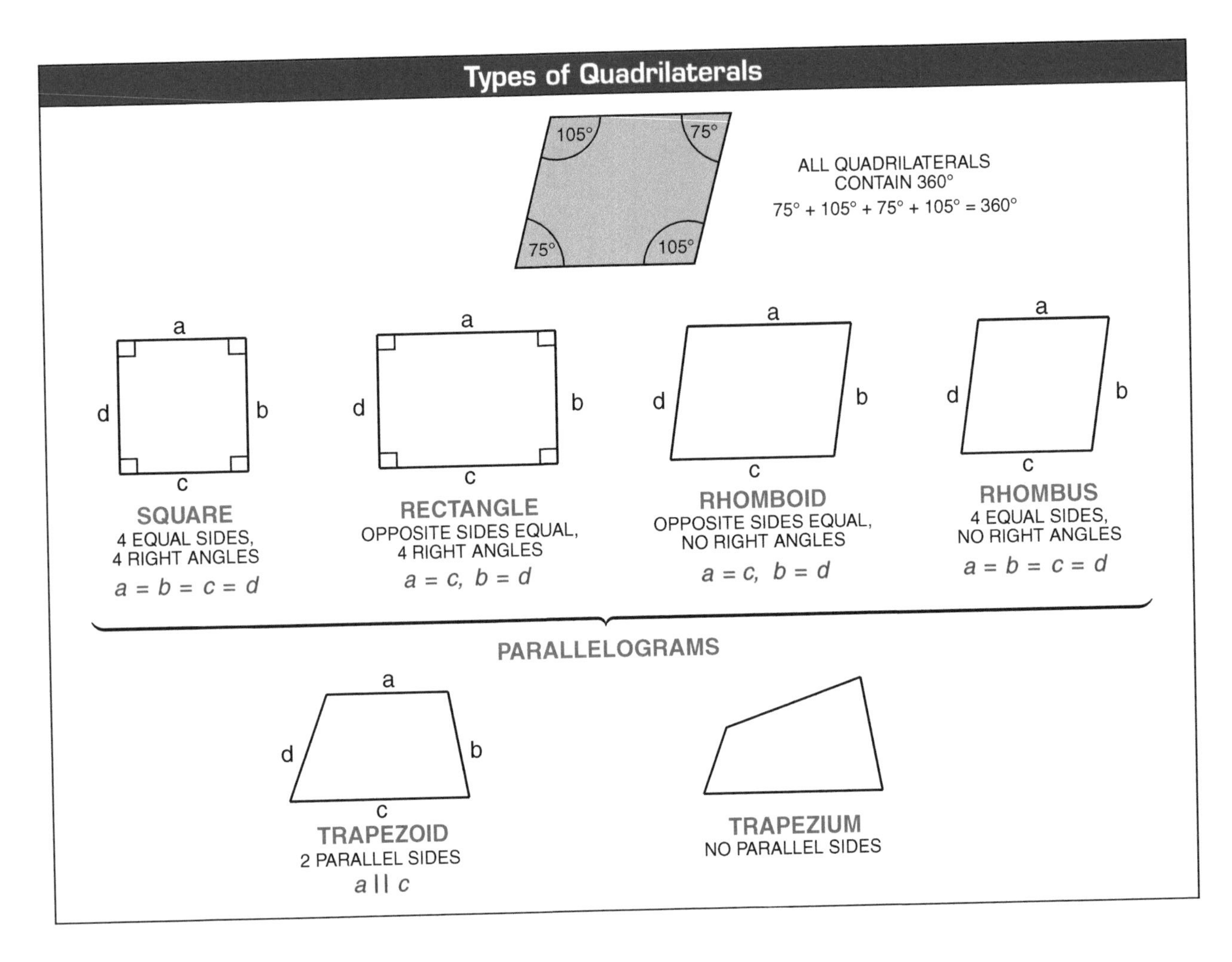

Calculating Area of Quadrilaterals

Squares	(sides a, a)	Area = a^2
Rectangles	(sides a, b)	Area = ab
Parallelograms	(side a, base b, height h)	Area = bh
Rhombuses	(side a, base a, height h)	Area = ah
Trapezoids	(top a, base b, height h)	Area = $\frac{(a+b)h}{2}$

Types of Triangles

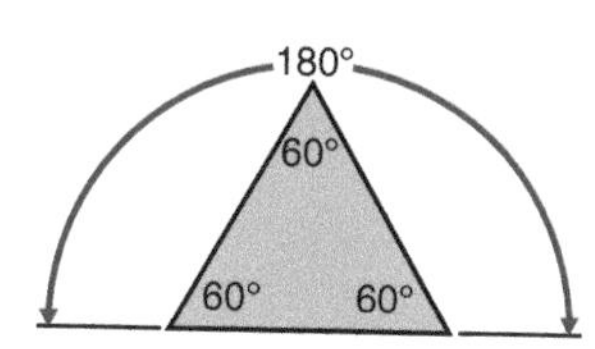

ALL TRIANGLES CONTAIN 180°
60° + 60° + 60° = 180°

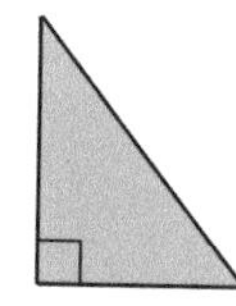

RIGHT
NO EQUAL SIDES,
1 RIGHT ANGLE

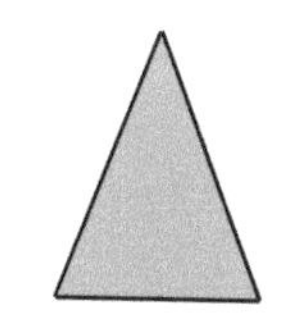

ISOSCELES
2 EQUAL SIDES,
2 EQUAL ANGLES

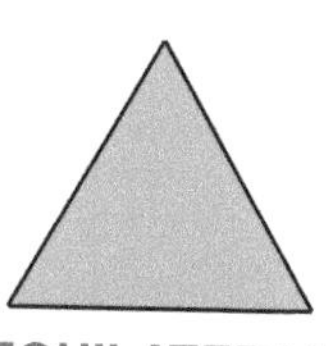

EQUILATERAL
3 EQUAL SIDES,
3 EQUAL ANGLES

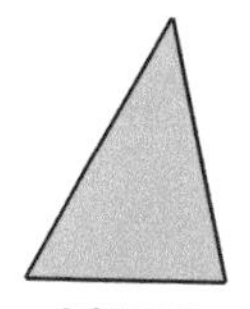

ACUTE
EACH ANGLE
LESS THAN 90°

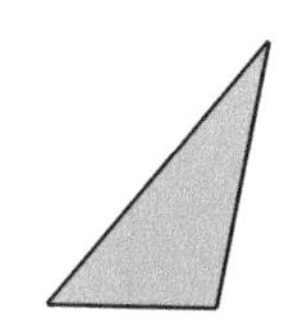

OBTUSE
1 ANGLE
GREATER THAN 90°

SCALENE
NO EQUAL SIDES, NO EQUAL ANGLES

Pythagorean Theorem

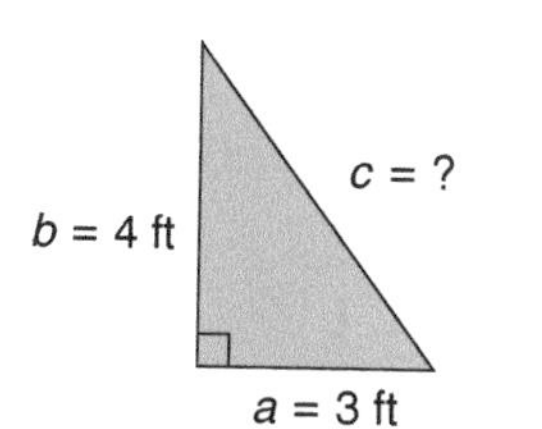

$c^2 = a^2 + b^2$

$c = \sqrt{a^2 + b^2}$

$c = \sqrt{(3\times3) + (4\times4)}$

$c = \sqrt{9 + 16}$

$c = \sqrt{25}$

$c = \textbf{5 ft}$

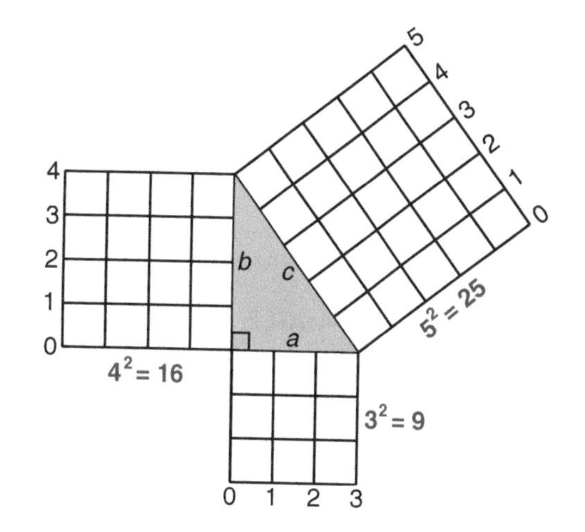

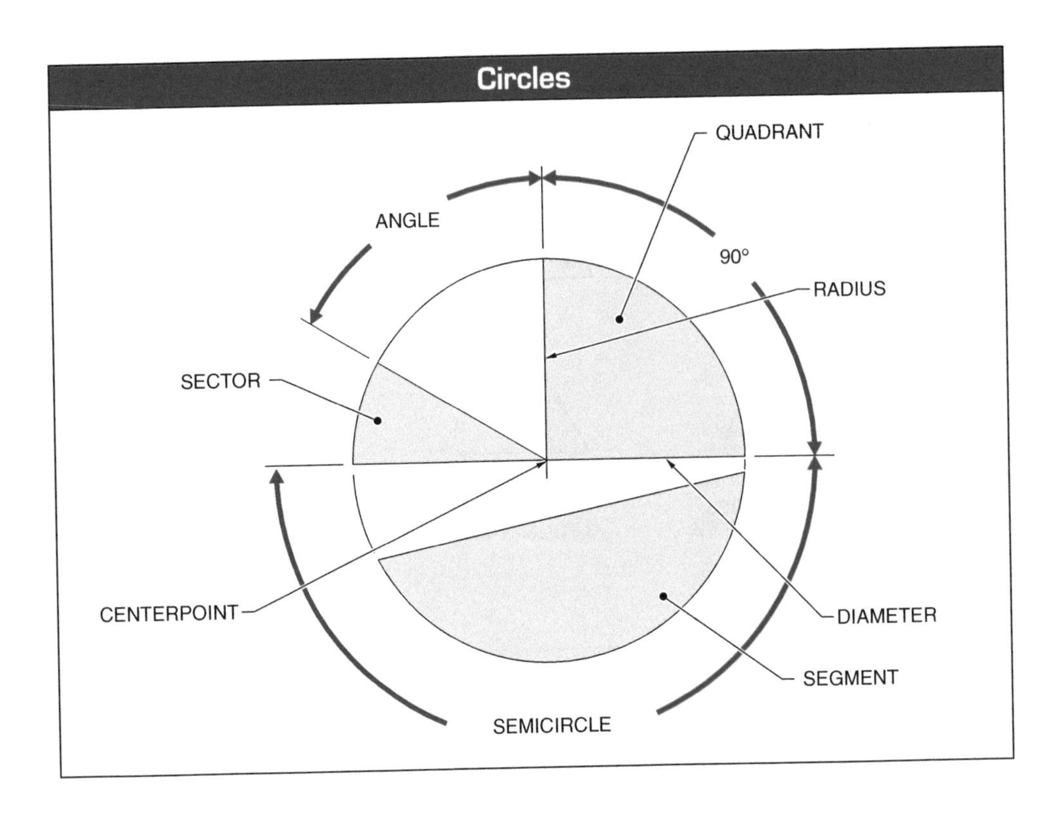
Circles
QUADRANT
ANGLE
90°
RADIUS
SECTOR
CENTERPOINT
DIAMETER
SEGMENT
SEMICIRCLE

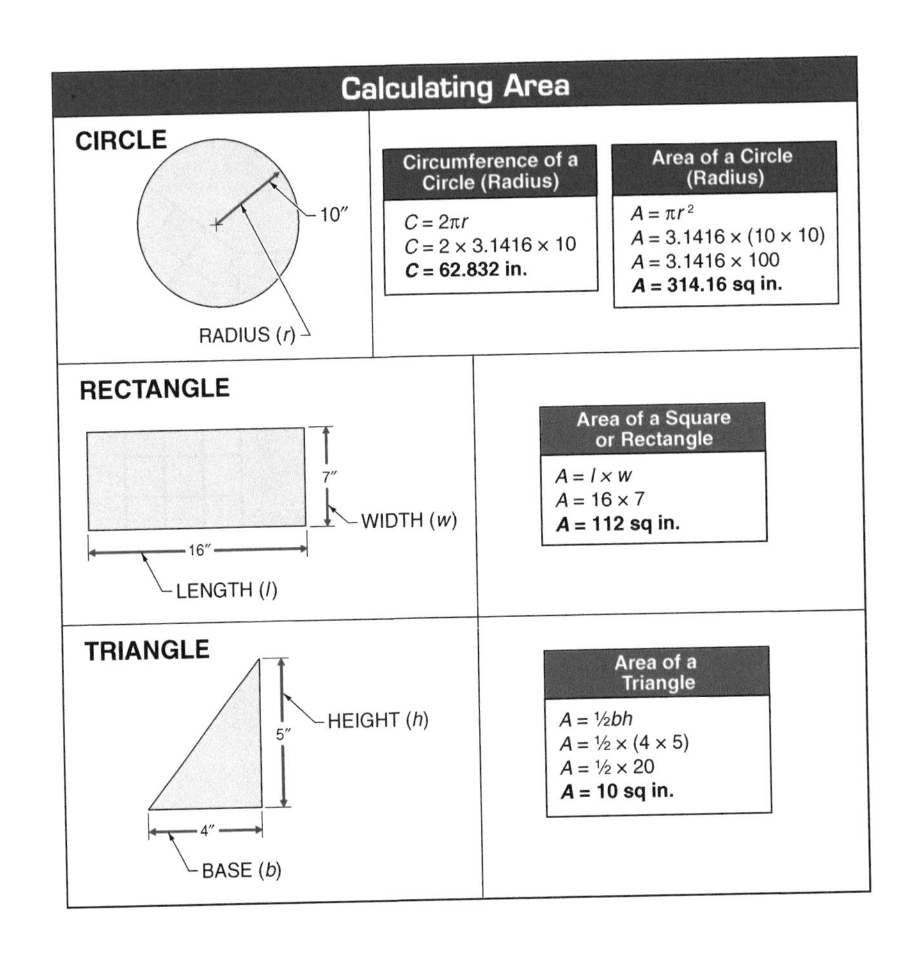
Calculating Area
CIRCLE
10″
RADIUS (r)
Circumference of a Circle (Radius)
C = 2πr
C = 2 × 3.1416 × 10
C = 62.832 in.
Area of a Circle (Radius)
A = πr²
A = 3.1416 × (10 × 10)
A = 3.1416 × 100
A = 314.16 sq in.
RECTANGLE
7″
WIDTH (w)
16″
LENGTH (l)
Area of a Square or Rectangle
A = l × w
A = 16 × 7
A = 112 sq in.
TRIANGLE
HEIGHT (h)
5″
4″
BASE (b)
Area of a Triangle
A = ½bh
A = ½ × (4 × 5)
A = ½ × 20
A = 10 sq in.

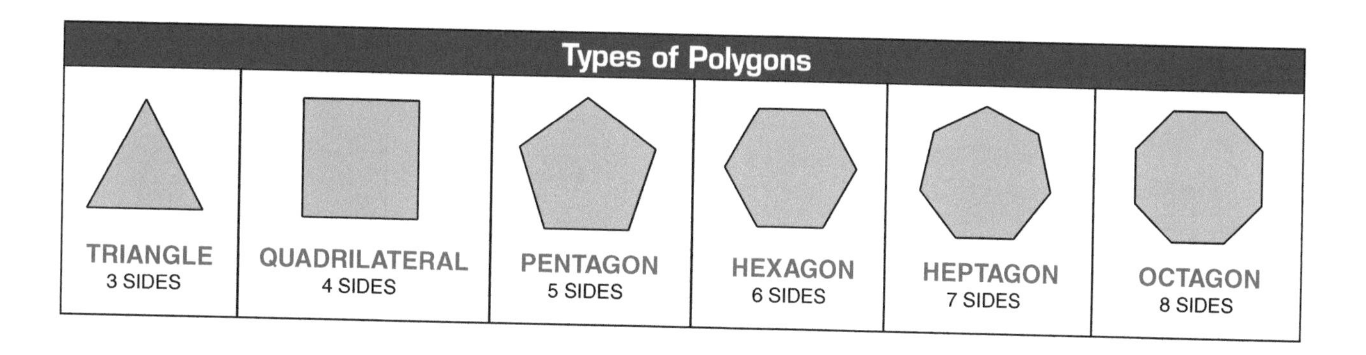
Types of Polygons
TRIANGLE
3 SIDES
QUADRILATERAL
4 SIDES
PENTAGON
5 SIDES
HEXAGON
6 SIDES
HEPTAGON
7 SIDES
OCTAGON
8 SIDES

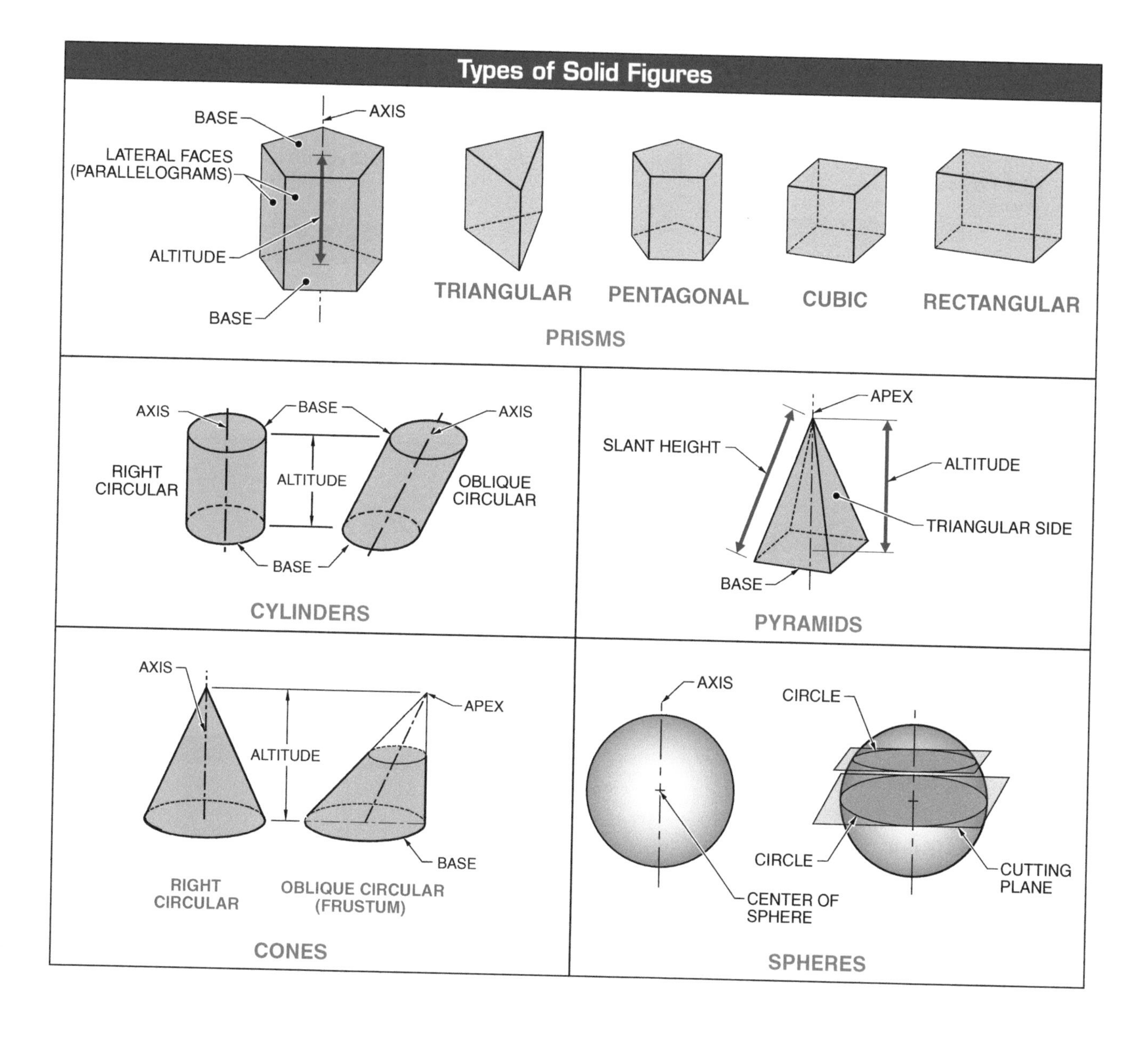
Types of Solid Figures
BASE
AXIS
LATERAL FACES
(PARALLELOGRAMS)
ALTITUDE
BASE
TRIANGULAR
PENTAGONAL
CUBIC
RECTANGULAR
PRISMS
AXIS
BASE
AXIS
RIGHT
CIRCULAR
ALTITUDE
OBLIQUE
CIRCULAR
BASE
CYLINDERS
APEX
SLANT HEIGHT
ALTITUDE
TRIANGULAR SIDE
BASE
PYRAMIDS
AXIS
APEX
ALTITUDE
BASE
RIGHT
CIRCULAR
OBLIQUE CIRCULAR
(FRUSTUM)
CONES
AXIS
CIRCLE
CIRCLE
CENTER OF
SPHERE
CUTTING
PLANE
SPHERES

Calculating Volume and Surface Area

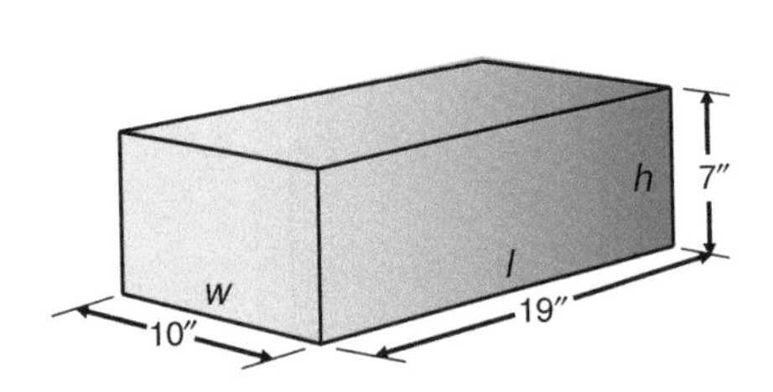

Surface Area of a Rectangular Solid

$A = 2(lw + wh + lh)$

$A = 2[(19 \times 10) + (10 \times 7) + (19 \times 7)]$

$A = 2(190 + 70 + 133)$

$A = 2(393)$

$\mathbf{A = 786}$ **sq in.**

Volume of a Rectangular Solid

$V = l \times w \times h$

$V = 19 \times 10 \times 7$

$\mathbf{V = 1330}$ **cu in.**

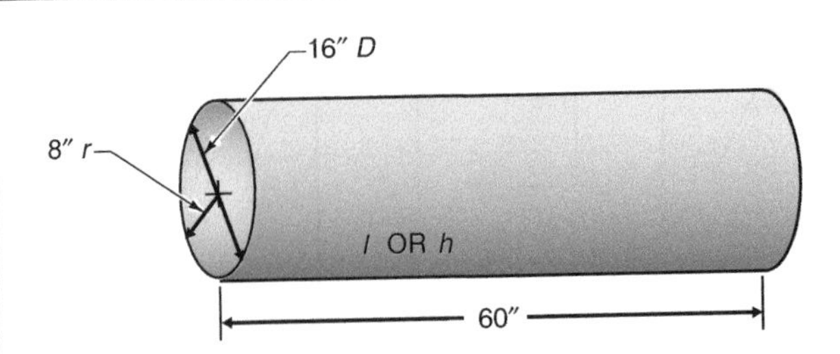

Surface Area of a Cylinder

$A = 2\pi r(r + h)$

$A = 2 \times 3.1416 \times 8(8 + 60)$

$A = 2 \times 3.1416 \times 8(68)$

$\mathbf{A = 3418.06}$ **sq in.**

Volume of a Cylinder

$V = \pi r^2 h$

$V = 3.1416 \times (8 \times 8) \times 60$

$V = 3.1416 \times 64 \times 60$

$\mathbf{V = 12{,}063.744}$ **cu in.**

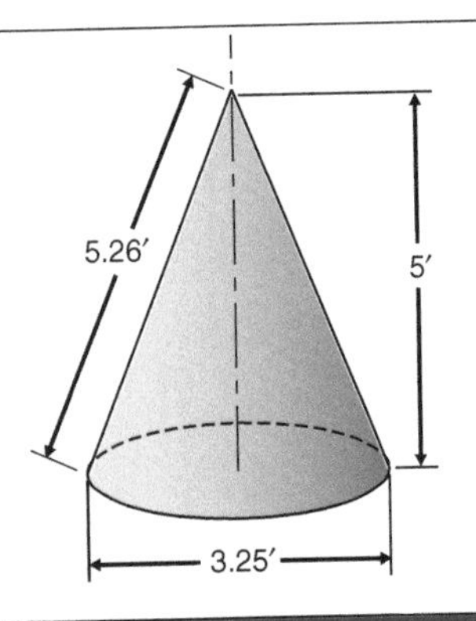

Surface Area of a Cone

$A = \pi r(r + s)$

$A = \pi \times 1.625(1.625 + 5.26)$

$A = 3.1416 \times 1.625(6.885)$

$\mathbf{A = 35.15}$ **sq ft**

Volume of a Cone

$V = \frac{1}{3}\pi r^2 h$

$V = \frac{1}{3} \times 3.1416 \times 1.625^2 \times 5$

$V = \frac{1}{3} \times 3.1416 \times 2.640625 \times 5$

$\mathbf{V = 13.83}$ **cu ft**

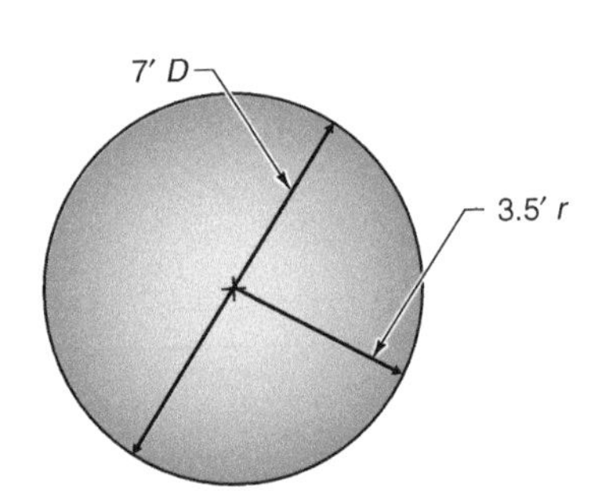

Surface Area of a Sphere

$A = 4\pi r^2$

$A = 4 \times \pi \times 3.5^2$

$A = 4 \times \pi \times 12.25$

$A = 4 \times 3.1416 \times 12.25$

$\mathbf{A = 153.94}$ **sq ft**

Volume of a Sphere

$V = \frac{4}{3}\pi r^3$

$V = \frac{4}{3} \times \pi \times 3.5^2$

$V = \frac{4}{3} \times \pi \times 42.875$

$V = \frac{4}{3} \times 3.1416 \times 42.875$

$\mathbf{V = 179.59}$ **cu ft**

JOB ESTIMATE SHEET

Project: ____________
Estimator: ____________

Sheet No.: ____________
Date: ____________
Checked: ____________

No.	Description	Dimensions			Quantity		Material		Labor		Equipment		Total	
						Unit	Unit Cost	Total	Unit Cost	Total	Unit Cost	Total	Unit Cost	Total
Total														

Glossary

A

absolute zero: A theoretical condition where no heat is present.

acre: An area of land containing 43,560 sq ft.

acute angle: An angle that measures less than 90°.

addition: The process of combining two or more numbers into a sum.

altitude: 1. The perpendicular distance between the two bases of a prism or cylinder. **2.** The perpendicular distance from the base to the apex of a pyramid or cone.

angle: A figure created by two intersecting lines.

apex: The point opposite the base of a pyramid or cone.

Arabic numerals: Numerals expressed using ten digits—0, 1, 2, 3, 4, 5, 6, 7, 8, and 9.

area: Space as expressed in square units.

B

bar graph: A graph in which the values of variables are represented by bars.

base: 1. The bottom of a polygon. **2.** One of the two parallel polygons of a prism. **3.** One of the two parallel circles of a cylinder. **4.** The bottom of a pyramid or cone.

base amount: A whole amount before a percentage is calculated in a percentage problem.

base number: A factor that is multiplied a given number of times according to an exponent.

borrowing: The process of moving "10" from the next higher column so that a difference will be positive.

C

cancellation: A method of removing common factors from both a numerator and denominator.

capacity: The maximum volume that a container can hold.

carrying: The process of moving a digit from one column into the column to its left.

circle: A round plane figure made of a curved line in which all points on the curve are the same distance from a centerpoint.

circumference: The boundary of a circle.

common denominator: A denominator that is the same among a group of fractions.

common factor: A factor that is common to two or more numbers.

complex fraction: A fraction that has a fraction, an improper fraction, a mixed number, or a mathematical process in its numerator, its denominator, or both.

compound ratio: The product of two or more ratios.

cone: A solid figure with a circular base and a curved surface that tapers from the base to the apex.

conversion factor: A number that translates one unit of measure into another unit of measure of the same value.

coordinates: Two numbers written in parentheses used to indicate points on a graph.

cylinder: A solid figure with two circular bases.

D

decimal: A number expressed with 10 as its base.

decimal fraction: A fraction with a denominator of 10 or a multiple of 10.

denominator: The number below a fraction bar that shows how many parts a whole number has been divided into.

diameter: The length of a line from one edge of a circle, through the centerpoint, and to the opposite edge.

difference: The number produced as a result of subtraction.

direct proportion: A proportion where an increase in one quantity leads to a proportional increase in the related quantity, and a decrease leads to a proportional decrease.
discount: A reduction of a marked price.
dividend: The number being divided in a division problem.
division: The process of determining the number of times one number is contained in another number.
divisor: The number a dividend is "divided by."

E

equilateral triangle: A triangle that contains three equal angles and three equal sides.
even number: Any number, except 0, that ends in 2, 4, 6, 8 or 0 and can be divided by 2 an exact number of times.
exponent: A number that indicates how many times a base number is to be multiplied.

F

factor: A number used as a multiplier.
fraction: A part of a whole number.

G

graph: A diagram that shows the relationship between two or more variables.
greatest common factor: The highest number in a group of factors.
grouped-bar graph: A bar graph where more than one bar is used to illustrate the variables in a group.

H

hectare: An area of land containing 10,000 m^2.
height: The length measurement from the base of a polygon to its other base or to the vertex on the opposite side.
horizontal line: A line that is parallel to the horizon.
hypotenuse: The side of a right triangle opposite the right angle.

I

improper fraction: A fraction with a numerator larger than its denominator, such as $\frac{7}{5}$, $\frac{8}{3}$, $\frac{4}{3}$, $\frac{10}{4}$, and $\frac{21}{7}$.
inclined line: A line that is neither horizontal nor vertical and is inclined in any direction. Also known as a slanted line.
integer: A negative whole number, a positive whole number, or 0.
inverse proportion: A proportion where an increase in one quantity leads to a proportional decrease in the related quantity, and a decrease leads to a proportional increase.
inverse ratio: The ratio of the reciprocals of two quantities.
isosceles triangle: A triangle with two equal angles and two equal sides.

L

length: Distance expressed in linear measure.
line: A one-dimensional figure that appears as a long, narrow band.
line graph: A graph in which points representing variables are connected by a line.
lowest common denominator (LCD): The smallest number into which the denominators of two or more fractions can divide an exact number of times.

M

marked price: The retail price of an item.
mass: The amount of matter contained in an object.
mixed number: A combination of a whole number and a proper fraction, such as $1\frac{1}{8}$, $2\frac{1}{4}$, $4\frac{1}{2}$, $6\frac{9}{16}$, and $11\frac{63}{64}$.
multiple discount: The total discount that results by applying more than one discount to the price of an item.
multiplication: The process of adding one factor as many times as indicated by another factor.

N

net price: The price of an item after a discount has been applied.
numerator: The number above a fraction bar that shows the number of parts taken from the denominator.

O

obtuse angle: An angle that measures more than 90° but less than 180°.

odd number: Any number that ends in 1, 3, 5, 7, or 9 and cannot be divided by 2 an exact number of times.

origin: The point where the two axes of a graph intersect.

P

parallel lines: Two or more lines that never intersect.

percentage: A number that represents part of a whole and is expressed as part of 100.

percent discount: The amount of a discount given as a percent.

percent rate: A percent value of an amount in a percentage problem and is the quantity found before the percent sign.

perfect square: A number whose square root is a whole number.

perimeter: The sum of the lengths of the sides of a closed plane figure.

period: A group of three places in a number that is separated from other periods by a comma.

perpendicular lines: Two lines that form a 90° angle.

pie chart: *See* pie graph.

pie graph: A graph in which a circle represents 100% of a variable and the sectors of the circle represent parts of the total. Also known as a pie chart.

plane figure: A two-dimensional figure.

point: A specific location in space.

polygon: A multiple-sided plane figure that has a perimeter of straight lines.

prime factor: A factor that is a prime number.

prime number: A whole number that can only be divided an exact number of times by itself and the number 1.

prism: A solid figure with two identical bases and lateral faces (sides) which are parallelograms.

product: A number produced as a result of multiplication.

proper fraction: A fraction with a numerator smaller than the denominator, such as $\frac{1}{8}$, $\frac{1}{4}$, $\frac{1}{2}$, $\frac{9}{16}$, and $\frac{25}{32}$.

proportion: An expression of equality between two ratios.

pyramid: A solid figure with a base that is a polygon and sides that are triangles.

Q

quadrilateral: A four-sided polygon with four interior angles.

quotient: A number produced as a result of division.

R

radical sign: A symbol (√) used to indicate the square root of a number.

radius: The distance from the centerpoint of a circle to its edge.

ratio: A mathematical way to represent the relationship between two or more numbers, or terms.

rectangle: A quadrilateral with opposite sides of equal length and and four 90° angles.

remainder: The undivided part of a quotient.

repeating decimal: A decimal that has a repeating number or group of numbers that repeat infinitely.

right angle: An angle that measures 90°.

right triangle: A triangle with one 90° angle.

Roman numerals: Numerals expressed by the letters I, V, X, L, C, D, and M.

rounding: The process of reducing the number of places in a decimal.

S

scalene triangle: A triangle with no equal angles or equal sides.

slanted line: *See* inclined line.

slant height: The distance from the side of the base to the apex of a pyramid or cone.

solid figure: A three-dimensional figure that has length, width, and height.

sphere: A solid figure with all points equidistant from its centerpoint.

square: A quadrilateral with four equal sides and four 90° angles.

square root: One of two equal factors, or roots, used to obtain a number.

stacked-bar graph: A bar graph where each bar is divided into more than one variable.

straight angle: An angle that measures 180° and forms a straight line.

subtraction: The process of taking one number away from another number.

sum: The number produced as a result of addition.

T

tax: A charge paid on income, products, and services.

temperature: A measurement of an amount of heat expressed in degrees.

triangle: A three-sided polygon with three interior angles.

U

unit of measure: A standard by which a quantity is measured.

V

vertex: A point of intersection of two or more lines.

vertical line: A line that is perpendicular to the horizon.

volume: Space as expressed in cubic measure.

W

weight: A measurement that indicates the heaviness of an object.

whole number: Any positive number that has no fractional or decimal parts.

X

x-axis: A horizontal line serving as a reference for the values of variables.

Y

y-axis: A vertical line serving as a reference for the values of variables.

Index

Page numbers in italic refer to figures.

T

U

V

W

X

Y